METHODS IN MOLECULAR BIOLOGY™

Series Editor
John M. Walker
School of Life Sciences
University of Hertfordshire
Hatfield, Hertfordshire, AL10 9AB, UK

For further volumes:
http://www.springer.com/series/7651

Pheromone Signaling

Methods and Protocols

Edited by

Kazushige Touhara

Department of Applied Biological Chemistry, Graduate School of Agricultural and Life Sciences, The University of Tokyo, Tokyo, Japan

Humana Press

Editor
Kazushige Touhara
Department of Applied Biological Chemistry
Graduate School of Agricultural and Life Sciences
The University of Tokyo
Tokyo, Japan

ISSN 1064-3745 ISSN 1940-6029 (electronic)
ISBN 978-1-4939-6012-5 ISBN 978-1-62703-619-1 (eBook)
DOI 10.1007/978-1-62703-619-1
Springer New York Heidelberg Dordrecht London

Softcover reprint of the hardcover 1st edition 2013

Printed on acid-free paper

Humana Press is a brand of Springer
Springer is part of Springer Science+Business Media (www.springer.com)

Preface

Many animal species utilize chemosensory systems to detect a variety of chemical substances in the external environment. When these substances convey information regarding food, predators, species, or sex, it is considered chemical communication between organisms. The notion of chemical communication was first proposed by Charles Darwin in the late nineteenth century, in conjunction with the physical signal-mediated communication described for the visual and auditory systems. At approximately the same time, Jean Henri Fabre described the attraction of male moths by conspecific females, and proposed that the cue was not a visual signal but some kind of smell. The concept of chemical interaction was scientifically recognized when Albrecht Beth proposed in 1932 the term "ectohormone" for substances that functioned like a hormone but were secreted into the external environment.

The first molecular evidence for a substance that was able to mediate chemical communication was obtained when Butenandt discovered bombykol, the sex pheromone of the silk moth *Bombyx mori*, in 1959, which is released by females and elicits the full sequence of sexual behavior in male moths [1]. The term "pheromone" was then defined by Karlson and Luscher as "substances which are secreted to the outside by an individual and received by a second individual of the same species, in which they release a specific reaction, for example, a definite behavior or developmental process" [2]. Pheromone is a combined term from the Greek words *pherein* for "to bear or transfer" and *hormao* for "to excite." Specific reactions by pheromones are generally categorized into two types: one that releases a visible stereotyped behavior such as attraction or avoidance and a second that elicits endocrinological changes such as puberty acceleration and estrous cycle regulation.

A pheromone appears to be a type of smell recognized by some species; however, the definition specifically avoids any use of the terms "odor" or "odorant." It does not even define pheromones as being volatile. Karlson and Luscher were shown to be insightful in this regard, as future research was to find that pheromones can be nonvolatile substances including relatively large organic compounds, peptides, and proteins. Recently, there has been some contention in the field as to whether the definition of pheromones should be extended, as pheromones appear to play far more diverse functional roles than previously defined. Possible criteria for a more inclusive definition of a pheromone may be that (1) pheromones are released by one individual and received by conspecifics; (2) pheromones themselves may convey meaningful information including sex, strain, and species to the receiver; and (3) pheromonal effects are stereotyped or innate. It is of note that the same molecule may be utilized as a pheromone in different species. It is also important to note that pheromones are distinct from individual chemical information that may be learned [3].

The mechanisms underlying pheromone action are also of significance. A receptor that senses insect sex pheromone was first discovered in the silk moth. The bombykol receptor was identified as a male-specific seven transmembrane protein that appeared to belong to the olfactory receptor superfamily, and that which showed a highly specific and sensitive response to bombykol [4]. It was later found that the mechanism of signal transduction was distinct from the one observed in vertebrates where the insect pheromone receptors formed a heteromeric ligand-gated cation channel complex [5, 6]. The electrical signal is transmitted to specific glomeruli in the antennal lobe, the first center of the olfactory sensory pathway, and integrated in the higher brain to elicit sexual behavior.

In vertebrates, volatile pheromones are detected by the main olfactory system in addition to the secondary olfactory system called the vomeronasal organ located at the bottom of the nasal cavity [7]. The vomeronasal system was identified in amphibians, rodents, and some primate species and detects not only volatile cues but also nonvolatile pheromones. The old-world monkeys and humans do not possess a vomeronasal organ, but rather detect pheromones in the main olfactory system, if at all. The receptors belong to the G protein-coupled receptor superfamily including the olfactory receptor (OR), trace amine-associated receptor (TAAR), and vomeronasal receptor (VR). The signal is further transmitted to the main or accessory olfactory bulb and subsequently conveyed to the higher brain areas responsible for specific behavioral and neuroendocrinological output.

The development of molecular biology and neuroscience techniques has permitted the precise method of pheromone action from the molecule to the receptor to be defined [7]. The next aim of pheromone research is to further our understanding of how pheromone information is integrated in the brain, and how specific behavioral output is regulated by the neural circuitry. The pheromone system is an ideal model for understanding the neural networks that lead to various adaptive behaviors. Another important area to investigate is whether humans also impart chemical communication via pheromones. Given humans have evolved visual and auditory senses as their major cue, it seems appropriate to suggest that meaningful chemical communication does not exist in humans today. However, it is also true that humans often feel appeased by the smell of infants. Menstrual synchrony in females living closely together is also another example of potential chemical communication within the human species [8]. In contrast, rodents appear to heavily rely on chemical communication throughout life. A full list of required compounds and their receptors, however, is yet to be fully revealed. The mechanism underlying how a selective pressure occurred on sensory systems so that each animal developed a sophisticated and individually tailored chemical communication strategy remains unknown. The strategy is thought to be optimal for the survival of each organism under their living environment. In this regard, the evolutionary history of changes in pheromone molecules and genome related to pheromone reception will tell us where we came from and in which direction we are headed.

Over the last decade, a great deal of progress has been made into understanding the molecular mechanisms underlying pheromone action, largely due to the discovery of receptor genes and the advancement of imaging techniques. The major goal of *Pheromone Signaling* is to provide experimental methods and protocols that allow us to perform pheromone research in a variety of organisms ranging from invertebrates to vertebrates. The book will cover a wide spectrum of experimental approaches necessary for handling pheromone molecules, measuring receptor response and neural activation, and analyzing behavioral output (Figure 1). Pheromone research requires multidisciplinary approaches including aspects of organic chemistry, biochemistry, molecular biology, electrophysiology, and behavioral science. On first glance, the transdisciplinary aspects of pheromone research may

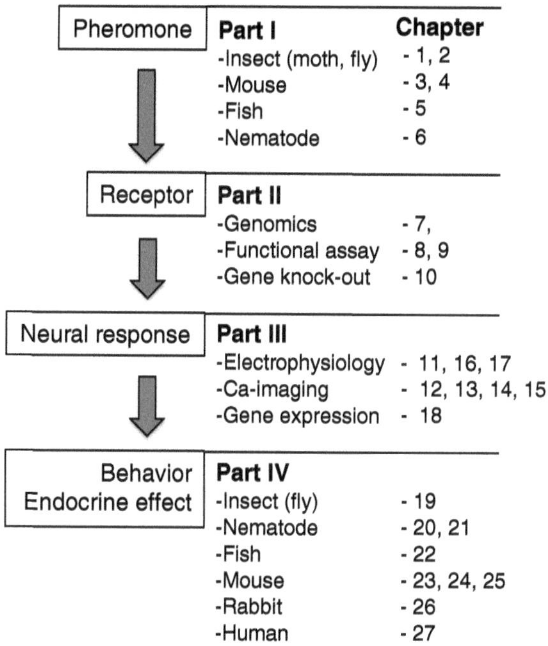

Fig. 1 The theme and a scheme for the book

seem overwhelming. This should not sidetrack investigators from delving into this field of research. This book will describe in detail the methodologies and techniques utilized in laboratories all over the world, making them accessible to those who want to begin investigation in the area of pheromone research.

Tokyo, Japan *Kazushige Touhara*

References

1. Butenandt A, Beckmann R, Stamm D, Hecker E (1959) Uber den Sexuallockstoff des Seidenspinners Bombyx mori, Reindarstellung und Konstitution. Z Naturforsch 14b:283–284
2. Karlson P, Luscher M (1959) Pheromones: a new term for a class of biologically active substances. Nature 183:55–56
3. Wyatt TD (2010) Pheromones and signature mixtures: defining species-wide signals and variable cues for identity in both invertebrates and vertebrates. J Comp Physiol A 196:685–700
4. Nakagawa T, Sakurai T, Nishioka T, Touhara K (2005) Insect sex-pheromone signals mediated by specific combinations of olfactory receptors. Science 307:1638–1642
5. Sato K, Pellegrino M, Nakagawa T, Nakagawa T, Vosshall LB, Touhara K (2008) Insect olfactory receptors are heteromeric ligand-gated ion channels. Nature 452:1002–1006
6. Wicher D, Schafer R, Bauernfeind R, Stensmyr MC, Heller R et al (2008) Drosophila odorant receptors are both ligand-gated and cyclic-nucleotide-activated cation channels. Nature 452:1007–1011
7. Touhara K, Vosshall LB (2009) Sensing odorants and pheromones with chemosensory receptors. Annu Rev Physiol 71:307–332
8. McClintock MK (1971) Menstrual synchorony and suppression. Nature 229:244–245

Contents

Contributors

Syrina Al Aïn • *Centre des Sciences du Goût, CNRS (UMR 6265), Université de Bourgogne, Dijon, France*
Tetsu Ando • *Graduate School of Bio-Applications and Systems Engineering, Tokyo University of Agriculture and Technology, Tokyo, Japan*
Cindy F. Baker • *National Institute of Water & Atmospheric Research Ltd., Hamilton, New Zealand*
Julie Bakker • *Groupe Interdisciplinaire de Genoproteomique Appliquee Neurosciences, University of Liege, Liege, Belgium*
Cornelia I. Bargmann • *Howard Hughes Medical Institute, The Rockefeller University, New York, NY, USA*
Michael J. Baum • *Department of Biology, Boston University, Boston, MA, USA*
Olivier Brock • *Netherlands Institute of Neuroscience, Amsterdam, The Netherlands*
Rebecca A. Butcher • *Department of Chemistry, University of Florida, Gainesville, FL, USA*
Jacqueline S.R. Chin • *Temasek Life Sciences Laboratories, Singapore, Singapore; Department of Biological Sciences, National University of Singapore, Singapore*
Sandeepa Dey • *Department of Molecular Genetics and Microbiology, Duke University Medical Center, Durham, NC, USA; Department of Cell Biology, The Scripps Research Institute, La Jolla, CA, USA*
Idan Frumin • *Department of Neurobiology, Weizmann Institute of Science, Rehovot, Israel*
Sachiko Haga-Yamanaka • *Stowers Institute for Medical Research, Kansas City, MO, USA*
Bill S. Hansson • *Department of Evolutionary Neuroethology, Max Planck Institute for Chemical Ecology, Jena, Germany*
Timothy E. Holy • *Department of Anatomy and Neurobiology, Washington University in St. Louis, St. Louis, MO, USA*
Tomohiro Ishii • *Department of Cell Biology, Graduate School of Medical and Dental Science, Tokyo Medical and Dental University, Tokyo, Japan*
Heeun Jang • *The Rockefeller University, New York, NY, USA*
Angeldeep Kaur • *Department of Cell Biology, The Scripps Research Institute, La Jolla, CA, USA*
Takefumi Kikusui • *School of Veterinary Medical Sciences, Azabu University, Sagamihara, Japan*
Kyuhyung Kim • *Department of Brain Science, Daegu Gyeongbuk Institute of Science and Technology (DGIST), Daegu, South Korea*
Susy M. Kim • *Neurobiology Section, Division of Biological Sciences, University of California San Diego, La Jolla, CA, USA*
Hiroko Kimoto • *Biological Research Laboratories, R&D Division, Daiichi Sankyo Co., LTD., Tokyo, Japan*
Masayuki Koganezawa • *Department of Developmental Biology and Neurosciences, Tohoku University Graduate School of Life Sciences, Sendai, Japan*

Soh Kohatsu • *Department of Developmental Biology and Neurosciences, Tohoku University Graduate School of Life Sciences, Sendai, Japan*
Kumi O. Kuroda • *Unit for Affiliative Social Behavior, RIKEN Brain Science Institute, Saitama, Japan*
Trese Leinders-Zufall • *Department of Physiology, School of Medicine, University of Saarland, Homburg, Germany*
Hiroaki Matsunami • *Department of Molecular Genetics and Microbiology, Duke Institute for Brain Sciences, Duke University Medical Center, Durham, NC, USA; Department of Neurobiology, Duke Institute for Brain Sciences, Duke University Medical Center, Durham, NC, USA*
Julian P. Meeks • *Department of Neuroscience, UT Southwestern Medical Center, Dallas, TX, USA*
Tatsuro Nakagawa • *Biological/Pharmacological Research Laboratories, JAPAN TOBACCO INC., Central Pharmaceutical Research Institute, Osaka, Japan*
Scott J. Neal • *Department of Biology, Brandeis University, Waltham, MA, USA; National Center for Behavioral Genomics, Brandeis University, Waltham, MA, USA*
Yoshihito Niimura • *Department of Bioinformatics, Medical Research Institute, Tokyo Medical and Dental University, Tokyo, Japan*
Jaime H. Noguez • *Department of Chemistry, University of Florida, Gainesville, FL, USA*
Milos V. Novotny • *Department of Chemistry, Institute for Pheromone Research, Indiana University, Bloomington, IN, USA*
Shannon B. Olsson • *Department of Evolutionary Neuroethology, Max Planck Institute for Chemical Ecology, Jena, Germany*
Bruno Patris • *Centre des Sciences du Goût, CNRS (UMR 6265), Université de Bourgogne, Dijon, France*
Benoist Schaal • *Centre des Sciences du Goût, CNRS (UMR 6265), Université de Bourgogne, Dijon, France*
Piali Sengupta • *Department of Biology, Brandeis University, Waltham, MA, USA; National Center for Behavioral Genomics, Brandeis University, Waltham, MA, USA*
Noam Sobel • *Department of Neurobiology, Weizmann Institute of Science, Rehovot, Israel*
Helena A. Soini • *Department of Chemistry, Institute for Pheromone Research, Indiana University, Bloomington, IN, USA*
Peter W. Sorensen • *Department of Fisheries, Wildlife, & Conservation Biology, University of Minnesota, St. Paul, MN, USA*
Michael Stewart • *National Institute of Water & Atmospheric Research Ltd., Hamilton, New Zealand*
Lisa Stowers • *Department of Cell Biology, The Scripps Research Institute, La Jolla, CA, USA*
Kazushige Touhara • *Department of Applied Biological Chemistry, Graduate School of Agricultural and Life Sciences, The University of Tokyo, Tokyo, Japan; ERATO Touhara Chemosensory Signal Project, JST, The University of Tokyo, Tokyo, Japan*
Yousuke Tsuneoka • *Unit for Affiliative Social Behavior, RIKEN Brain Science Institute, Saitama, Japan*
Jing W. Wang • *Neurobiology Section, Division of Biological Sciences, University of California San Diego, La Jolla, CA, USA*
Pei Sabrina Xu • *Department of Anatomy and Neurobiology, Washington University in St. Louis, St. Louis, MO, USA*

DAISUKE YAMAMOTO • *Department of Developmental Biology and Neurosciences, Tohoku University Graduate School of Life Sciences, Sendai, Japan*
JOANNE Y. YEW • *Temasek Life Sciences Laboratories, Singapore, Singapore; Department of Biological Sciences, National University of Singapore, Singapore*
C. RON YU • *Stowers Institute for Medical Research, Kansas City, MO, USA*
SENMIAO ZHAN • *Department of Molecular Genetics and Microbiology, Duke University Medical Center, Durham, NC, USA; Boston University School of Medicine, Boston, MA, USA*
XINXING ZHANG • *Department of Chemistry, University of Florida, Gainesville, FL, USA*
YUE ZHOU • *Department of Chemistry, University of Florida, Gainesville, FL, USA*
FRANK ZUFALL • *Department of Physiology, School of Medicine, University of Saarland, Homburg, Germany*

Part I

Sampling, Chromatographic Purification, and Chemical Analysis of Pheromone Molecules

Chapter 1

Female Moth Pheromones

Tetsu Ando

Abstract

The sex pheromone, a volatile secreted by a female moth, is stored in the pheromone gland and can be easily extracted with hexane. The extract is effectively analyzed using a gas chromatography combined with an electro-antennogram detector (GC-EAD) and a mass spectrometry (GC-MS), both of which are equipped with a capillary column. GC-EAD analysis indicates the number of pheromone components that have a different chromatographic behavior. The mass spectrum measured by GC-MS suggests the outline of the chemical structure. In addition to a comparison with chemical data of authentic synthetic compounds, micro-chemical reactions reveal a precise structure of the natural pheromone. Finally, the chemical structure is confirmed by field evaluation of the synthetic pheromone.

Key words Insect, Lepidoptera, Electro-antennogram, GC-MS, Diagnostic ion, Chiral HPLC, Mating communication, Male attractant

1 Introduction

Lepidoptera is the second largest insect group, including about 160,000 described species, which are currently divided into 47 supper families and further approximately 130 families (*see* **Note 1**). As a consequence of the diversity of Lepidoptera, interesting species-specific sex pheromone systems are exhibited in this insect group. The quite varied pheromones, which have been identified from female moths of nearly 630 species around the world [1, 2], are classified into groups of Type I (75 %), Type II (15 %), and miscellaneous (10 %) according to their chemical structures [3]. Figure 1 shows the structures of typical pheromone components. Type I pheromones are composed of unsaturated compounds with a C_{10}–C_{18} straight chain and a terminal functional group, such as a hydroxyl, an acetoxyl, or a formyl group. On the other hand, Type II pheromones are composed of unsaturated hydrocarbons and epoxy derivatives with a C_{17}–C_{23} straight chain. The biosynthetic pathways of pheromones in these two types are quite different (*see* **Note 2**).

Kazushige Touhara (ed.), *Pheromone Signaling: Methods and Protocols*, Methods in Molecular Biology, vol. 1068,
DOI 10.1007/978-1-62703-619-1_1,

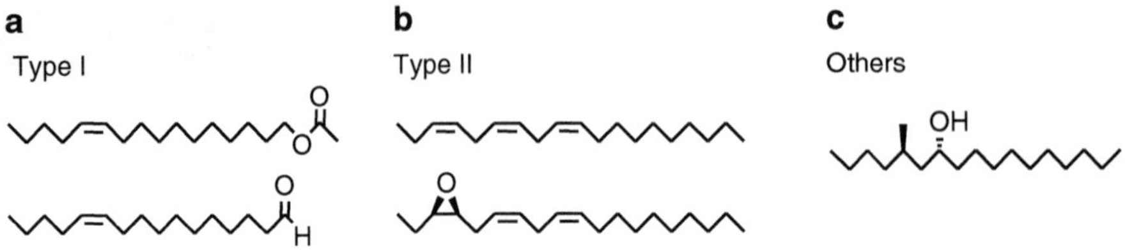

Fig. 1 Representative lepidopteran sex pheromones: (**a**) Type I pheromone of the diamond back moth (*Plutella xylostella*, Plutellidae), (**b**) Type II pheromone of the Japanese giant looper (*Ascotis selenaria cretacea*, Geometridae), and (**c**) a methyl-branched pheromone of a lichen moth (*Miltochrista calamine*, Arctiidae)

Type I pheromones have been identified from the species in various families, such as Crambidae, Tortricidae, and Noctuidae, which include many agricultural pest insects. For example, females of the diamond back moth (*Plutella xylostella*, Plutellidae) secrete a mixture of (*Z*)-11-hexadecenyl acetate and (*Z*)-11-hexadecenal [4]. Each lepidopteran species produces a species-specific pheromone with a difference of the chemical structure, i.e., functional group, carbon chain length, number of double bond(s), unsaturated position and configuration. The diversity of lepidopteran sex pheromones is also generated by blending multiple components. Type II pheromones are mainly produced by the species in highly evolved insect groups, such as Geometridae, Lymantriidae, and Arctiidae. For example, females of the Japanese giant looper (*Ascotis selenaria selenaria*, Geometridae) secrete a mixture of (3*Z*,6*Z*,9*Z*)-3,6,9-nonadecatriene and (3*R*,4*S*)-3,4-epoxy-(6*Z*,9*Z*)-6,9-nonadecadiene [5]. In addition to the unbranched compounds, several novel pheromones with a methyl branch have been identified. Females of a lichen moth (*Miltochrista calamine*, Arctiidae) secrete a (5*R*,7*R*)-5-methylheptadecan-7-ol [6]. Furthermore, sex attractants have been reported for the male moths of about 1,200 species, which were found by screening tests of known synthetic pheromones and their analogues in the field [2].

Since we have extensive knowledge of the chemical communication systems developed in lepidopteran insects, new identification should be planned with reference to previous works carried out with a species taxonomically related to a new target moth. Although sex pheromones play an important role in reproductive isolation, closely related species possibly produce similar pheromone components (*see* **Note 3**). Generally, a pheromone titer is very low (<100 ng/female) (*see* **Note 4**), and mass rearing of pest insects is not easy. Additionally, a considerable number of species are univoltine. In these cases, every experiment has limitations. Male antennae, however, respond to a pheromone quite sensitively, and, thus, not only a major component but also a minor component can be detected using an electrophysiological technique, namely, by measurement with an electro-antennogram (EAG) [7]. Gas chromatography combined with an EAG detector (GC-EAD) is the most useful instrument to identify pheromone candidates in a crude extract of pheromone glands [8]. The sensitivity of the EAD against a pheromone component is higher than that of a flame ionization detector (FID), as shown in Fig. 2. The chemical structure of each EAG-active component can be estimated by the analysis of the mass spectrum measured by GC combined with mass spectrometry (GC-MS). Epoxy components in Type II pheromones include chiral centers. Even in an unsuccessful case of optical resolution by chiral GC columns, enantiomeric separation of the epoxides has been accomplished by high-performance liquid chromatography (HPLC) with a chiral column [3]. Finally, field evaluation of synthetic pheromone candidates reveals the pheromone with an attractive activity for males of the target species. This chapter deals with the above techniques and some related experimental procedures.

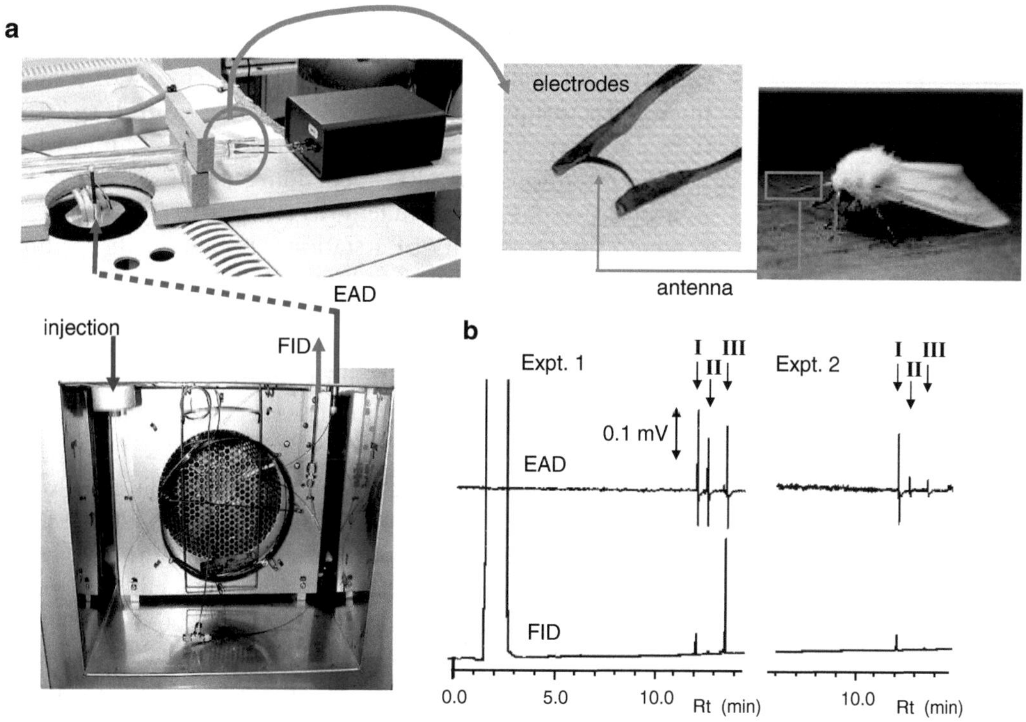

Fig. 2 GC-EAD analysis: (**a**) an instrument equipped with a capillary column and a handmade EAD (**b**) data of three synthetic epoxides stimulating a male antenna of the mulberry looper (*Menophra atrilineata*, Geometridae). Compounds; **I** = 9,10-epoxy-(*Z*)-6-octadecene (a minor pheromone component), **II** = 9,10-epoxy-(3*Z*,6*Z*)-3,6-octadecadiene (a major pheromone component), and **III** = 9,10-epoxy-(3*Z*,6*Z*)-3,6-nonadecadiene (a pheromone homologue). Expt. 1; **I** (10 ng), **II** (1 ng), and **III** (100 ng). Expt. 2; **I** (10 ng), **II** (0.001 ng), and **III** (1 ng)

2 Materials

2.1 Collection

1. Organic solvents: Hexane and diethyl ether.
2. Solid phase micro-extraction (SPME) fiber: A fused-silica fiber coated with poly(dimethylsiloxane).
3. Porous polymer beads: Porapak Q or Tenax TM.

2.2 Bioassay

1. Flask (100 ml).
2. Filter paper.
3. Red light.
4. Wind tunnel: An acrylic tube (0.3 m ID × 2 m) with a steady airflow of 0.4 m/s.
5. Sticky board trap with a 30 × 27 cm bottom plate with a roof.
6. Dispenser of synthetic pheromone: Rubber septum (OD 8 mm).

2.3 GC

1. GC instrument: Agilent 7890A GC system equipped with FID.
2. Gas: He (carrier gas, >99.9995 % purity), H_2 (>99.9995 % purity) and dry air (>99.999 % purity) for FID, and N_2 (make-up gas, >99.9995 % purity).
3. Capillary columns: A low polar column such as DB-5 and a high polar column such as DB-23 (0.25 mm ID × 30 m, 0.25 μm film). A compound is eluted from the column with a specific retention time (Rt), depending on its volatility (i.e., molecular weight and chain length) and polarity. The Rt is also changed by the polarity of a stationary phase in the column. Therefore, analyses with both low and high polar columns are informative. By comparing the Rts measured by two columns, the number of double bonds and conjugation of two double bonds are indicated. Separation of geometrical and positional isomers of unsaturated compounds is successfully accomplished by the high polar column.
4. EAD: The effluent from a capillary column is split equally into two lines, one leading to FID, and the other, to the EAD (Fig. 2a). An antenna is excised at the base from the male moth, and a few distal segments are cut off. Insect saline is used for the electrodes. Each end of the antenna is attached to a droplet of a saline solution on an electrode of the EAD device so that the sensilla faces the airflow from the GC. The rate of the airflow carrying eluted compounds to the antenna is about 8 cm/s.

2.4 GC-MS

1. GC-MS instrument: Electron impact GC-MS is conducted with a quadrupole mass spectrometer (Agilent 5973 mass spectrometer system) or a magnetic sector mass spectrometer. The carrier gas is helium, and the ionization voltage is 70 eV.
2. He gas: Carrier gas (>99.9995 % purity).

2.5 Chiral HPLC

1. HPLC instrument: Jasco LC-2000 series equipped with ultraviolet (UV) and refractive index (RI) detectors.
2. Solvents: Hexane, 2-propanol, methanol, and water.
3. Chiral columns: A normal-phase column such as Chiralpak AD and Chiralpak AS, and a reversed-phase column such as Chiralcel OJ-R (4.6 mm ID×25 cm). In the case of epoxy pheromones, no chiral columns have the ability to separate enantiomers of many epoxides universally [3]. A reasonable resolution has been accomplished by selection of the best chiral column after trial and error.

2.6 Reagents for Purification and Chemical Reaction

1. Florisil.
2. Organic solvents: Hexane, diethyl ether, CH_2Cl_2, ethanol, diethylene glycol, THF.
3. Dimethyl disulfide (DMDS).
4. 4-Methyl-1,2,4-triazoline-3,5-dione (MTAD).
5. Reduction: Hydrazine hydrate, H_2O_2, KOH, $LiAlD_4$.
6. Mesylation: Methanesulfonyl chloride, triethylamine, *N*,*N*-dimethyl-4-aminopyridine.

3 Methods

3.1 Pheromone Collection

1. Pheromone gland extraction: Sex pheromone components are secreted from a pheromone gland, which is located at an abdominal tip (eighth and ninth abdominal segments) of a virgin female in many lepidopteran species. With reference to observation of a mating behavior, the gland of a target species is cut off at a calling time and then immersed in hexane (50 μl/tip) for 15 min to extract the pheromone components that are stored in the gland. It is important to avoid contamination of obstructive lipids extracted with a polar solvent for a long-time immersion. This solvent extraction is the most convenient method for pheromone collection.
2. Absorption by SPME: SPME is used to collect airborne pheromone components emitted by a female. The fiber is placed a few mm from the abdominal gland of a calling female for 2–3 h [9]. Collection is also accomplished by gently rubbing the fiber on a tegument of a glandular area for 5 min [10]. SPME samples are injected by exposing the fiber in the hot injector of GC for 5 min before the start of a chromatogram.
3. Collection by an adsorbent tube: A small glass tube is filled with adsorbent (porous polymer beads, 200 mg), which is sufficiently washed with diethyl ether [11]. To collect an airborne pheromone by the adsorbent, the tube is connected to a gas

scrubber bottle including female moths and supplied air through an activated charcoal filter. The adsorbed materials are eluted from the tube with diethyl ether.

3.2 Bioassay

1. Laboratory bioassays: In the case of nocturnal species, responses of the males placed in a flask are observed in darkness under a red light (10 lx). A piece of filter paper is loaded with a natural pheromone (less than one female equivalent (1 FE)) or other test materials dissolved in hexane, and the biological activity is examined after evaporation of the solvent. Mating behavior is divided into four categories: vigorous antennal swing, orientation flight, mating dance, and abdominal contact with the pheromone source [12].
2. Wind tunnel: Male responses are also observed in a wind tunnel. The filter paper with a test material is hung 15 cm above the floor and 15 cm upwind of the wind tunnel. Males are released at the leeward side, and the following male behaviors are recorded for 3 min: upwind flight, approach to the source, contact with the source, and attempt of copulation [11].
3. GC-EAD: Differently from the case of mating behavior, a male antenna responds to each constituent of a multicomponent pheromone. Injection of a crude pheromone extract (1 FE) indicates the number of included components and approximate length of their carbon chains. A natural component (>10 pg) can be sensitively detected by a male antenna, and structure-related analogues of a pheromone also stimulate the antenna as shown in Fig. 2b. The double-bond positions of a dienyl pheromone can be presumed by the EAG-activities of a series of synthetic monoenyl compounds [13].

3.3 Partial Purification by a Mini-Column

If a crude extract includes many impurities, column chromatography of the extract (100 FE) is conducted on Florisil (500 mg) using 5, 10, 20, and 30 % ether in hexane (3 ml each) as an eluent. Acetates and aldehydes of Type I pheromones are recovered in the 10 % ether fraction, and alcohols are recovered in the 30 % ether fraction [14].

3.4 GC-MS Analysis

A mass spectrum of each EAG-active component (>10 ng/gland) is obtained by an injection of the crude extract (1–10 FE) (*see* **Note 4**). The EI-MS analysis usually shows a molecular ion (M^+) and characteristic fragment ions indicating a summary of the chemical structure, i.e., chain length, functional group, and degree of unsaturation [15]. M^+ of monoenyl compounds of the Type I pheromone is scarcely detected, but ions at m/z M-18 of alcohols and aldehydes and ions at m/z M-60 and 61 of acetates indicate the structure. It is not easy to determine a double-bond position by the mass spectrum, except for some conjugated dienyl compounds that show diagnostic fragment ions reflecting the positions (*see* Fig. 3 and **Note 5**). After determination of the position (see next section), configuration of the double bond is determined by comparing with Rts of authentic geometrical isomers. The positions of the epoxy ring and double bond(s) of Type II pheromones can be estimated by some diagnostic fragment ions, which have been published with many spectral data (*see* **Note 6**). Methyl-branched hydrocarbons show characteristic secondary fragment ions indicating the position of the branch [3]. In the case of branched alkenes, ketones, and alcohols, positional isomers show similar mass spectra [15].

3.5 Determination of Double-Bond Positions

1. DMDS derivatization (Fig. 4a): The double-bond position of a monounsaturated Type I pheromone is revealed by derivatization with DMDS [16]. A crude extract including a monoene (about 250 ng, more than 5 FE) is treated with DMDS (50 μl) and a diethyl ether solution of iodine (60 mg/ml, 50 μl) after removal of the solvent. The mixture is warmed at 40 °C overnight, and a 5 % sodium thiosulfate solution (0.5 ml) is added to it. Products are extracted with hexane and analyzed by GC-MS. The DMDS adduct shows two characteristic fragment ions produced by cleavage of the bond between the sulfur-substituted carbons indicating the position of the original double bond.

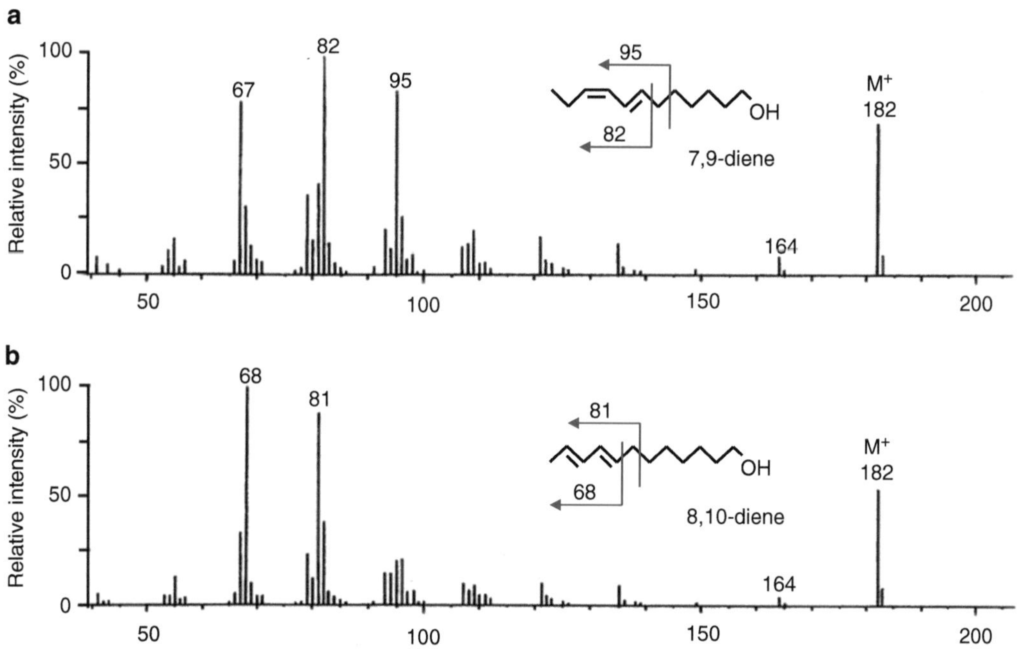

Fig. 3 Mass spectra of C_{12} dienyl pheromones indicating two characteristic fragmentations: (**a**) (7*E*,9*Z*)-7,9-dodecadien-1-ol of the grapevine moth (*Lobesia botrana*, Olethreutinae) and (**b**) (8*E*,10*E*)-8,10-dodecadien-1-ol of the codling moth (*Cydia pomonella*, Olethreutinae)

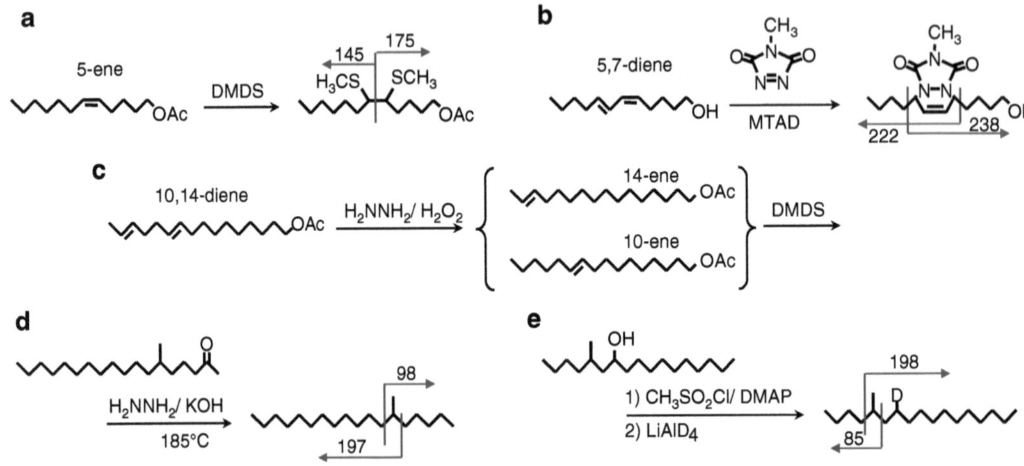

Fig. 4 Chemical derivatization of moth pheromones and characteristic fragment ions of the derivatives: (**a**) a DMDS adduct of 5-dodecenyl acetate, (**b**) an MTAD adduct of 5,7-doecadien-1-ol, (**c**) partial diimide reduction, (**d**) Wolff–Kishner reduction of a methyl-branched 2-ketone, and (**e**) mesylation and $LiAlD_4$ reduction of a methyl-branched secondary alcohol

2. MTAD derivatization (Fig. 4b): The double-bond positions of a conjugated diene are determined by derivatization with MTAD [17]. A crude pheromone (about 0.5 μg) dissolved in CH_2Cl_2 (10 μl) is treated with a CH_2Cl_2 solution of MTAD (10 mg/ml, 40 μl) for 30 min at room temperature. After changing the solvent to hexane, products are analyzed by GC-MS. The MTAD adduct, which is formed by Diels–Alder cycloaddition between the conjugated diene and the dienophile, gives mass spectrum including abundant fragment ions that are diagnostic for the unsaturated positions in the parent compound.
3. Diimide reduction (Fig. 4c): The unsaturated positions of an unconjugated polyene are confirmed by combination of a partial diimide reduction and DMDS derivatization [18]. A crude pheromone (about 1 μg) dissolved in hexane (10 μl) is treated with 10 μl each of a hydrazine and a hydrogen peroxide solution in ethanol (N_2H_4: 0.3 ml of hydrazine hydrate in 10 ml of ethanol, H_2O_2: 0.04 ml of 31 % H_2O_2 in 10 ml of ethanol). The mixture is maintained at 55 °C for 1 h for partial hydrogenation. Since the double bonds of a polyene are unselectively reduced to make a mixture of monoenyl derivatives, the crude products are directly treated with DMDS and adducts are analyze by GC-MS to determine the double-bond positions.

3.6 Determination of Branched-Methyl Positions

1. Wolff–Kishner reduction (Fig. 4d): Carbonyl compounds are converted into hydrocarbons to reveal carbon skeletons [19]. A crude pheromone (about 1 μg) dissolved in diethylene glycol (0.2 ml) is treated with hydrazine hydrate (1 μg) and KOH

(0.2 mg) and heated at 185 °C for 4 h. After cooling, water (1 ml) is added to the mixture. Crude products are extracted with hexane and analyzed by GC-MS.

2. $LiAlD_4$ reduction (Fig. 4e): Secondary alcohols are also converted into hydrocarbons [6]. A crude pheromone (about 1 μg) dissolved in CH_2Cl_2 (0.1 ml) is mixed with triethylamine (10 mg), *N,N*-dimethyl-4-aminopyridine (1 mg), and methanesulfonyl chloride (20 mg) and stirred at 0 °C for 2 h and at room temperature for 1 day. After the usual work-up, the produced mesylate is dissolved in dry THF (0.1 ml) and treated with $LiAlD_4$ (5 mg) under an argon atmosphere. The reaction mixture is acidified with 1N HCl (2 ml) after 3 h of stirring and then extracted with hexane. The extract is analyzed by GC-MS.

3.7 Determination of Stereochemistry

Absolute configuration of the pheromone with a chiral center is assigned by the biological activities of both enantiomers, which have been synthesized by starting from a chiral compound or by a chiral reaction. The stereochemistry of a chiral natural pheromone is determined by comparison with the chromatographic behaviors of synthetic enantiomers on a chiral GC or HPLC column. In the case of HPLC analysis, compounds with a conjugated system are highly sensitive to a UV detector. Otherwise Rts of the two synthetic enantiomers must be measured by an RI detector, and then the natural pheromone is analyzed. After the injection of the pheromone, eluted materials are collected every 30 s around the Rt and quantitatively analyzed by GC-MS.

3.8 Field Evaluation of Synthetic Pheromone

Each synthetic pheromone candidate (1 mg) dissolved in hexane (100 μl) is incorporated into a rubber septum, which is used as a dispenser. The lure is placed at the center of a sticky board trap. Traps are hung separately in the field at intervals of 10 and 1.5 m height. Captured males are counted weekly.

4 Notes

1. (Taxonomy of Moths): Nearly 99 % of lepidopteran species belong to Ditrysia, which has two separate genital openings (a copulatory orifice and an ovipore). Ditrysia includes two superfamilies of butterflies and 31 superfamilies of moths. For the most part, female sex pheromones have been identified from the species of the following large superfamilies; Tortricoidea (167 species), Pyraloidea (87 species), Geometroidea (48 species), and Noctuoidea (174 species) [2].
2. (Biosynthesis of Moth Pheromones): The variation of the chemical structures results from differences in both a starting

material and a biosynthetic enzyme system [3]. The Type I compounds are biosynthesized from saturated fatty acids produced by de novo synthesis starting from acetyl-CoA. The Type II compounds are biosynthesized starting from dietary linoleic and linolenic acids, which are changed into unsaturated hydrocarbons (6,9-dienes and 3,6,9-trienes) outside of the pheromone gland (in oenocytes). Some females are able to introduce an additional double bond and produce further modified polyenes, such as 4,6,9-trienes and 1,3,6,9-tetraenes. The hydrocarbons are transported to the gland via hemolymph and emitted into the atmosphere directly and/or after epoxidation.

3. (Common Pheromone Structures in Related Species): It is expected that taxonomically related species that have developed from a common ancestor also exhibit similarity at the pheromone level [3]. The family of Tortricidae in Tortricoidea is divided into subfamilies, such as Olethreutinae and Tortricinae. The species in Olethreutinae usually secrete C_{12} chain compounds unsaturated at the 8- or 9-position. On the other hand, the species in Tortricinae utilize C_{14} chain compounds mainly unsaturated at the 11-position. Noctuoidea, the largest and highly evolved superfamily, is divided into several families, such as Noctuidae, Lymantridae, and Arctiidae. In the family of Noctuidae, species in one group (Trifinae) produce Type I pheromones, while species in the other group (Quadrifinae) produce Type II pheromones. Many species of Lymantridae and Arctiidae also utilize Type II pheromones. Although no sex pheromone has been reported from female butterflies, a chemical cue is an important factor for the mating communication of diurnal moths, such as clearwing moths in Sesiidae. Almost all of the Sesiidae pheromones are composed of 2,13- and 3,13-dienyl compounds with a C_{18} chain.
4. (Pheromone Titer): The titer fluctuates with a circadian rhythm, and the most abundant pheromone is usually detected at a calling time. The pheromone content is also different depending on the species. Although a quantitative correlation between the size of a female moth and its pheromone titer is not always observed, the gland of a large insect, such as a Noctuidae species, usually contains around 10–100 ng of a main pheromone component. On the contrary, the pheromone content in micro-lepidopteran species, such as leafminer moths, is very limited, sometimes lower than a 1-pg order.
5. (Diagnostic Ions of Conjugated Dienes): Compounds unsaturated at a side of the terminal methyl group [$\omega n,\omega(n+2)$-diene, $n=1$–5] show characteristic fragment ions, $[CH_2CH{=}CHCH{=}CHC_{n-1}H_{2n-1}]^+$ and $[CH_2{=}CHCH{=}CHC_{n-1}H_{2n-1}]^+$, reflecting the position of unsaturations [15]. For example, the positional

isomers of C_{12} dienyl compounds are distinguished by the following fragment ions: *m/z* 123 and 110 of 5,7-dienes (ω5,ω7-diene), *m/z* 109 and 96 of 6,8-dienes (ω4,ω6-diene), *m/z* 95 and 82 of 7,9-dienes (ω3,ω5-diene), *m/z* 81 and 68 of 8,10-dienes (ω2,ω4-diene), and *m/z* 67 and 54 of 9,11-dienes (ω1,ω3-dienes) (Fig. 3). In particular, ions at *m/z* 67 of ω1,ω3-dienes, *m/z* 81 or 68 of ω2,ω4-dienes, and *m/z* 95 or 82 of ω3,ω5-dienes are recorded as the characteristic base peaks. Although it is noteworthy that 4,6-dienal shows a characteristic base peak at *m/z* 84 [12], the spectral patterns of the dienes unsaturated at a side of the functional group are similar.

6. (Diagnostic Ions of Type II Pheromones): While a characteristic fragment ion at *m/z* 79, $[H(CH{=}CH)_3]^+$, is detected for many compounds with a homoconjugated dienyl structure, each pheromone component can be identified by some diagnostic ions, which reflect the positions of a double bond and an epoxy ring as follows (underlined numbers mean base peaks) [15]; ions at *m/z* M-56, 108 and <u>79</u> for 3,6,9-trienes, *m/z* M-85, 163, 110 and <u>79</u> for 6,9,11-trienes, *m/z* M-98, 150 and <u>79</u> for 6,9,12-trienes, *m/z* M-54, 106, 91 and <u>79</u> for 1,3,6,9-tetraenes, *m/z* M-82, 108 and <u>79</u> for 3,6,9,11-tetraenes, *m/z* M-96, 148 and <u>79</u> for 3,6,9,12-tetraenes, *m/z* M-58, M-72 and <u>79</u> for 3,4-epoxy-6,9-dienes, *m/z* M-69, M-83, 111, 97 and <u>67</u> for 6,7-epoxy-3,9-dienes, *m/z* M-109, M-123, 122, 108 and <u>79</u> for 9,10-epoxy-3,6-dienes, and *m/z* M-107, M-121, 120, <u>106</u> and 79 for 9,10-epoxy-1,3,6-trienes.

References

1. El-Sayed AM. Internet database 2012. http://www.pherobase.com/
2. Ando T. Internet database 2012. http://www.tuat.ac.jp/~antetsu/LepiPheroList.htm
3. Ando T, Inomata S, Yamamoto M (2004) Lepidopteran sex pheromones. Top Curr Chem 239:51–96
4. Tamaki Y, Kawasaki K, Yamada H, Koshihara T, Osaki N, Ando T, Yoshida S, Kakinohana H (1977) (*Z*)-11-Hexadecenal and (*Z*)-11-hexadecenyl acetates: sex-pheromone components of the diamondback moth (Lepidoptera: Plutellidae). Appl Entomol Zool 12:208–210
5. Ando T, Ohtani K, Yamamoto M, Miyamoto T, Qin X-R, Witjaksono (1997) Sex pheromone of Japanese giant looper, *Ascotis selenaria cretacea*: identification and field tests. J Chem Ecol 23:2413–2423
6. Yamakawa R, Kiyota R, Taguri T, Ando T (2011) (5*R*,7*R*)-5-Methylheptadecan-7-ol: a novel sex pheromone component produced by a female lichen moth, *Miltachrista calamine*, in the family Arctiidae. Tetrahedron Lett 52:5808–5811
7. Schnaider D (1969) Insect olfaction: deciphering system for chemical messages. Science 163:1031–1037
8. Arn H, Stadler E, Rauscher S (1975) The electroantennographic detector—a selective and sensitive tool in the gas chromatographic analysis of insect pheromone. Z Naturforsch 30c:722–725
9. Mozuraitis R, Borg-Karlson A-K, Buda V, Ivinskis P (1999) Sex pheromone of the spotted tentiform leafminer moth *Phyllonorycter blancardella* (Fabr.) (Lep., Gracillariidae). J Appl Entomol 123:603–606
10. Frérot B, Malosse C, Cain A-H (1997) Solid-phase microextraction (SPME): a new tool in pheromone identification in Lepidoptera. J High Resolut Chromatogr 20:340–342
11. Fujii T, Nakano R, Takubo Y, Qian S, Yamakawa R, Ando T, Ishikawa Y (2010) Female sex pheromone of a lichen moth *Eilema japonica* (Arctiidae, Lithosiinae). J Insect Physiol 56:1986–1991
12. Naka H, Vang LV, Inomata S, Ando T, Kimura T, Honda H, Tsuchida K, Sakurai H (2003)

Sex pheromone of the persimmon fruit moth, *Stathmopoda masinissa*: identification and laboratory bioassay of (4*E*,6*Z*)-4,6-hexadecadien-1-ol derivatives. J Chem Ecol 29:2447–2459
13. Roelofs W, Comeau A, Hill GM (1971) Sex attractant of the codling moth: characterization with electroantennogram technique. Science 174:297–299
14. Do ND, Ohbayashi K, Naka H, Nakada K, Ando T (2011) Identification and field evaluation of sex pheromone components of the pear barkminer moth, *Spulerina astaurota*. J Chem Ecol 37:1222–1230
15. Ando T, Yamakawa R (2011) Analyses of lepidopteran sex pheromones by mass spectrometry. Trends Anal Chem 30:990–1002
16. Buser H-R, Arn H, Guerin P, Rauscher S (1983) Determination of double bond position in mono-unsaturated acetates by mass spectrometry of dimethyl disulfide adducts. Anal Chem 55:818–822
17. Young DC, Vouros P, Holick MF (1990) Gas chromatography-mass spectrometry of conjugated dienes by derivatization with 4-methyl-1,2,4-triazoline-3,5-dione. J Chromatogr 522:295–302
18. Wakamura S, Ohno S, Arakaki N, Kohama T, Haraguchi D, Yasui H (2010) Identification and field activity of the sex pheromone component of the sweetpotato vine borer moth *Omphisa anastomosalis* (Lepidoptera: Crambidae). Appl Entomol Zool 45:635–640
19. Yamamoto M, Kamata T, Do ND, Adachi Y, Kinjo M, Ando T (2007) A novel lepidopteran sex pheromone produced by females of a Lithosiinae species, *Lyclene dharma dharma*, in the family of Arctiidae. Biosci Biotechnol Biochem 71:2860–2863

Chapter 2

Pheromones in the Fruit Fly

Jacqueline S.R. Chin and Joanne Y. Yew

Abstract

The identification of pheromones (chemical communication cues) is critical to our understanding of complex social behavior in insects and other animals. In this chapter, we describe analytical methods for the purification of lipid pheromones by thin layer chromatography and the quantification and determination of their elemental composition by mass spectrometry.

Key words Pheromones, *Drosophila melanogaster*, Lipids, Hydrocarbons, Thin layer chromatography, Mass spectrometry

1 Introduction

Insect pheromones can consist of a mixture of compounds. Many of these signaling molecules are lipids composed of branched and straight-chain alkanes and alkenes and oxygen-containing hydrocarbons. Lipid pheromones are usually found on the surface of the insect cuticle or are secreted by specialized glands [1, 2]. In order to study pheromone profiles, the cuticle or glands are chemically extracted and the extract is analyzed using mass spectrometry or nuclear magnetic resonance spectroscopy. One disadvantage of whole extract analysis is that less abundant compounds are usually masked by more abundant components. To address this shortcoming, chromatographic methods are used to separate complex mixtures into simpler fractions and allow for a more thorough analysis.

One chromatographic method, thin layer chromatography (TLC), has been proven to be useful in identifying novel lipid pheromones. TLC plates consist of glass or aluminum coated with an adsorbent layer of silica gel. Using TLC, components of the chemical extract can be separated into discrete fractions according to hydrophobicity. Different molecules will have different degrees of polarity depending on their molecular weight and functional groups. The *Drosophila melanogaster* male sex pheromone, *cis-vaccenyl* acetate (cVA), was discovered by Brieger and Butterworth

Kazushige Touhara (ed.), *Pheromone Signaling: Methods and Protocols*, Methods in Molecular Biology, vol. 1068,
DOI 10.1007/978-1-62703-619-1_2, © Springer Science+Business Media, LLC 2013

[3, 4] using TLC together with gas-chromatography mass spectrometry (GCMS). Comparing *D. melanogaster* extracts from males and virgin females, a male-specific spot on the TLC plate was identified. A similar strategy was also used to identify a female aphrodisiac, heptacosadiene [5]. In this case, a behavioral assay was used to test individual fractions for their ability to induce high levels of courtship from males.

Recent technological advances have been useful for detecting compounds directly from the insect cuticle, forgoing chemical extraction altogether. These techniques allow for spatial expression of a compound to be preserved and in some cases, for the insect to remain alive during sampling. One method makes use of solid phase micro-extraction (SPME), wherein flies are probed with a thin fiber coated with adsorbent compounds and the fiber is analyzed with GCMS [6, 7]. Recent work using SPME led to the identification of 17 new *Drosophila* hydrocarbons [8].

In addition to improved methods of sampling, developments in mass spectrometry instrumentation have also been important in revealing new chemical cues. Ultraviolet laser desorption/ionization orthogonal time-of-flight mass spectrometry (UV-LDI-o-TOF-MS) allows detection of novel polar compounds from the cuticles of individual insects [9]. Use of this method led to the discovery of a male *D. melanogaster* pheromone, CH503, a long-lived inhibitor of male courtship [10]. In addition, a number of novel triacylglyceride molecules were found in the anogenital region of male *Drosophila mojavensis* and *Drosophila arizonae* [11]. These newly identified compounds were undetected by previous GCMS analysis due to the high molecular weight and polarity of the molecules. Direct Analysis in Real Time (DART) MS is another recently introduced method that has been useful for the rapid analysis of cuticular lipids [12, 13]. Both UV-LDI MS and DART MS provide high sensitivity detection of polar and apolar lipids with higher molecular weight (m/z 500–1,000). However, unlike GCMS, neither method is capable of detecting saturated hydrocarbons nor elucidating branched vs. unbranched structures under standard conditions. For the most comprehensive profile of chemical compounds on the cuticular surface, multiple analytical methods are necessary.

2 Materials

2.1 *Drosophila* Cuticular Hydrocarbon Extraction and Semiquantitative Analysis

1. Hexane spiked with 10 μg/mL hexacosane (Sigma-Aldrich; St. Louis, Missouri, USA).
2. Gas-tight glass syringe (Hamilton, Reno, Nevada, USA).
3. Wash solution: chloroform–methanol (66:33; per vol).

4. 1.8 mL glass vials with Teflon-lined caps (s/n 224740; Wheaton, Millville, New Jersey, USA).
5. Nitrogen evaporator or rotary evaporator.

2.2 Separation of Cuticular Lipids by Thin Layer Chromatography

1. Silica gel 60 TLC plates (Merck, Darmstadt, Germany).
2. Flat bottom TLC chamber for 20 × 20 cm plates, with glass lid (Camag, Muttenz, Switzerland).
3. TLC solvent systems:
   (a) For neutral lipid analysis (triacylglycerides, alkanes, alkenes, fatty acid methyl esters, acetate esters such as cVA): hexane–diethyl ether–acetic acid (90:9:1; per vol). *See* Fig. 1a. Alternative system: petroleum/ether (70:30; per vol).
   (b) For polar lipid analysis (secondary and primary alcohols, oxoaldehydes, free fatty acids, acetoxy esters such as CH503): hexane–diethyl ether–acetic acid (66:33:1; per vol). *See* Fig. 1b. Alternative solvents: hexane–methanol–diethyl ether (85:10:5; per vol) or ether–petroleum (70:30; per vol).

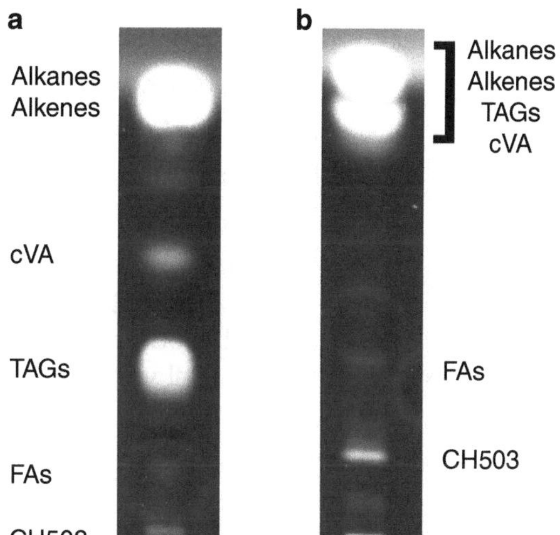

Fig. 1 A comparison of solvent systems for TLC separation of cuticular extract from Drosophila. Both plates were stained with the neutral lipid dye primuline (0.02 % in 20 % acetone) and visualized under UV illumination. Extract from approximately 50 flies (a mixture of males and females) is loaded onto each lane. (**a**) Hexane–diethyl ether–acetic acid (90:9:1; per vol) provides separation of neutral lipids. Alkanes, alkenes, triacylglycerides (TAGs), and the male sex pheromone cVA are clearly resolved, while more polar components of the cuticle such as fatty acids (FAs) and CH503 only moved a short distance from the origin. (**b**) Hexane–diethyl ether–acetic acid (66:33:1; per vol) provides separation of polar lipids. Alkanes, alkenes, TAGs, and cVA are no longer separated, but CH503 and FAs are clearly resolved

4. Lipid and pheromone standards can be obtained from Sigma-Aldrich, Cayman Chemical Company (Ann Arbor, Michigan, USA), Pherobank (Wageningen, Netherlands), and Avanti Polar Lipids (Alabaster, Alabama, USA).
5. Chromatography paper (Whatman, Springfield Mill, UK).
6. Sample applicator (Linomat IV; Camag, Muttenz, Switzerland) or microsyringe (Hamilton).
7. Staining reagents:
   (a) Primuline (Sigma-Aldrich): 0.02 % in 20 % acetone in aerosol spray or shallow glass staining dish.
   (b) Sulfuric acid charring: 10 % sulfuric acid in aerosol spray.
   (c) Iodine crystals (Sigma-Aldrich): sealed glass container containing a few crystals of iodine.
   (d) Water: Milli Q-purified water in aerosol spray.
8. UV-light for visualization of primuline stain.
9. Phosphorimager imager for quantification of lipid signals from TLC plates.

2.3 Purification of Compounds from TLC Plates

1. Clean razor or scalpel blade.
2. Glass cutter or diamond knife.
3. Weighing paper.
4. Narrow glass funnel.
5. Chloroform–methanol (66:33; per vol).
6. Flash chromatography column:
   (a) Stuff a small amount of glass wool fiber (Pall Corporation, Ann Arbor, Michigan, USA) into the tapered end of a disposable borosilicate glass Pasteur pipette (15 cm length). The wool will serve as a stopper.
   (b) Suspend the pipette vertically using a ring stand and clamp.
7. Nitrogen evaporator.

2.4 Analysis of Extracts by GCMS

1. 0.1 mL micro inserts with spring bottom for GCMS vials (Supelco, St. Louis, Missouri, USA).
2. Glass syringe (Hamilton).
3. Lipid and pheromone standards (*see* Subheading 2.2).
4. Gas chromatography mass spectrometer. We use a GCMS QP2010 system from Shimadzu (Kyoto, Japan).

2.5 Analysis of Intact Insects by UV-LDI MS

1. Double-sided tape. We use self-adhesive tabs from Plano (Plano GmbH, Wetzlar, Germany).
2. Glass coverslips (22 × 22 mm, VFM, Wales, UK).
3. Fine stainless steel forceps.

4. External calibrants consisting of synthetic peptide standards such as Substance P, Bradykinin, Renin substrate (all are available from Sigma-Aldrich). Calibrants are used in 1 mM concentration, dissolved in ddH_2O.
5. Matrix for analysis of calibrants: 10 mg/mL 2,5-Dihydroxybenzoic acid (DHB; Sigma-Aldrich) in 70 % methanol–water.
6. Time-of-flight mass spectrometer with a MALDI source and equipped with a UV-laser (either N_2 or Nd:YAG). The ion source must be able to provide a 1–3 mbar buffer gas environment. We use an orthogonal time-of-flight MS with a modified oMALDI2 source (ABI Qstar Elite, Framingham, MA, USA).

2.6 Analysis of TLC Plates by UV-LDI MS

1. Double-sided tape. We use self-adhesive tabs from Plano (Plano GmbH, Wetzlar, Germany).
2. Aluminum-backed TLC plates (Sigma-Aldrich).
3. External calibrants consisting of synthetic peptide standards such as Substance P, Bradykinin, Renin substrate (all are available from Sigma-Aldrich). Calibrants are used in 1 mM concentration, dissolved in ddH_2O.
4. Matrix for analysis of calibrants and TLC plates:
   (a) 10 mg/mL 2,5-Dihydroxybenzoic acid (DHB; Sigma-Aldrich) in 70 % methanol–water.
   (b) Graphite pencil.
5. Time-of-flight mass spectrometer with a MALDI source and equipped with a UV-laser (either N_2 or Nd:YAG). The ion source must be able to provide a 1–3 mbar buffer gas environment. We use an orthogonal time-of-flight MS with a modified oMALDI2 source (ABI Qstar Elite, Framingham, MA, USA).

2.7 Analysis of Intact Insects by DART MS

1. Fine stainless steel forceps.
2. Undiluted polyethylene glycol (Sigma-Aldrich).
3. 5 M ammonium acetate solution in water.
4. Melting point tubes, 1.0–2.0 mm OD (World Precision Instruments, Sarasota, Florida, USA).
5. Atmospheric pressure ionization time-of-flight mass spectrometer (AccuTOF-DART; JEOL USA, Inc., Peabody, MA, USA) equipped with a DART interface (IonSense Inc., Saugus, MA, USA).

3 Methods

Carry out all procedures at room temperature unless otherwise stated.

Use gas-tight glass syringes for transfer of solvents.

3.1 *Drosophila* Cuticular Lipid Extraction and Semiquantitative Analysis

1. Anesthetize flies with cold or CO_2 and place in a glass vial with a Teflon cap. For quantitative analysis by GCMS, prepare five replicate sets for each genotype (*see* **Notes 1** and **2**). Place vials on ice so that the flies remain anesthetized.
2. Using a glass syringe, add enough hexane to cover the flies (*see* **Note 3**). For quantitative analysis, add 120 μL of hexane with hexacosane standard into each vial.
3. Cap the vial and incubate at room temperature for 20 min.
4. Transfer all of the solvent into a new vial (*see* **Notes 4** and **5**). For quantitative analysis, transfer 100 μL of the solvent from each vial. Only a portion of the total extraction solvent is removed in order to ensure consistent volumes from sample to sample.
5. Add additional hexane to quickly rinse residual lipids and pool together with the first extraction. This step is omitted for quantitative analysis.
6. Evaporate the extract under a gentle stream of nitrogen (*see* **Note 6**). If a nitrogen stream is not available, the solvent can be evaporated by leaving the vials uncapped in the fume hood for 4–6 h at RT.
7. Store extract at −20 °C until analysis.

3.2 Separation of Cuticular Lipids by Thin Layer Chromatography

1. A tank with dimensions (17.5 cm × 16.0 cm × 6.2 cm) can fit two 10 × 10 cm TLC plates at once.
2. Clean the inside of the tank with methanol and allow the tank to dry.
3. Prepare 100 mL of the solvent system.
4. Place a sheet of filter paper approximately 30 cm × 15 cm into the tank in order to allow the tank's environment to equilibrate. Seal the tank lid with silicone gel in order to prevent evaporation of solvent. Place a heavy weight on the lid. Allow equilibration for at least 5 h (*see* **Note 7**).
5. While waiting for the tank to equilibrate, prepare the TLC plate: Redissolve the extract by adding 10 μL of hexane for every 50 flies into the vial. To load the extract on the TLC plate, use an automated system such as the Linomat (Camag). Alternatively, use a pencil to draw a line across about 1.5 cm above the bottom of the plate. Use a ruler to measure and mark points at even intervals with a pencil. Using a capillary or glass syringe, dot 10 μL of the extract on a marked point, allowing the hexane to dry at intervals in order to prevent the size of the dot from spreading too much (*see* **Note 8**).
6. Place the plate into solvent in tank, taking care that the edges of the plate touch the bottom of the tank evenly. Cover tank and replace the weight. Allow solvent front to reach approximately

1 cm from top of the plate. This should take about 14–20 min (*see* **Note 9**).

7. Remove plate. Allow the plate to dry in the fumehood for at least 15 min (*see* **Note 10**).
8. Stain the plate with any of the following methods:
   (a) Spray with primuline using aerosol spraying device or by dipping into a solution of primuline for approximately 20 s. Let dry for 1 h. Bands can be visualized under UV light (*see* **Note 11**). Relative quantification can be performed using a phosphorimager to measure the brightness of bands.
   (b) Spray with sulfuric acid. Visualize by placing on a thermal plate heated to 100 °C for 10 min. Bands will appear as dark brown or black spots.
   (c) Iodine crystals can also be used to visualize lipid bands and is generally considered to be nondestructive. Place the TLC plate in a sealed chamber with a few crystals of potassium iodine. After 30–60 min, dark brown lipid stains can be observed. Once the plate is removed from the container, the position of these bands should be marked quickly with a pencil since the stain fades rapidly.
   (d) If lipids are highly concentrated, spraying water on the plate will allow visualization of the lipid bands. This is the least destructive method of detection but also the least sensitive.

3.3 Purification of Compounds from TLC Plates

1. Run a test plate with one lane in order to determine the optimal solvent conditions and optimal loading amount (*see* **Note 8** regarding signs of overloading). Scale up to multiple plates with ten lanes each. The number of plates will depend on the natural abundance of the target compound and the amount of product desired.
2. After developing the TLC plates, use a glass cutter or diamond knife to cut out the first lane. Stain only this lane with the method of choice. If nondestructive dyes such as iodine vapor or water are used, the entire plate can be used for purification.
3. After drying, align the stained, cut lane with the rest of the plate. Using a clean razor blade, draw two lines across marking the top and the bottom of the band of interest. Extend these lines onto the unstained plate. Scrape the silica gel onto weighing paper.
4. Prepare a flash chromatography column. Wash the column with 10 volumes of chloroform methanol in order to remove contaminants.

5. Place a clean glass vial under the column for collection of the solvent. Make sure the volume of the vial will be sufficient for at least 5× the volume of the scraped silica gel.
6. Carefully add the scraped gel from the TLC plate to the column using a glass funnel.
7. To elute compounds from the silica gel, add solvent slowly into the column using a glass syringe (*see* **Note 12**). Use the syringe to wash residual silica dust from the funnel and walls of the pipette. To ensure thorough elution, use a volume of solvent 5× the volume of the silica gel.
8. Push out residual solvent by aspiration using a rubber bulb placed over the top opening of the column.
9. Evaporate the eluant under a gentle stream of nitrogen.

3.4 Analysis of Extracts by GCMS

1. Prior to analysis of the experimental samples, establish the minimal concentration and sample volume that is suitable for the instrument. These parameters will vary depending on the instrument (*see* **Notes 13** and **14**).
2. Establish the retention times of compounds of interest by first analyzing synthetic standards. The retention time will change depending on the instrument gradient conditions and column.
3. Standard gradient conditions for analysis of *D. melanogaster* hydrocarbons are as follow:
   (a) Temperature gradient: column temperature begins at 50 °C for 2 min and increases to 300 °C at a rate of 15 °C/min.
   (b) Column: 5 % phenyl-methylpolysiloxane (DB-5, 30 m length, 0.32 i.d., 0.25 μm film thickness, Agilent, Santa Clara, CA, USA).
4. Redissolve the prepared extracts in hexane (*without* hexacosane standard) and transfer to the appropriate sample vials specific for the GCMS instrument (*see* **Notes 15** and **16**).
5. For quantification, prepare a dilution series of synthetic standards in the appropriate concentration range (*see* **Note 17**).
6. Run all samples (e.g., experimental genotypes vs. control) under the same instrument conditions.
7. Align retention times of peaks in sample with standards. If electron ionization (EI) information is available, match the EI spectrum with the National Institute of Standards and Technology (NIST) Mass Spectral Library (*see* **Note 18**) or with the EI spectrum obtained from analysis of the synthetic version.
8. Quantification can be done by normalizing the area under the compounds of interest to the area under the peak of hexacosane

or to the total area under all of the peaks (the total ion chromatogram). The relative abundance amount of the compound is the average of the three to five replicates (*see* **Note 19**).

3.5 Analysis of Intact Insects Using UV-LDI MS

1. We operate under 2 mbar buffer gas in the ion source (*see* **Note 20**). Under these conditions, mostly potassiated molecules ($[M+K]^+$) and on occasion, sodiated molecules ($[M+Na]^+$) are observed.
2. Anesthetize flies with cold or CO_2. Mount flies on adhesive tape on glass cover slips. Mount the cover slip on the customized target plate, making sure the area to be analyzed is parallel to the same plane on which the laser beam is focused. Stack adhesive tape and glass cover slips to achieve desired height.
3. Add a calibrant to the sample plate by mixing 0.5 μL of a synthetic peptide standard with 0.5 μL of DHB matrix directly on the edge of the target plate. Allow to dry.
4. Calibrate the instrument using the monoisotopic peak for the peptide standard and the signal for DHB. Internal calibrants that can be detected from the surface of the fly include dihexose ($[M+K]^+$ 381.07) and a triacylglyceride ($[M+K]^+$ 815.65).
5. Raster the laser beam over areas of interest such as the anogenital region and the dorsal surface of the fly. Parameters such as laser pulse duration, firing rate, fluence, and irradiation time should be kept consistent from sample to sample.
6. Use an elemental composition calculator to determine all possible chemical formula for a given mass, instrument accuracy, and charge (*see* **Note 21**).
7. Relative quantitation of compound abundance can be performed by normalizing individual signal intensities to the total area underneath all detected signals (*see* **Note 19**).

3.6 Analysis of TLC Plates Using UV-LDI MS

1. For detection of compounds from TLC plate without extraction, use a TLC plate with aluminum backing. Run the plate using the same method and conditions as would be used for a glass-backed TLC plate (*see* Subheading 3.3) making sure to include a duplicate lane that will be used for staining.
2. Excise one lane with a pair of scissors and stain the lane with a lipophilic dye (*see* Subheading 3.3).
3. Mark corresponding spots/bands of interest on the unstained portion of the plate with a graphite pencil. Color over a portion of the band with the pencil. Overlay the colored band with 0.5 μL of DHB solution (*see* **Note 22**).
4. Apply the graphite/DHB matrix to a portion of the TLC plate that does not contain any bands. This region will serve as the negative control.

5. Adhere the TLC plate to the sample plate using sticky tabs or double-sided tape, making sure that the area to be analyzed is parallel to the same plane on which the laser beam is focused.
6. Add a calibrant to the sample plate by mixing 0.5 μL of a synthetic peptide standard with 0.5 μL of DHB matrix directly on the edge of the target plate. Allow to dry.
7. Calibrate the instrument using the signal for the peptide standard and the signal for DHB.
8. Raster the laser over the individual bands, making note of which signals are detected in which position on the plate.
9. Obtain a mass spectrum from the spot serving as a negative control. DHB and graphite will produce signals as well. This control will facilitate identification of signals belonging to the biological sample vs. signals arising from the matrix or silica.

3.7 Analysis of Intact Insects Using DART MS

1. DART MS analysis is performed using the following ion source settings: the gas heater is set to 200 °C, the glow discharge needle is set at 3.5 kV. Electrode 1 is set to +150 V and electrode 2 is set to +250 V. Helium gas flow is set to 2.5 L/min. Under these conditions, mostly protonated ($[M+H]^+$) molecules are observed (*see* **Note 23**).
2. Anesthetize flies with cold or CO_2.
3. Using forceps that have been cleaned with 70 % ethanol, pick up a fly by both its wings, making sure not to poke or squish the fly (*see* **Note 24**).
4. Put the fly in the stream of charged helium gas. Hold the fly in the stream until peaks of triacylglycerides start to appear (*see* **Note 25**).
5. Hold the sample in the DART stream in approximately the same location for the same amount of time each time in order to produce reproducible spectra from sample to sample.
6. Remove the fly and immediately after, place a capillary tube dipped into PEG into the stream (*see* **Note 26**).
7. Calibrate mass spectra using PEG as a standard.
8. Relative quantitation of compound abundance can be performed by normalizing individual signal intensities to the total area underneath all detected signals (*see* **Note 19**).
9. Use an elemental composition calculator to determine all possible chemical formulae for a given mass, instrument accuracy, and charge (*see* **Note 21**).

4 Notes

1. The number of flies needed will depend on the sensitivity of the instrument used for detection. Each adult *D. melanogaster* contains approximately 1–1.5 μg of cuticular lipids. For GCMS, six to ten flies is adequate. The limit of detection of the instrument should be assessed prior to experiments by testing extractions using different quantities of flies, e.g., from 1 to 10.
2. Flies are isolated as pupae to prevent mating and hydrocarbon exchange from contact. Flies will adsorb cuticular hydrocarbons from other flies even in the absence of mating. Extraction should be performed around the same time each day since quantities of cuticular hydrocarbon varies with circadian rhythm.
3. Hexane is the standard solvent used for extraction. A chloroform–methanol (66:33; per vol) solution will extract more lipids, particularly triacylglycerides. However, because this solvent seems to break the cuticle down faster, we recommend a shorter incubation time of 5 min. Ethanol may be used when selectively extracting polar compounds such as sugars.
4. Be careful not to damage the fly tissue by piercing the cuticle with the syringe tip since doing so will result in internal stores of triacylglycerides and other lipids to be released.
5. In between each transfer, clean the glass syringe by aspirating from a solution of chloroform–methanol (66:33; per vol) and dispensing into a waste container. Repeat three times. Repeat this process with a second bottle of chloroform–methanol (66:33; per vol). Hereafter, all cleaning of syringes should proceed by using each bottle of washing solvent in the same order.
6. When evaporating volumes above 15 mL, it is more efficient to use a rotary evaporator. Vacuum evaporators such as a SpeedVac should be avoided since more volatile compounds such as fatty acid methyl esters are likely to evaporate.
7. Waiting for too short a time causes the solvent front to be curved and bands to be misaligned.
8. Usually a maximum of ten lanes can be plated for a 10 × 10 cm TLC plate. With each lane containing a concentration corresponding to approximately 100 flies, the maximum number of flies that can be extracted for each plate is 1,000. Having too high a concentration for each lane causes obstruction to the separation of the compounds where the larger compounds such as the triacylglycerides block the smooth migration of small compounds such as dienes.
9. The solvent in the tank can be reused for ten plates though the retention index (distance from the origin) will change with

each repeated use. Other factors such as temperature, humidity can also affect separation. Therefore, it is important to use one lane for staining or to have a standard on each plate in order to determine the retention index of the target compound.

10. Fumes from the TLC tank are hazardous. Avoid inhaling and use tanks only in the fumehood.
11. Wear goggles to prevent UV light from going into the eyes.
12. Elute the column with the same solvent that was used for extraction.
13. Use the minimal volume allowed for the GCMS instrument. By using smaller volumes, the extract will be more concentrated and fewer flies may be needed for analysis.
14. The volume of sample solvent can be reduced by using a glass micro insert into the sample vial. A sample volume of 100–200 μL is normally needed for standard GCMS autosamplers. The insert allows volumes as small as 20–50 μL to be used.
15. Make sure the sample vials are compatible with the brand of instrument. Using the wrong vials can damage the autosampler syringe.
16. If amounts are scarce, the extract can be dissolved in small volumes (less than 10 μL) and directly injected into the instrument.
17. When performing absolute quantification, it is necessary to generate a standard curve using use the synthetic version of the compound of interest rather than a generic hydrocarbon. The signal intensity for a compound depends on both its abundance and ionization efficiency.
18. The NIST MS library is available by subscription and can be found on most GCMS instruments.
19. The total area under a subset of peaks can be selected for normalization rather than the total area under all detected peaks. This method is preferred if there is variation in the number or intensity of background or contamination peaks from sample to sample. It is important to note that these methods of normalization will provide only relative quantification and do not indicate an absolute increase or decrease in amounts of a particular compound.
20. Optimal conditions for ion source pressure, extraction voltages, and quadrupole voltages will differ for each instrument [10].
21. A freeware elemental composition calculator can be downloaded here: http://www.wsearch.com.au/Tools/elemental_composition_calculator.htm. Typically, we limit the elements to C, H, O, Na, and K.

22. In our experience, combining both DHB and graphite results in more intense signals when analyzed by UV-LDI MS. Graphite and pencil lead have been shown to be effective matrices for MS analysis of small molecules and lipids [14, 15].
23. Placing a vial of concentrated ammonium acetate beneath the DART stream can facilitate the formation of NH_4 adducts. This modification is useful for triacylglycerides and higher molecular weight lipids.
24. Damaging the fly tissue will cause compounds from internal stores to be detected.
25. Hold the fly tightly to prevent it from being sucked into the instrument. Detection of triacylglyceride peaks indicates that the cuticle has been broken and compounds from internal stores are detected.
26. Only a small amount of PEG is needed for detection. Be careful not to touch any part of the inlet or DART source with the capillary tube containing PEG. Doing so can result in contamination peaks for subsequent analyses.

References

1. Wyatt T (2003) Pheromones and animal behaviour. Cambridge University Press, Cambridge, MA
2. Howard RW, Blomquist GJ (2005) Ecological, behavioral, and biochemical aspects of insect hydrocarbons. Annu Rev Entomol 50:371–393
3. Butterworth FM (1969) Lipids of Drosophila: a newly detected lipid in the male. Science 163:1356–1357
4. Brieger G, Butterworth FM (1970) Drosophila melanogaster: identity of male lipid in reproductive system. Science 167:1262
5. Antony C, Jallon J (1982) The chemical basis for sex recognition in Drosophila melanogaster. J Insect Physiol 28:873–880
6. Tentschert J, Kolmer K, Hölldobler B et al (2001) Chemical profiles, division of labor and social status in Pachycondyla queens (Hymenoptera: formicidae). Naturwissenschaften 88:175–178
7. Bland JM, Osbrink WL, Cornelius ML et al (2001) Solid-phase microextraction for the detection of termite cuticular hydrocarbons. J Chromatogr A 932:119–127
8. Everaerts C, Farine JP, Cobb M et al (2010) Drosophila cuticular hydrocarbons revisited: mating status alters cuticular profiles. PLoS One 5:e9607
9. Yew JY, Soltwisch J, Pirkl A et al (2011) Direct laser desorption ionization of endogenous and exogenous compounds from insect cuticles: practical and methodologic aspects. J Am Soc Mass Spectrom 22:1273–1284
10. Yew JY, Dreisewerd K, Luftmann H et al (2009) A new male sex pheromone and novel cuticular cues for chemical communication in Drosophila. Curr Biol 19:1245–1254
11. Yew JY, Dreisewerd K, de Oliveira CC et al (2011) Male-specific transfer and fine scale spatial differences of newly identified cuticular hydrocarbons and triacylglycerides in a Drosophila species pair. PLoS One 6:e16898
12. Cody RB, Laramée JA, Durst HD (2005) Versatile new ion source for the analysis of materials in open air under ambient conditions. Anal Chem 77:2297–2302
13. Yew JY, Cody RB, Kravitz EA (2008) Cuticular hydrocarbon analysis of an awake behaving fly using direct analysis in real-time time-of-flight mass spectrometry. Proc Natl Acad Sci U S A 105:7135–7140
14. Black C, Poile C, Langley J et al (2006) The use of pencil lead as a matrix and calibrant for matrix-assisted laser desorption/ionisation. Rapid Commun Mass Spectrom 20: 1053–1060
15. Langley GJ, Herniman JM, Townell MS (2007) 2B or not 2B, that is the question: further investigations into the use of pencil as a matrix for matrix-assisted laser desorption/ionisation. Rapid Commun Mass Spectrom 21:180–190

Chapter 3

Analysis of Volatile Mouse Pheromones by Gas Chromatography Mass Spectrometry

Milos V. Novotny and Helena A. Soini

Abstract

High-precision quantitative profiling of volatile organic constituents in rodent physiological fluids and glandular secretions is needed to relate olfactory signals to physiology and behavior. Whereas capillary gas chromatography-mass spectrometry (GC-MS) analysis has become the most widely applied in such investigations, the extraction and preconcentration of volatile organics is arguably the most critical step in the overall analytical task. In this chapter, we describe technical details of two main sample extraction procedures used in our laboratory: dynamic headspace trapping, and stir bar sorptive extraction (SBSE). They have been demonstrated here for the chromatographic analysis of mouse urine, serum, saliva, and preputial gland specimens.

Key words Dynamic headspace trapping, Stir bar sorptive extraction, Gas chromatography-mass spectrometry, Mouse urine, Mouse serum, Mouse saliva, Mouse preputial glands

1 Introduction

1.1 General Considerations

Among the rodent species, the house mouse (*Mus domesticus*) has become a most popular and widely studied mammal in terms of chemical (olfactory) communication. Although olfactory communication is ubiquitous in nature, ranging from microbial systems to insects, lizards, birds, etc., to mammals including primates, the house mouse has evolutionarily developed a most sophisticated system of chemical messengers. Some of the structurally identified substances in this species truly deserve denotation as "pheromones", the term initially and primarily coined for the insect world [1]. The unique scientific opportunities to study chemical communication systems of rodents, in general, and the house mouse, in particular, were noted already in the 1950s and 1960s during the first observations of the importance of olfactory signals in the Lee-Boot effect for estrus suppression [2, 3] and the Whitten effect in estrus synchronization [4, 5]. Many years after these basic biological discoveries, the interest of scientific community has further

Kazushige Touhara (ed.), *Pheromone Signaling: Methods and Protocols*, Methods in Molecular Biology, vol. 1068,
DOI 10.1007/978-1-62703-619-1_3, © Springer Science+Business Media, LLC 2013

been reinforced through the structural identification of the first primer mouse pheromones such as dehydro-*exo*-brevicomin and 2-*sec*-butyl-4,5-dihydrothiazole [6], 2,5-dimethylpyrazine [7], and the farnesene isomers that are dominance-signaling factors [8] in male mice.

Following the groundbreaking discovery by Buck and Axel [9] in decoding the basic mechanisms in mammalian olfaction, the rodent species have become increasingly popular as appropriate models for pheromone-receptor investigations. This, in turn, has increased the needs for structural identification of additional constituents of rodents' biological fluids and glandular secretions as the potential "ligands" for neurobiological studies in the vomeronasal organ [10–12] and the major olfactory epithelium [13–15]. Additionally, the knowledge of chemical structure of such "olfactants" has been important in elucidating the relevant biochemical mechanisms such as protein-pheromone binding (complex formation and its relevant structural and functional attributes at the molecular level) [16–22].

The unique role of laboratory-bred mice for numerous biological, biomedical, and pharmacological investigations will further necessitate improved knowledge of the composition of complex metabolite and protein mixtures excreted by these most valuable experimental animals. Simultaneously, new bioanalytical techniques in the fields of genomics, proteomics, and metabolomics are now capable yielding a nearly complete knowledge of the systems biology of the house mouse from which some parallels can be drawn concerning more ecologically significant attributes of the wild mice. The pertinent aspects of bioanalytical quantitative profiling of volatile metabolites (relatively small molecules) will be detailed below.

While chemical communication in mice can be principally mediated by either volatile or nonvolatile molecules, the mouse has on its body a variety of known and potential signal sources including urine, saliva, preputial glands, plantar eccrine glands, Harderian glands, exorbital lacrimal glands, coagulating glands, submaxillary glands, and non-specialized sebaceous glands [23]. The best studied sources of volatile chemosignals seem urine and preputial glands, where a variety of both primer and releaser pheromone signals were initially located. These two sources are also easily available for the profiling studies by modern analytical procedures such as capillary gas chromatography-mass spectrometry (GC-MS) and liquid chromatography-mass spectrometry (LC-MS).

Whereas our methodological descriptions in this chapter relate extensively to volatile organic compounds as chemical messengers, the unique capability of the house mouse and some other rodents has been to excrete into their environments a series of lipocalin proteins [24, 25], which can interact with the volatile pheromones and facilitate their ecological impact. The best known of these biomolecules are so-called major urinary proteins (MUPs), which had been noticed as endocrinologically dependent urinary constituents for a number of decades [26], but could be studied in their molecular details only recently through the availability of modern structural tools, such as mass spectrometry [27, 28], NMR spectrometry [18, 19, 29] and X-ray crystallography [30]. The current studies seem to endorse the hypothesis that the associations between the volatile molecules and such proteins are at the heart of the complex chemical communication system that the rodent uniquely utilize in their environment, although the MUPs may have unique roles of their own in the chemical communication events [31, 32].

1.2 Methodological Considerations

The current success of the volatile pheromone profiling not only reflects the instrumental capabilities of today's GC-MS in terms of measurement sensitivity and resolution of complex biological mixtures, but it is also dependent on the modern techniques of solute preconcentration and sampling. Different sampling conditions will be described in this article, representing primarily the procedures practiced in the authors' laboratory, while other modified or optimal conditions may need to be sought for other specialized investigations of volatile constituents of rodent urines and other related biological materials.

The metabolomic profiling is usually related to a series of quantitative measurements where precision (repeatability) is of a primary concern, such as the analytical runs aiming at a comparison between male and female animals, the experiments involving endocrinological or behavioral manipulations, genetic studies, etc. [7, 33–39]. As an example, the biological relevance of such comparative analyses is represented in Fig. 1. Here, the "urinary output" of the male *Mus domesticus* and the male (monogamous) *Mus spicilegus* is being compared with the possible correlation of their different chemistry [39] and behavioral traits [40, 41].

Sampling pheromones from different biological materials for the GC-MS analysis requires efficient enrichment of the compounds of interest, as certain biologically active components may act at trace levels. Enrichment techniques, such as solvent extraction [42–44], solid-phase microextraction, SPME [45–48], adsorption methods using porous polymers [34, 37, 49–51], and the sorptive stir bar extraction (SBSE) [38, 39, 52, 53] are among the commonly used approaches. Solvent extraction is more applicable to less volatile compounds due to the potential losses of volatiles during the solvent evaporation steps. Volatile compounds, which are suitable for solvent extraction need to be present at relatively high levels to be detected using this method, since only a fraction of the final sample volume will be introduced to the analytical instrument. Solid-phase microextraction (SPME) has relatively low sample capacity, while displacement of some low-level

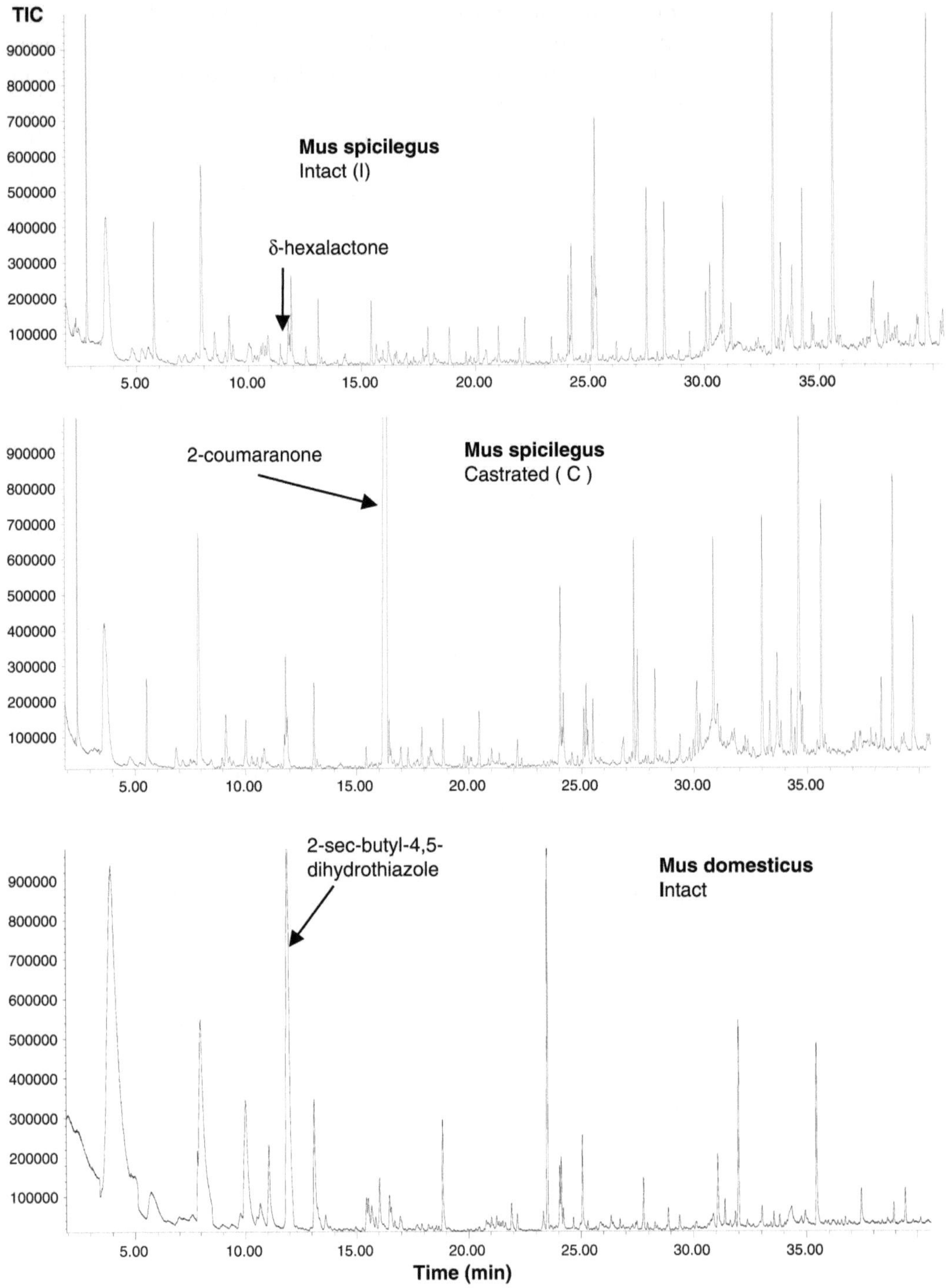

Fig. 1 Comparative urinary profiles (GC-MS total ion current) of intact and castrated male *Mus spicilegus* and intact *Mus domesticus*. Reprinted with permission from ref. [39]. Method as described in Subheading 3.2.1, except the urine sample volume was 200 μl

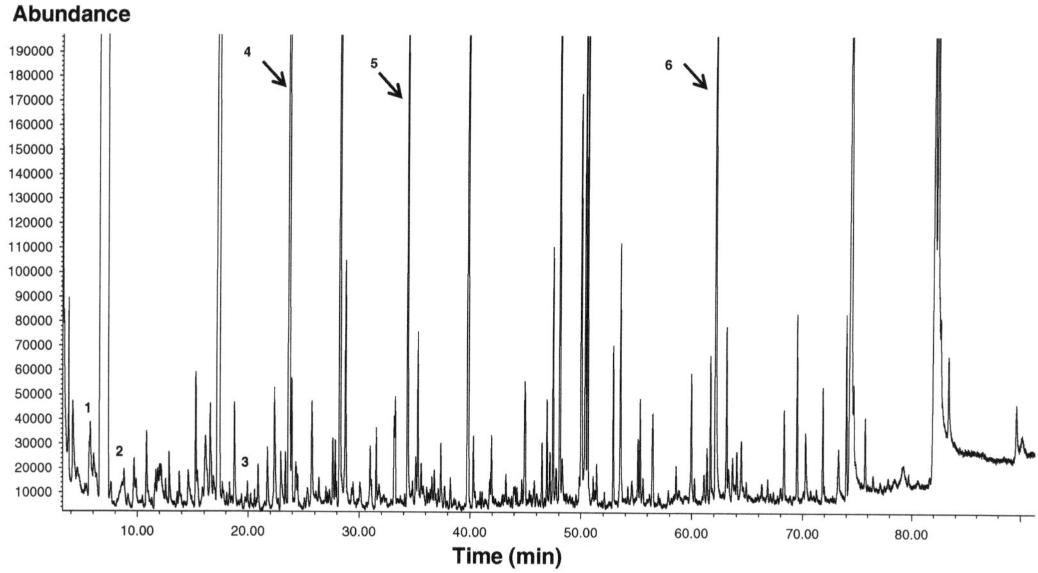

Fig. 2 Volatile compound profile (GC-MS total ion current) measured from male ICR mouse serum as described in Subheading 3.2.3 below. *Arrows* indicate 1: and 2: dihydrofuran decomposition products of 6-hydroxy-6-methyl-3-heptanone; 3: dehydro-*exo*-brevicomin; 4: 2-*sec*-butyl-4,5-dihydrothiazole; 5: geraniol; 6: farnesol. Modified and reprinted from ref. [38]

or low molecular weight compounds by the more concentrated sample constituents or higher molecular weight compounds can take place within a limited polymer volume of the SPME fiber [54]. Therefore, we generally prefer the sorptive stir bar extraction (SBSE) with higher PDMS (polydimethylsiloxane) sorptive polymer volume (24 μl) as compared with 0.6 μl polymer volumes in typical SPME fibers [54].

In this chapter, we will emphasize two methods being actively used in our laboratory: (1) Tenax sorbent-aided dynamic headspace procedure used for the mouse urine samples; and (2) the sorptive stir bar extraction (SBSE) method used for mouse urine, preputial glands, serum and saliva samples. In addition to urine, investigation of the volatile compounds in the preputial glands are important for the mouse chemical communication, since these glands are the original source of important pheromone constituents in urine, such as the farnesenes, which are linked to dominant and aggressive behavior in males [8]. Measuring volatile compound levels in serum, on the other hand, provides information on the metabolic routes involved in the biosynthetic pathways leading to the chemical signaling compounds [55]. Figure 2 (Subheading 3.2.3 described below) illustrates the volatile compound profile extracted from mouse serum [38].

Saliva, like urine, can be collected without sacrificing the animal. Anesthetized mice can produce enough saliva to prove the presence of some of the key male mouse pheromones in this material,

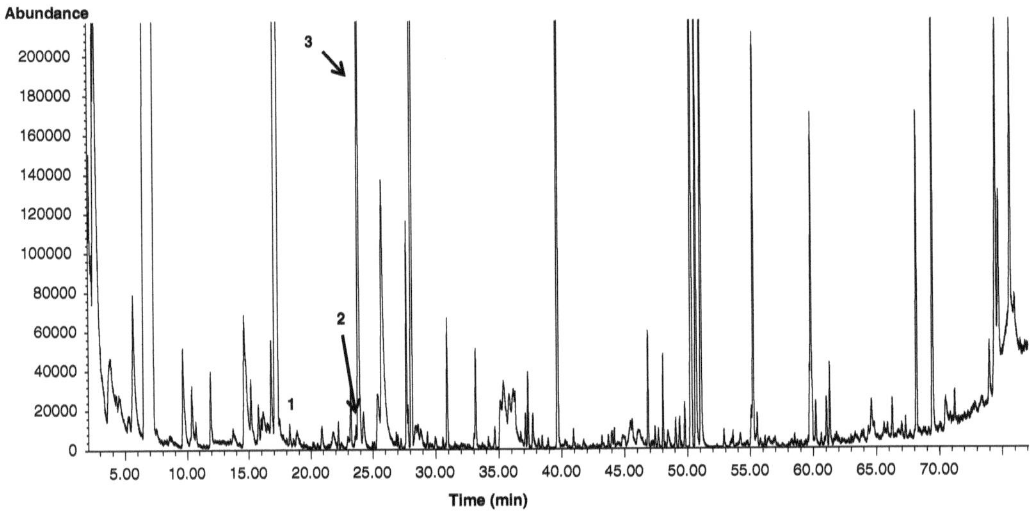

Fig. 3 Volatile compound profile (GC-MS total ion current) from male ICR mouse saliva as described in Subheading 3.2.4 below. *Arrows* indicate selected identified compounds. 1: limonene; 2: 2-*sec*-butyl-4,5-dihydrothiazole; 3: nonanal [56]

which are also present in serum and urine, as shown in Fig. 3 (Subheading 3.2.4 described below) [56].

While this article is confined to documentation of the analytical strategies to investigations on the mouse chemosignals, similar strategies have served our laboratory in other studies in the field of chemical ecology, albeit sometimes with procedural modifications.

2 Materials

All used materials need to be at the highest purity to avoid co-extracting exogenous impurities that can interfere with the compounds of biological interest. Rinsing glassware and ceramics with acetone and drying them at 80–95 °C in the oven is highly recommended. When choosing the plastic materials as sample vials, inert polypropylene is our recommended choice.

2.1 Dynamic Headspace Trapping of Urinary Volatiles

1. Ultrapure water (OmniSolv® EM Science, Gibbstown, NJ).
2. Methanol (J.T. Baker, Baker Analyzed® HPLC Solvent, Mallinckrodt Baker, Phillipsburg, NJ).
3. Internal standard, 7-tridecanone (Aldrich Chemical Company, Milwaukee, WI) (*see* **Note 1**).
4. Helium gas (Ultrahigh purity, Airgas, Inc., Radnor, PA).
5. Glass condenser with the ice water cooling loop (*see* Fig. 4 and **Note 2**).

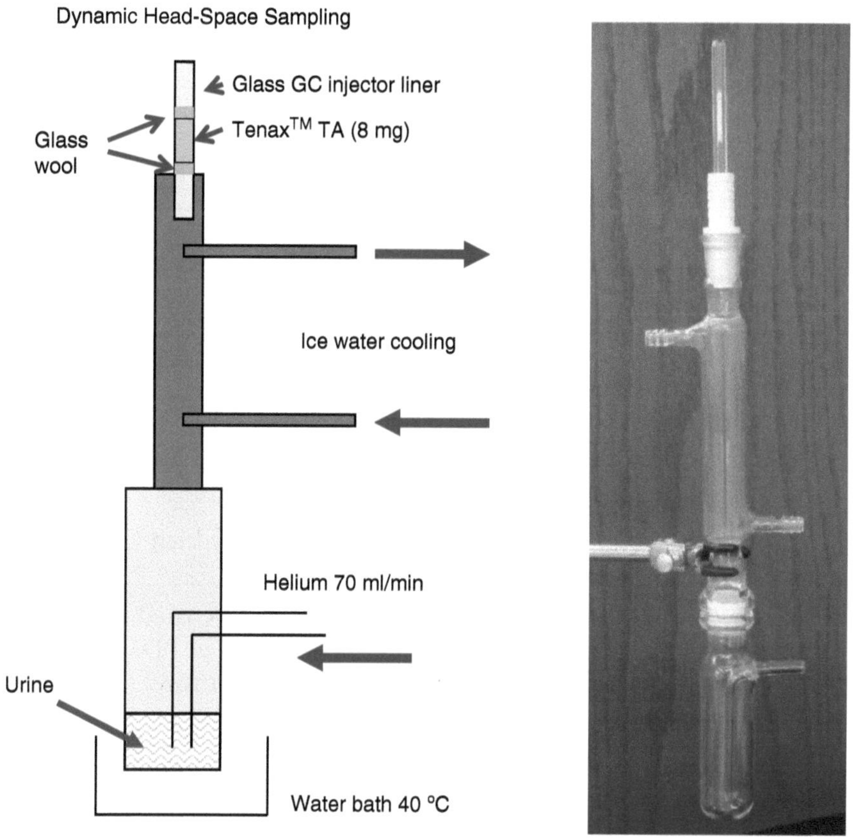

Fig. 4 Schematic of the dynamic head space sampling setup and a photograph of the used glassware

6. Tenax™ sorbent (TA, 35/65 mesh, Alltech, Deerfield, IL).
7. Splitless GC injector liner 78.5×6.0 mm, 2 mm, i.d. (Grace Davison Discovery Sciences, Deerfield, IL), to pack Tenax polymer between glass wool plugs (glass wool, DMCS treated, Alltech Associates, Deerfiled, IL) (*see* Fig. 4).
8. A modified GC injector port.
9. A loop of uncoated deactivated silica tubing (30 cm×0.25 mm, i.d., Restek) attached with the universal Press-Tight Connector (Restek Corporation, Bellefonte, PA) to the separation column.
10. Plastic 200-ml liquid nitrogen container.
11. The separation capillary column DB-5MS (30 m×0.25 mm, i.d., 0.25 μm film thickness) (Agilent, J&W Scientific, Folsom, CA).
12. Finnigan MAT Magnum ion trap GC-MS system (Finnigan MAT, San Jose, CA).

2.2 Stir Bar Extraction of Urine

1. Ultrapure water (OmniSolv® EM Science, Gibbstown, NJ).
2. Methanol (J.T. Baker, Baker Analyzed® HPLC Solvent, Mallinckrodt Baker, Phillipsburg, NJ).

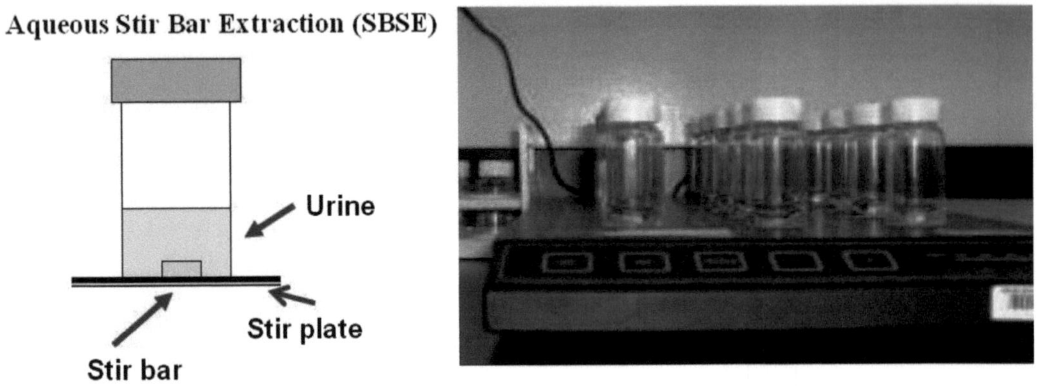

Fig. 5 Schematic of the stir bar sorptive extraction (SBSE) setup. In the photograph, samples are being extracted in a batch mode

3. Internal standard, 7-tridecanone (Aldrich Chemical Company, Milwaukee, WI).
4. Stir bars (Twister™ PDMS polymer-coated stir bar, dimensions 10 mm, 0.5 mm film thickness, 24 μl PDMS volume, Gerstel GmbH, Mülheim an der Ruhr, Germany) (*see* **Note 3**).
5. Wheaton disposable glass scintillation vials, 20-ml (VWR International, West Chester, PA) (*see* Fig. 5 and **Note 4**).
6. Stir plate (15-place) Variomag Multipoint HP 15 stirplate (H+P Labortechnic, Oberschleissheim, Germany).
7. Kimtech Kimwipes (Kimberly Clark, Roswell, GA).
8. TDS Desorption tubes (Gerstel GmbH) (*see* **Note 5**).
9. The gas chromatograph-mass spectrometer (GC-MS) instrument: Agilent 6890N gas chromatograph connected to 5973i MSD mass spectrometer (Agilent Technologies, Inc., Wilmington, DE).
10. The separation capillary column DB-5MS (30 m × 0.25 mm, i.d., 0.25 μm film thickness) (Agilent, J&W Scientific, Folsom, CA).
11. Thermal Desorption Autosampler and Cooled Injection System (TDSA–CIS4) (Gerstel GmbH, Mülheim an der Ruhr, Germany).

2.3 Stir Bar Extraction of Preputial Glands

1. Miltex disposable safety scalpel (VWR International, West Chester, PA).
2. CoorsTek porcelain 65-ml mortar and pestle (VWR International, West Chester, PA).
3. Liquid nitrogen (*see* **Note 6**).
4. Ultrapure water (OmniSolv® EM Science, Gibbstown, NJ).

5. Methanol (J.T. Baker, Baker Analyzed® HPLC Solvent, Mallinckrodt Baker, Phillipsburg, NJ).
6. Internal standard, 7-tridecanone (Aldrich Chemical Company, Milwaukee, WI).
7. Ammonium sulfate (99.95 %, Aldrich Chemical Company, Milwaukee, Wisconsin) (*see* **Note 7**).
8. *See*, **items 8–11** in Subheading 2.2 above.

2.4 Stir Bar Extraction of Serum

1. Nonspecific protease enzyme (type XIV), Pronase E from *Streptomyces griseus*, 5.3 units/mg (Sigma-Aldrich, St. Louis, MO).
2. 50 mM ammonium bicarbonate (NH_4HCO_3), pH 8.2 (J.T. Baker, Phillipsburg, NJ).
3. Ultrapure water (OmniSolv® EM Science, Gibbstown, NJ).
4. Methanol (J.T. Baker, Baker Analyzed® HPLC Solvent, Mallinckrodt Baker, Phillipsburg, NJ).
5. Internal standard, 7-tridecanone (Aldrich Chemical Company, Milwaukee, WI).
6. *See* **items 8–11** in Subheading 2.2 above.

2.5 Stir Bar Extraction of Saliva

1. Enzyme α-amylase from *Bacillus licheniformis* (Sigma-Aldrich, St. Louis, MO).
2. Enzyme Glyko Sialidase A from *Arthobacter ureafaciens* (Prozyme, Hayward, CA).
3. Ultrasonication bath FS30 (Fisher Scientific, Waltham, MA).
4. Ammonium sulfate (99.95 %, Aldrich Chemical Company, Milwaukee, Wisconsin).
5. 30 mM sodium phosphate (Na_2HPO_4) (Mallincrodt Chemicals, Austin, TX), pH to 6.5 with o-phosphoric acid (Sigma-Aldrich, St. Louis, MO).
6. Methanol (J.T. Baker, Baker Analyzed® HPLC Solvent, Mallinckrodt Baker, Phillipsburg, NJ).
7. Internal standard, 7-tridecanone (Aldrich Chemical Company, Milwaukee, WI).
8. *See* **items 8–11** in Subheading 2.2 above.

3 Methods

The used sample preparation and analysis methods aim to release maximally small organic volatile compounds from the matrix compounds and to do so reproducibly and quantitatively. These attributes are extremely important in the biological studies comparing compound levels in different study subject groups in the exact statistical manner. These methodological attributes should also be applicable at the low-concentration levels, since some of the pheromonally active sample components may exist at low levels. Different matrix materials are known to influence markedly the success in isolating small organic molecules from different media. This is taken into consideration by using appropriate additives such as ammonium sulfate, digestive enzymes, co-solvents, etc., to improve recoveries of the analytes of interest. Verification of the compound identities is generally performed using either commercially available or synthesized standard materials, with a subsequent recording of retention time index data [57], through high-resolution mass spectrometry measurements [55], and element-specific detector responses for verifying the presence of nitrogen- and sulfur-containing compounds in our samples [38].

3.1 Dynamic Headspace Trapping of Urinary Volatiles

The current procedure used in our laboratory has its origin in the analytical studies of the 1970s [58–60], in which various volatile trace constituents of aqueous media were purged and successfully preconcentrated on a small precolumn (capsule) containing a thermostable porous polymer. Following analytically adequate preconcentration time (optimized during the method development), the trapped volatiles are released at high temperature. Typically, this takes place at the modified hot injection port of a gas chromatograph. Released (desorbed) volatiles are directed to a cryogenically cooled portion of the capillary precolumn [34, 37, 52] to cryofocus the compound mixture into a narrow "frozen" band. Actual injection onto the separation column occurs when the frozen section of the column is rapidly heated and the compounds from the cryofocused band transfers back into the gas phase and the mixture is separated in the capillary column, followed by detection in the MS for quantitative evaluations. As described in the original studies of the analytical aspects of this preconcentration procedure [58–60], the purpose of the preconcentration step and the subsequent recording of the profiles of volatiles is the capability of *comparative studies* of different samples. Thus, the reproducibility of trapping and desorption, rather than a complete solute adsorption, is the primary objective. Our dynamic headspace sampling apparatus is shown in Fig. 4. As an alternative, there are available commercial purge-and-trap autosamplers dedicated to GC-MS instruments, such as an autosampler device from Teledyne-Tekmar (Mason, OH).

3.1.1 Dynamic Head Space Extraction of Urine

1. Place 1.0 ml of urine and 2 ml of ultrapure water in a glass vessel connected to a helium flow to purge the volatiles from the urine sample.
2. Place the glass vessel in the water bath at 40 °C.

3. Place the condenser with the ice-water cooling flow on the top of urine vessel.
4. Pack 8 mg of Tenax TA in the 7.8-cm long glass injector liner and close Tenax section ends with a plug of glass wool.
5. Place the injector liner containing Tenax section on the top of the condenser.
6. Turn cold water stream on to reduce the amount of condensed water in the gaseous volatile mixture and turn helium flow on (70 ml/min) for collecting the sample headspace as shown in Fig. 4 (*see* **Note 2**).
7. Let the helium flow purge (bubble) into the sample for 30 min, while volatiles are being trapped in the Tenax adsorbent.
8. Remove the injector liner with the Tenax section and place it inside the GC injector (carrier gas flow disconnected) and follow the GC-MS instructions in Subheading 3.3 below.

3.2 Stir Bar Extraction of Urine, Preputial Glands, Serum and Saliva

During the late 1990s, Sandra and coworkers found the polydimethylsiloxane (PDMS) polymer that was well suited for preconcentrating organic molecules from dilute aqueous and gaseous environmental samples [61, 62]. When the PDMS film (thickness 0.5 or 1.0 mm) is placed around a magnetic stir bar (1-cm length), sorptive extraction of small organic molecules into the PDMS layer can be performed in the aqueous phase, while the sample placed in a flat bottom vial is stirred with the immersed PDMS stir bar, with the vial being placed on the top of the magnetic stir plate (*see* Fig. 5). The PDMS polymer action is stated as "sorptive" rather than "adsorptive". Small organic molecules "dissolve" inside the polymer instead of adsorbing on the surface. While using 45–60 min extraction time, the sorptive process reaches an equilibrium, which improves the method's reproducibility greatly. In early 2000, Twister® stir bars based on this idea became commercially available from the Gerstel Company (Germany). Precise manufacturing of stir bars of identical dimensions allows to extract multiple samples simultaneously as illustrated in Fig. 5 with excellent reproducibility [38, 54, 63].

3.2.1 SBSE of Urine

1. Place 0.5 ml mouse urine in a disposable 20-ml glass scintillation vial (*see* Fig. 5).
2. Add 2 ml of ultrapure water.
3. Pipet 5 μl of internal standard 7-tridecanone (8 ng in 5 μl methanol).
4. Insert clean stir bar into the vial (*see* **Note 8**).
5. Stir for 60 min at 850 rpm.
6. Take the stir bar out from the sample, on the top of clean lint-free paper tissue.

7. Pipet 2 × 1 ml of ultrapure water on the top of the stir bar.
8. Dry the stir bar with clean dry Kimwipes sheet.
9. Place the stir bar in the desorption tube.
10. *See* Subheading 3.4 for thermal desorption, cryofocusing, and the GC-MS analysis.

3.2.2 SBSE of Preputial Glands

1. Weigh the frozen preputial gland.
2. While still frozen, cut the gland into thin slices.
3. Place the gland slices on the clean ceramic mortar vessel.
4. Pour about 50 ml of liquid nitrogen in the ceramic vessel.
5. Crush with the pestle the frozen tissue quickly into powder. If necessary, add more liquid nitrogen.
6. Rinse the contents of the ceramic vessel into a clean 20-ml glass scintillation vial with 2 ml of ultrapure water.
7. Add 100 mg of ammonium sulfate (*see* **Note 9**).
8. Add 50 μl of 7-tridecanone (80 ng in 50 μl methanol) (*see* **Note 10**).
9. Stir for 60 min at 850 rpm.
10. Take the stir bar out from the sample vial, on the top of clean lint-free paper tissue.
11. Pipet 2 × 1 ml of ultrapure water on the top of the stir bar.
12. Dry the stir bar with clean dry Kimwipes.
13. Place the stir bar in the desorption tube.
14. *See* Subheading 3.4 for thermal desorption, cryofocusing and the GC-MS analysis.

3.2.3 SBSE of Serum Samples

1. Pipet 10 ml of ultrapure water and 10 ml of 50 mM buffer into a clean 20-ml scintillation glass vial.
2. Add 200 μl serum.
3. Add 2 mg of Pronase E protease.
4. Wrap vial in foil, and allow enzyme digestion to take place overnight at room temperature (*see* **Note 11**).
5. Insert a clean stir bar in a vial and add 20 μl 7-tridecanone (40 ng in 20 μl of methanol).
6. Stir for 2 h at 850 rpm.
7. Take the stir bar on the top of clean lint-free paper tissue.
8. Pipet 2 × 1 ml of ultrapure water on the top of the stir bar.
9. Dry the stir bar with clean dry Kimwipes.
10. Place the stir bar in the desorption tube.
11. *See* Subheading 3.4 for thermal desorption, cryofocusing and the GC-MS analysis.

3.2.4 SBSE of Saliva Samples

1. Pipet 1.8 ml of 30 mM sodium phosphate buffer adjusted to pH 6.5 with *o*-phosphoric acid and 150 μl of saliva into a 20-ml scintillation vial.
2. Add enzymes α-amylase and sialidase A (5 μl of each) (*see* **Note 12**).
3. Wrap the vial in foil and allow to digest overnight at room temperature (*see* **Notes 10** and **11**).
4. Place a clean stir bar, 20 μl of the internal standard 7-tridecanone in methanol (16 ng in 20 μl) and 300 mg of ammonium sulfate $(NH_4)_2SO_4$ in a vial.
5. Sonicate the sample vial for 60 min at room temperature.
6. Take the stir bar on the top of clean lint-free paper tissue.
7. Pipet 2× 1 ml of ultrapure water on the top of the stir bar.
8. Dry the stir bar with clean dry Kimwipes.
9. Place the stir bar in the desorption tube.
10. *See* Subheading 3.4 below for thermal desorption, cryofocusing and the GC-MS analysis.

3.3 Hot Injector Desorption-Cooled Capillary Loop: GC-MS Analysis

1. Injector temperature is set at 250 °C.
2. Helium carrier gas flow is stopped.
3. Injector liner containing Tenax adsorbent is placed inside the injector.
4. Place the precolumn uncoated capillary loop inside the 200-ml liquid nitrogen container in the GC oven, keep door open.
5. Resume carrier gas flow (1.0 ml/min).
6. Let helium flow thermal desorption to release the volatiles from Tenax and cryofocus them in the deactivated precolumn loop immersed in liquid nitrogen (−150 °C) for 15 min.
7. Remove the liquid nitrogen container from the GC oven and close the oven door.
8. Start the GC oven temperature program: 40 °C (hold 5 min), an increase 2 °C/min to 200 °C (hold 10 min), the helium carrier gas head pressure of 12.0 psi, flow rate of 1.0 ml/min.
9. Start manually the MS acquisition: The manifold and transfer line temperatures are set at 220 °C and 300 °C, respectively, the ion trap in the positive electron ionization mode (70 eV), spectra scanning from 40 to 350 msu (1 scan/s).

3.4 TDSA–CIS/GC-MS Analysis

1. TDSA is in a splitless mode for thermal desorption: TDSA temperature program for desorption 20 °C (hold 0.5 min), a 60 °C/min ramp to 270 °C (final hold 5 min), temperature of the transfer line to the CIS, at 280 °C.
2. CIS (cooled injection system) is used for cryofocusing, injection in the solvent vent mode: The CIS-4 is cooled with liquid

nitrogen to −80 °C, after desorption and cryotrapping, CIS-4 is heated at 12 °C/s to 280 °C (hold 10 min) for injection, vent pressure 9.0 psi, a vent flow 50 ml/min, and a purge flow at 50 ml/min.

3. The GC temperature program is 40 °C (hold 5 min), an increase 2 °C/min to 200 °C (hold 10 min), the helium carrier gas head pressure of 9.0 psi, flow rate of 1.0 ml/min at constant flow mode.
4. The MS instrument is at positive electron ionization (EI) mode at 70 eV, the scanning rate of 2.41 scans/s over the mass range of 40–350 amu, the MSD transfer line temperature is set at 280 °C, the ion source and quadrupole temperatures are set at 230 °C and 150 °C, respectively.

4 Notes

1. Internal standard, 7-tridecanone (dihexylketone), is stored in a silica desiccator after opening the bottle due to hygroscopic nature of 7-tridecanone crystals. Methanol solutions can be stored at +4 °C.
2. Helium flow (carrying volatiles) passes through an ice-water-chilled section to minimize water excess before the volatiles are trapped on the Tenax polymer.
3. Stir bars are cleaned and reconditioned after use in the TC-2 Tube Conditioner with AUX 163 Controller (Gerstel GmbH, Mülheim an der Ruhr, Germany). Stir bars are placed in Gerstel TDS desorption tubes at 300 °C under the high flow of ultrapure helium (120 ml/min in each desorption tube).
4. Disposable glass scintillation vials are rinsed with acetone and dried at 95 °C prior to use. Vial caps with metal foil lining are used. Caps do not need acetone wash. As the alternative, Gerstel (Gerstel GmbH, Mülheim an der Ruhr, Germany) 10 or 20-ml headspace glass vials can be used with screw caps provided with blue silicone/PTFE septa (low bleed). The vials and caps are ultraclean, not requiring acetone wash prior to use.
5. TDS tubes are (1) washed with acetone, (2) dried in the oven at 95 °C, (3) conditioned in the TC-2 Tube Conditioner with AUX 163 Controller (Gerstel GmbH, Mülheim an der Ruhr, Germany) at 300 °C under the high flow of ultrapure helium (120 ml/min in each desorption tube), (4) rinsed with acetone, and (5) dried at 95°. These steps were found necessary, especially when tissue extractions were performed with numerous samples.
6. Store liquid nitrogen in the small Dewar vessel for the use within 1–2 days.

7. A small container size, e.g., 50 g has been found optimal for ammonium sulfate to avoid exogenous volatile compounds entering inside the reagent bottle headspace over a long time.
8. Stir bar extraction can be also performed in the headspace mode. While the sample is placed in the bottom of the vial, the Twister® stir bar is hung in air space above the sample for an appropriate time (1–12 h). The sample can be kept at room temperature or heated to 37 °C.
9. Ammonium sulfate increases small-molecule recoveries from the matrix containing proteins due to the salting-out effects and protein denaturation [64].
10. Large amount of 7-tridecanone is necessary because of the presence of competing hydrophobic matrix material.
11. Enzyme digest takes place in dark for 12 h at room temperature, loosely covered with aluminum foil to allow oxygen flow into the sample. Protease enzyme is digesting macromolecular proteins into smaller peptide fragments, while releasing small organic volatile molecules that have affinity to the protein surfaces.
12. Sialidase enzyme breaks very large mucin glycoprotein structures while reducing viscosity in the saliva sample and subsequently improving the extraction recoveries for small organic target compounds.

Acknowledgments

This work was jointly sponsored by the METACyt Initiative of Indiana University, a major grant from the Lilly Endowment, Inc., and the Lilly Chemistry Alumni Chair funds (to M.V.N.).

References

1. Karlson P, Lüscher M (1959) "Pheromones": a new term for a class of biologically active substances. Nature 183:55–56
2. Lee S, van der Boot LM (1955) Spontaneous pseudopregnancy in mice. Acta Physiol Pharmacol Neerl 4:442–443
3. Lee S, van der Boot LM (1956) Spontaneous pseudopregnancy in mice II. Acta Physiol Pharmacol Neerl 5:213–214
4. Whitten WK (1956) Modification of the oestrus cycle of the mouse by external stimuli associated with the male. J Endocrinol 13:399–404
5. Whitten WK, Bronson FH, Greenstein JA (1968) Estrus-inducing pheromone of male mice: transport by movement of air. Science 161:584–585
6. Jemiolo B, Harvey S, Novotny M (1986) Promotion of the Whitten effect in female mice by synthetic analogs of male urinary constituents. Proc Natl Acad Sci USA 83:4576–4579
7. Novotny M, Jemiolo B, Harvey S, Wiesler D, Marchlewska-Koj A (1986) Adrenal-mediated endogenous metabolites inhibit puberty in female mice. Science 231:722–725
8. Novotny M, Harvey S, Novotny M (1990) Chemistry of male dominance in the house mouse, *Mus domesticus*. Experientia 46:109–113
9. Buck LB, Axel R (1991) A novel multigene family may encode odorant receptors: a molecular basis for odor recognition. Cell 65:175–187
10. Moss RL, Flynn RE, Shen X-M, Dudley C, Shi J, Novotny M (1997) Urine-derived compound

evokes membrane responses in mouse vomeronasal receptor neurons. J Neurophysiol 77: 2856–2862
11. Leinders-Zufall T, Lane AP, Puche AC, Ma W, Novotny MV, Shipley MT, Zufall F (2000) Ultrasensitive pheromone detection by mammalian vomeronasal neurons. Nature 405: 792–796
12. Holy TE, Dulac C, Meister M (2000) Responses of vomeronasal neurons to natural stimuli. Science 289:1569–1572
13. Sam M, Vora S, Malnic B, Ma W, Novotny MV, Buck LB (2001) Odorants may arouse instinctive behaviours. Nature 412:142
14. Boehm U, Zou Z, Buck LB (2005) Feedback loops link odor and pheromone signaling with reproduction. Cell 123:683–695
15. Soucy ER, Albeanu DF, Fantana AL, Murthy VN, Meister M (2009) Precision and diversity in an odor map on the olfactory bulb. Nat Neurosci 12:210–220
16. Bacchini A, Gaetani E, Cavaggioni A (1992) Pheromone binding proteins of the mouse, *Mus musculus*. Experientia 48:419–421
17. Hurst JL, Robertson DHL, Tolladay U, Beynon RJ (1998) Proteins in urine scent marks of male mouse extend the longevity of olfactory signals. Anim Behav 55:1289–1297
18. Zidek L, Stone MJ, Lato SM, Ellington AD, Novotny MV (1999) NMR Mapping of recombinant mouse major urinary protein I binding site occupied by a pheromone 2-sec-butyl-4,5-dihydrothiazole. Biochemistry 38:9850–9861
19. Novotny MV, Ma W, Wiesler D, Zidek L (1999) Positive identification of the puberty-accelerating pheromone of the house mouse: the volatiles associating with the major urinary protein. Proc R Soc Lond B 266: 2017–2022
20. Sharrow SD, Vaughn JL, Zidek L, Novotny MV (2002) Pheromone binding by polymorphic mouse major urinary proteins. Protein Sci 11:2247–2256
21. Cavaggioni A, Mucignat-Caretta C, Zagotto G (2003) Absolute configuration of 2-*sec*-butyl-4,5-dihydrothiazole in male mouse urine. Chem Senses 28:791–797
22. Armstrong SD, Robertson DHL, Cheetman SA, Hurst JL, Beynon RJ (2005) Structural and functional differences in isoforms of mouse major urinary proteins: a male-specific protein that preferentially binds a male pheromone. Biochem J 391:343–350
23. Bronson FH (1971) Rodent pheromones. Biol Reprod 4:344–357
24. Finlayson JS, Asofsky R, Potter M, Runner CC (1965) Major urinary protein complex of normal mice: origin. Science 149:981–982
25. Flower DR, North ACT, Sansom CE (2000) The lipocalin protein family: structural and sequence overview. Biochim Biophys Acta 1482:9–24
26. Parfentjer JA (1932) Calcium and nitrogen content in urine of normal and cancer mice. Proc Soc Exp Biol Med 29:1285–1287
27. Robertson DH, Cox KA, Gaskell SJ, Evershed RP, Beynon RJ (1996) Molecular heterogeneity in the major urinary proteins of the house mouse *Mus musculus*. Biochem J 316:265–272
28. Mechref Y, Zidek L, Ma W, Novotny MV (2000) Glycosylated major urinary protein of the house mouse: characterization of its N-linked oligosaccharides. Glycobiology 10:231–235
29. Zidek L, Novotny MV, Stone MJ (1999) Increased protein backbone conformational entropy upon hydrophobic ligand binding. Nat Struct Biol 6:1118–1121
30. Timm DE, Baker LJ, Mueller H, Zidek L, Novotny MV (2001) Structural basis of pheromone binding to mouse major urinary protein (MUP-I). Protein Sci 10:997–1004
31. Hurst JL, Payne CE, Nevison CM, Marie AD, Humphries RE, Robertson DHL, Cavaggioni A, Beynon RJ (2001) Individual recognition in mice mediated by major urinary proteins. Nature 414:631–634
32. Stopka P, Janotova K, Heyrovsky D (2007) The advertisement role of major urinary proteins in mice. Physiol Behav 91:667–670
33. Jemiolo B, Alberts J, Sochinski-Wiggins S, Harvey S, Novotny M (1985) Behavioural and endocrine responses of female mice to synthetic analogues of volatile compounds in male urine. Anim Behav 33:1114–1118
34. Schwende FJ, Wiesler D, Jorgenson JW, Carmack M, Novotny M (1986) Urinary volatile constituents of the house mouse, *Mus musculus*, and their endocrine dependency. J Chem Ecol 12:277–296
35. Andreolini F, Jemiolo B, Novotny M (1987) Dynamics of extraction of urinary chemosignals in the house mouse (*Mus musculus*) during the natural estrous cycle. Experientia 43:998–1002
36. Harvey S, Jemiolo B, Novotny M (1989) Pattern of volatile compounds in dominant and subordinate male mouse urine. J Chem Ecol 15:2061–2071
37. Ma W, Miao Z, Novotny MV (1998) The role of the adrenal gland and adrenal-mediated chemosignals in estrus suppression in the house mouse: the Lee-Boot effect revisited. Biol Reprod 59:1317–1320
38. Novotny MV, Soini HA, Koyama S, Bruce KE, Penn D (2007) Chemical identification of MHC-influenced volatile compounds in mouse

urine. I: Quantitative proportions of major chemosignals. J Chem Ecol 33:417–434
39. Soini HA, Wiesler D, Koyama S, Féron C, Baudoin C, Novotny MV (2009) Differences in urinary scents of two related mouse species, *Mus spicilegus* and *Mus domesticus.* J Chem Ecol 35:580–589
40. Patris B, Baudoin C (2000) A comparative study of parental care between two rodent species: implications for the mating system of the mound-building mouse *Mus spicilegus.* Behav Processes 51:35–43
41. Poteaux B, Gouat P, Jacquot C, Christophe N, Baudoin C (2008) Socio-genetic structure of mound-building mice, *Mus spicilegus*, in autumn and early spring. Biol J Linn Soc Lond 93:689–699
42. Nishimura K, Utsumi K, Yuhara M, Fujitani Y (1989) Identification of puberty-accelerating pheromones in male mouse urine. J Exp Zool 251:300–3005
43. Singer AG, Beauchamp GK, Yamazaki K (1997) Volatile signals of the major histocompatibility complex in male mouse urine. Proc Natl Acad Sci USA 94:2210–2214
44. Zhang JX, Rao XP, Sun L, Zhao CH, Qin XW (2007) Putative chemical signals about sex, individuality, and genetic background in the preputial gland and urine of the house mouse (*Mus musculus*). Chem Senses 32:293–303
45. Lin DY, Zhang SZ, Block E, Katz LC (2005) Encoding social signals in the mouse main olfactory bulb. Nature 434:470–477
46. Willse A, Kwak J, Yamazaki K, Preti G, Wahl JH, Beauchamp GK (2006) Individual odortypes: interaction of MHC and background genes. Immunogenetics 58:967–982
47. Osada K, Tashiro T, Mori K, Izumi H (2008) The identification of attractive volatiles in aged male mouse urine. Chem Senses 33:815–823
48. Schaefer ML, Wongravee K, Holmboe ME, Heinrich NM, Dixon SJ, Zeskind JE, Kulaga HM, Brereton RG, Reed RR, Trevejo JM (2010) Mouse urinary biomarkers provide signatures of maturation, diet, stress level and diurnal rhythm. Chem Senses 35:459–471
49. Schwende FJ, Jorgenson JW, Novotny M (1984) A possible chemical basis for the histocompatibility-related mating preference in mice. J Chem Ecol 10:1603–1615
50. Jemiolo B, Gubernick DJ, Yoder MC, Novotny M (1994) Chemical characterization of urinary volatile compounds of *Peromyscus californicus*, a monogamous biparental rodent. J Chem Ecol 20:2489–2500
51. Röck F, Mueller S, Weimar U, Rammensee HG, Overath P (2006) Comparative analysis of volatile constituents from mice and their urine. J Chem Ecol 32:1333–1346
52. Soini HA, Bruce KE, Wiesler D, David F, Sandra P, Novotny MV (2005) Stir bar sorptive extraction: a new quantitative and comprehensive sampling technique for determination of chemical signal profiles from biological media. J Chem Ecol 31:377–392
53. Röck F, Hadeler KP, Rammensee HG, Overath P (2007) Quantitative analysis of mouse urine volatiles: in search of MHC-dependent differences. PLoS One 5(e429):1–8
54. Baltussen E, David F, Sandra P, Janssen H-G, Cramers CA (2002) Sorptive sample preparation—a review. Anal Bioanal Chem 373:3–22
55. Novotny MV, Soini HA (2008) Volatile mammalian chemosignals: structural and quantitative aspects. In: Hurst JL, Beynon RJ, Roberts SC, Wyatt T (eds) Chemical signals in vertebrates 11. Springer, New York, pp 13–23
56. Soini HA, Koyama S, Karn RC, Novotny MV (2013) To be published
57. Kovats E (1965) Gas chromatographic characterization of organic substances in the retention index system. In: Giddings JC, Keller RA (eds) Advances in chromatography. Marcel Dekker, New York, pp 229–247
58. Novotny M, Lee ML, Bartle KD (1974) Some analytical aspects of the chromatographic headspace concentration method using a porous polymer. Chromatographia 7:333–338
59. Novotny M, McConnell ML, Lee ML, Farlow R (1974) High resolution gas-chromatographic analysis of the volatile constituents of body fluids with use of glass capillary columns. Clin Chem 20:1105–1110
60. Novotny M, McConnell ML, Lee ML (1974) Some aspects of high-resolution gas-chromatographic analysis of complex volatile samples. J Agric Food Chem 22:765–769
61. Baltussen E, David F, Sandra P, Janssen H-G, Cramers CA (1998) Sorption tubes packed with polydimethylsiloxane: a new and promising technique for the preconcentration of volatiles and semivolatiles from air and gaseous samples. J High Resolut Chromatogr 21: 332–340
62. Baltussen E, Sandra P, David F, Cramers C (1999) Stir bar sorptive extraction (SBSE), a novel extraction technique for aqueous samples: theory and principles. J Microcolumn Sep 11:737–747
63. Tienpont B, David F, Desmet K, Sandra P (2002) Stir bar sorptive extraction-thermal desorption-capillary GC-MS applied to biological fluids. Anal Bioanal Chem 373:46–55
64. Blanchard J (1981) Evaluation of the relative efficacy of various techniques for deproteinizing plasma samples prior to high-performance liquid chromatographic analysis. J Chromatogr 226:455–460

Chapter 4

Murine Nonvolatile Pheromones: Isolation of Exocrine-Gland Secreting Peptide 1

Hiroko Kimoto and Kazushige Touhara

Abstract

Our search for a substance recognized by the vomeronasal neurons revealed that the extra-orbital lacrimal gland (ELG) isolated from adult male mice produced the male-specific peptide pheromone exocrine gland-secreting peptide 1 (ESP1). The following protocol reveals how ESP1 may be extracted from the ELG, purified using anion-exchange and reverse-phase high-performance liquid chromatography (HPLC), and analyzed by mass spectrometry. This protocol has been specifically designed for the purification of ESP1, but may be modified to isolate a variety of peptides from the exocrine glands. Peptides purified in this manner may help further define the molecular mechanisms underlying pheromone communication in the vomeronasal system.

Key words Exocrine gland-secreting peptide 1, Extra-orbital lacrimal gland, Vomeronasal organ, Dialysis, High-performance liquid chromatography, Mass spectrometry

1 Introduction

The vomeronasal organ (VNO) plays an essential role in the detection of pheromones by mice [1]. The vomeronasal pumping system enables the uptake of not only volatile chemicals, but also nonvolatile peptides and proteins into the VNO [2, 3]. Whereas volatile chemical substances are transferred immediately, but transiently, and vary in accordance with the physiological conditions, peptides and proteins remain intact at scent marks for extended periods of time and convey definite information throughout their life [1]. Volatiles and nonvolatile peptides have been suggested to mediate distinct types of social and sexual information [4].

Several chemical compounds originating in the urine and the major urinary proteins have been reported to stimulate the vomeronasal sensory neurons [5, 6]. Neuronal firing in the vomeronasal system of behaving mice has also been shown to be modulated following physical contact with the facial area [7], implicating the existence of additional sources of pheromone secretion other than urine.

Kazushige Touhara (ed.), *Pheromone Signaling: Methods and Protocols*, Methods in Molecular Biology, vol. 1068,
DOI 10.1007/978-1-62703-619-1_4, © Springer Science+Business Media, LLC 2013

Whilst searching for the source of pheromone substances recognized by the vomeronasal neurons among extracts of exocrine glands, we discovered that the extra-orbital lacrimal gland (ELG), which lies subcutaneously at the base of the ear, secreted an active component [8]. Via a three-step purification process using anion-exchange and reverse-phase high-performance liquid chromatography (HPLC), we identified a male specific 7-kDa peptide that we termed exocrine gland-secreting peptide 1 (ESP1) [8]. ESP1 is a member of a multi-gene family that is clustered within close proximity to the class I major histocompatibility complex region [8, 9]. ESP1 induced c-Fos expression in the vomeronasal sensory neurons expressing the V2Rp5 receptor in female mice, and evoked a neural action potential [8–10]. In addition, ESP1 released into the tears of males exhibits an important role as a sex pheromone that enhances female sexual receptive behavior [10].

Here, we describe protocols for extraction, purification, and analysis of ESP1 from the ELG. The innate ESP1 isolated in this manner has proved useful for further defining the molecular basis underlying information transmission between individuals via the vomeronasal system. Recombinant ESP1 produced in *E. coli*, which has a c-Fos inducing activity practically equivalent to innate ESP1 [8], has also be utilized.

2 Materials

2.1 ELG Sampling Components

1. Adult male BALB/c mice.
2. Forceps and fine dissection scissors.
3. Tris buffer (20 mM Tris–HCl, pH 7.5).
4. 1.5-ml microcentrifuge tubes.
5. Pellet pestle for 1.5-ml microcentrifuge tube.
6. High speed refrigerated microcentrifuge.
7. Dialysis tubing (10,000 MWCO, Spectra/Por; Spectrum Laboratories, *see* **Note 1**).
8. Closure clips for dialysis tubing (*see* **Note 2**).
9. Magnetic stirrer.
10. Magnetic stir bar.

2.2 HPLC Components

All buffers and solvents used in the following method should be HPLC-grade or filtered using a 0.22-μm filter prior to use. Degassing mobile-phase solvents by sonication under pressure or a HPLC degasser is necessary to prevent air bubble formation in flow lines.

1. 0.45-μm filter (MILLEX-HV; Millipore).
2. Microcentrifuge tubes.

3. HPLC system (*see* **Note 3**).
4. Anion-exchange DEAE column (TSK-GEL DEAE-5PW, 7.5×75 mm or 22.5×150 mm; Tosoh, *see* **Note 4**).
5. Tris buffer (20 mM Tris–HCl, pH 7.5).
6. 0.5 M NaCl in Tris buffer.
7. 50-ml conical tube (BD Falcon; BD Biosciences).
8. Liquid nitrogen.
9. Vacuum freeze drying equipment.
10. Reverse-phase C4P column (PEGASIL-300, 4.6×250 mm; Senshu, *see* **Note 4**).
11. 10 % acetonitrile (ACN) in 0.1 % trifluoroacetic acid (TFA).
12. 60 % ACN in 0.1 % TFA.
13. Siliconized microcentrifuge tubes (*see* **Note 5**).
14. SpeedVac concentrator (GMI).
15. 30 % ACN in 0.1 % heptafluorobutyric acid (HFBA).
16. 60 % ACN in 0.1 % HFBA.

2.3 Mass Spectrometry Components

1. Matrix: 10 mg of α-cyano-4-hydroxycinnamic acid (CHCA) in 50 % ACN containing 0.1 % TFA (*see* **Note 6**).
2. Matrix-assisted laser desorption ionization-time of flight/mass spectrometer (Voyager-DE STR; Applied Biosystems).
3. Bovine insulin (Sigma).

3 Methods

3.1 ELG Sampling

1. Collect the intact ELGs from sacrificed adult male BALB/c mice. Rinse the collected samples with ice-cold Tris buffer and transfer to a 1.5-ml microcentrifuge tube (*see* **Note 7**).
2. Add ice-cold Tris buffer (10 μl/mg of gland tissue) to the tube and homogenize with a pellet pestle on ice.
3. Centrifuge at 14,000×*g* for 10 min at 4 °C.
4. During centrifugation, prepare dialysis tubing by rinsing in deionized water and seal one end of the tubing with a closure clip (*see* **Note 8**).
5. Transfer the supernatant into the dialysis tubing and close the open end using a second closure clip (*see* **Note 2**).
6. Place the dialysis tubing in a large volume of Tris buffer, and dialyze with stirring overnight at 4 °C with two to three buffer changes (*see* **Note 9**).
7. At the completion of dialysis, remove one closure clip and pipette the dialyzed sample into a clean microcentrifuge tube (*see* **Note 10**).

3.2 Purification of ESP1

Please note, that when not in use, columns and flow lines need to be filled with an organic solvent such as 80 % methanol or 90 % acetonitrile.

1. Filter the dialyzed sample through a 0.45-μm filter into a fresh microcentrifuge tube (*see* **Note 11**).
2. Prepare the mobile phase for anion-exchange HPLC by filling the apparatus with Tris buffer with or without 0.5 M NaCl. Set the flow rate to 0.5 ml/min.
3. Connect an anion-exchange DEAE column and rinse with Tris buffer (*see* **Note 12**).
4. Inject the filtered sample acquired at **step 1** onto the DEAE column and elute with Tris buffer. While monitoring the peptide effluent at 280 nm, collect the flow through into a 50 ml conical tube (*see* **Note 13**).
5. Freeze the flow through sample in liquid nitrogen and lyophilize using vacuum freeze drying equipment (*see* **Note 14**).
6. Set up the mobile phase for reverse-phase HPLC by changing the solution filling the apparatus to 10 % or 60 % ACN in 0.1 % TFA. Set the flow rate to 1 ml/min.
7. Connect a reverse-phase C4P column and wash using a linear gradient program of 10–60 % ACN in 0.1 % TFA over 10 min at a flow rate of 1 ml/min.
8. Dissolve the lyophilized flow through sample (acquired at **step 5**) in 10 % ACN in 0.1 % TFA.
9. Inject the dissolved sample onto the C4P column and run a linear gradient program of 10–60 % ACN in 0.1 % TFA over 50 min at a flow rate of 1 ml/min (Fig. 1a). Collect 1 ml fractions in siliconized microcentrifuge tubes.

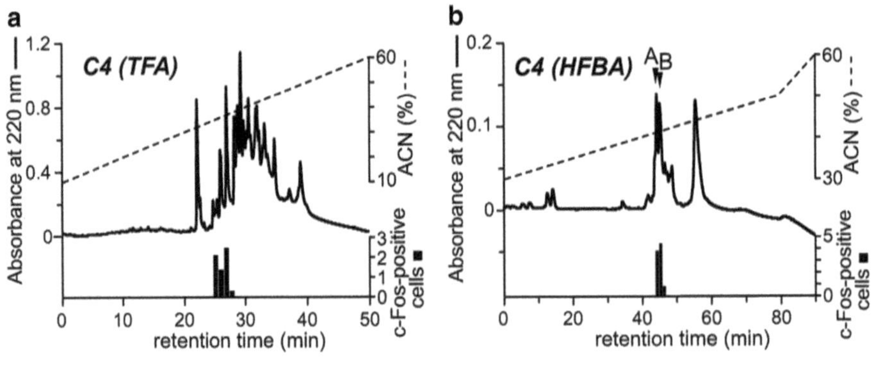

Fig. 1 Purification and analysis of ESP1 from the ELGs of male BALB/c mice. (**a**) The second step in HPLC purification; a C4 column and TFA were used. This elution profile was obtained from a sample that contained the equivalent of 48 ELGs. (**b**) The third step in the HPLC purification; a C4 column and HFBA were used. This elution profile was obtained from a sample that contained the equivalent of 43 ELGs. Fractions A and B had c-Fos-inducing activity. These figures were reproduced from [8]

10. Check the activity using a sample of each fraction equivalent to five ELGs. Transfer to a 1.5-ml microcentrifuge tube and dry in a SpeedVac concentrator. Dissolve peptide in 200 μl of Tris buffer, transfer all the peptides with 35–40 mg of a cotton swab, and repeat the drying process. Expose the cotton swab transfused with the fraction to a BALB/c female mouse and examine c-Fos induction in the vomeronasal sensory neurons (*see* **Note 15** and Chapter 19 for the detailed c-Fos induction methodology).
11. Set up the mobile phase for the second reverse-phase HPLC using the same C4P column. Transfer filled solution to 30 or 60 % ACN in 0.1 % HFBA (*see* **Note 16**). Set the flow rate to 1 ml/min.
12. Connect the C4P column and wash using a linear gradient program of 30–60 % ACN in 0.1 % HFBA over 10 min at a flow rate of 1 ml/min.
13. Dry the active fractions (acquired at **step 9**) in a SpeedVac concentrator and dissolve in 30 % ACN in 0.1 % HFBA.
14. Inject the dissolved active component onto the C4P column and run a linear gradient program of 30–50 % ACN in 0.1 % HFBA over 80 min followed by 50–60 % ACN in 0.1 % HFBA over 10 min at a flow rate of 1 ml/min (Fig. 1b). Collect 1-ml fractions in siliconized microcentrifuge tubes.
15. Check the activity using a portion of each fraction and the c-Fos assay (*see* **step 10** and **Note 17**).

3.3 Analysis by Mass Spectrometry

1. Mix a few microliters of the peptide fraction with 1 μl of the matrix.
2. Spot 1 μl of the mixture onto the MALDI plate and dry. Record the plate position numbers for each of the samples.
3. Prepare for mass spectrometry. Calibrate the mass scale by acquiring spectra of a known peptide such as bovine insulin.
4. Measure the molecular mass of the peptides in the active fractions following the manufacturer's instructions (Fig. 2a).

If you would like to confirm the amino acid sequence of the purified ESP1, we suggest using N-terminal protein sequencing (Fig. 2b). For N-terminal amino acid analysis, transfer the HPLC active fraction onto glass support disks pre-cycled with polybrene and run the sequencer. MS/MS experiments may also be utilized to identify the sequence by searching MS/MS spectra of peptides against sequences in known DNA or protein databases.

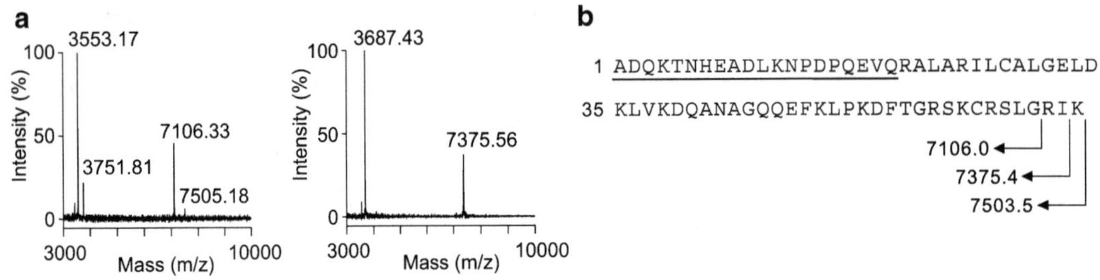

Fig. 2 (**a**) Mass spectra of the active fractions A and B from the final column. Fraction A exhibited a 7,106 Da main peak and a 7,505 Da minor peak (*left*). Fraction B exhibited a 7,376 Da peak (*right*). (**b**) The amino-acid sequence of ESP1. The N-terminal amino-acid sequence obtained from the peptides in fractions A and B is *underlined.* The molecular weights determined by mass spectrometry corresponded to three different C-terminal truncated forms of the peptide. The *numbers* show the theoretical molecular weights for each peptide. These figures were reproduced from [8]

4 Notes

1. We were able to use this dialysis tubing even though the size of ESP1 was under 10 kDa. We hypothesize that this was likely due to oligomerization of ESP1. Consider using other MWCO size tubing, depending on the size of a target peptide.
2. A knot tied in the tubing may also be used as a substitute for a closure clip. However, simply forming a knot risks leakage and alteration to pore size.
3. The HPLC system should be able to mix two mobile phases in a gradient manner, be set to manage a flow rate precisely and detect on a UV absorption spectrum. Although we collected each fraction manually, a fraction collector and an auto-sampler may be used.
4. HPLC columns produced by different manufacturers may be used successfully. However, results may vary due to column matrixes exhibiting slightly different characteristics.
5. We used siliconized microcentrifuge tubes to prevent sample loss. Other tubes which have a low binding surface may also be used for collection.
6. The matrix is used for improving crystallization. CHCA is appropriate for measuring peptides under 10 kDa.
7. The ELG is located under the skin at the base of the ear and should be easily isolated from the surrounding tissue. ELG samples may be stored at −80 °C until required.
8. Always follow the manufacturer's instructions when using dialysis tubing from different sources.
9. The volume of the dialysis buffer should be at least 200 times the sample volume. This step drastically decreased UV

absorbance at 280 nm of the flow-through sample at **step 4** in Subheading 3.2.

10. The tube containing the dialyzed extract may be stored at −80 °C until required.
11. This step protects the HPLC column and improves reproducibility of the analytical data. The sample may be stored at −80 °C for further use.
12. In this case, it is sufficient to wash the column with Tris buffer. When components absorbed to the column are to be collected, simply run a gradient program of 0–0.5 M NaCl in Tris buffer.
13. ESP1 was not bound to the DEAE column, probably a result of its high isoelectric point. Following the collection of flow-through sample, the DEAE column should be washed using a linear gradient of 0–0.5 M NaCl in Tris buffer.
14. The lyophilized sample may be stored at 4 °C or −20 °C.
15. The dried cotton swabs soaked with each fraction diluted in ACN buffer did not exhibit any c-Fos-inducing activity. The cotton swabs soaked with each fraction that was redissolved in Tris buffer demonstrated very weak activity. Therefore, we concluded that redissolving in Tris buffer and drying in the SpeedVac are required to ensure optimum activity.
16. A TMS column (PEGASIL-300 TMS, 4.6 × 250 mm; Senshu) did not improve the separation or reduce the ACN concentration in active fractions.
17. In our experiments, the active fractions including ESP1 were eluted at approximately 41–42 % ACN (Fig. 1b).

References

1. Touhara K (2008) Sexual communication via peptide and protein pheromones. Curr Opin Pharmacol 8(6):759–764
2. Wysocki CJ, Wellington JL, Beauchamp GK (1980) Access of urinary nonvolatiles to the mammalian vomeronasal organ. Science 207(4432):781–783
3. Meredith M, O'Connell RJ (1979) Efferent control of stimulus access to the hamster vomeronasal organ. J Physiol 286:301–316
4. Isogai Y, Si S, Pont-Lezica L, Tan T, Kapoor V, Murthy VN, Dulac C (2011) Molecular organization of vomeronasal chemoreception. Nature 478(7368):241–245
5. Leinders-Zufall T, Lane AP, Puche AC, Ma W, Novotny MV, Shipley MT, Zufall F (2000) Ultrasensitive pheromone detection by mammalian vomeronasal neurons. Nature 405(6788):792–796
6. Chamero P, Marton TF, Logan DW, Flanagan K, Cruz JR, Saghatelian A, Cravatt BF, Stowers L (2007) Identification of protein pheromones that promote aggressive behaviour. Nature 450(7171):899–902
7. Luo M, Fee MS, Katz LC (2003) Encoding pheromonal signals in the accessory olfactory bulb of behaving mice. Science 299(5610):1196–1201
8. Kimoto H, Haga S, Sato K, Touhara K (2005) Sex-specific peptides from exocrine glands stimulate mouse vomeronasal sensory neurons. Nature 437(7060):898–901
9. Kimoto H, Sato K, Nodari F, Haga S, Holy TE, Touhara K (2007) Sex- and strain-specific expression and vomeronasal activity of mouse ESP family peptides. Curr Biol 17(21):1879–1884
10. Haga S, Hattori T, Sato T, Sato K, Matsuda S, Kobayakawa R, Sakano H, Yoshihara Y, Kikusui T, Touhara K (2010) The male mouse pheromone ESP1 enhances female sexual receptive behaviour through a specific vomeronasal receptor. Nature 466(7302):118–122

Chapter 5

Chemical Analysis of Aquatic Pheromones in Fish

Michael Stewart, Cindy F. Baker, and Peter W. Sorensen

Abstract

Pheromones are chemicals that pass between members of the same species that have inherent meaning. In the case of fish, pheromones are water-soluble and found in low concentrations. As such, sensitive and selective methods are needed to separate and analyze these pheromones from an environmental matrix that may contain many other chemicals. This chapter describes a generic method used to concentrate and identify these chemicals and two extremely sensitive and selective methods for analysis, namely, mass spectrometry and enzyme-linked immunosorbent assay.

Key words Fish pheromone, Liquid chromatography, Solid phase extraction chromatography, Mass spectrometry, Enzyme-linked immunosorbent assay, ELISA

1 Introduction

Pheromones, or chemical signals that pass between members of the same species, are used by fish to mediate a variety of physiological responses and behaviors including endocrinological synchrony, migratory orientation, sexual arousal, attraction, and aggression. While most fish pheromones are mixtures of relatively unspecialised bodily metabolites, some can function as independent entities [1]. Approximately a dozen fish pheromones have now been identified, meaning that their release, detection and biological function have all been elucidated and their chemical structures identified [2]. There are also many chemicals utilized by fish that are not strictly pheromones, but these compounds still play an important role in the life cycle of fish, such as predator detection, locating food, shoaling, and homing. This chapter only covers techniques used for the detection and analysis of chemicals utilized by fish and does not address their biological importance.

1.1 Pheromone Classes

Steroids and prostaglandins are the predominant classes of compounds that have been identified as pheromones used by fish. These are usually modified with polar functional groups, thereby

Kazushige Touhara (ed.), *Pheromone Signaling: Methods and Protocols*, Methods in Molecular Biology, vol. 1068,
DOI 10.1007/978-1-62703-619-1_5, © Springer Science+Business Media, LLC 2013

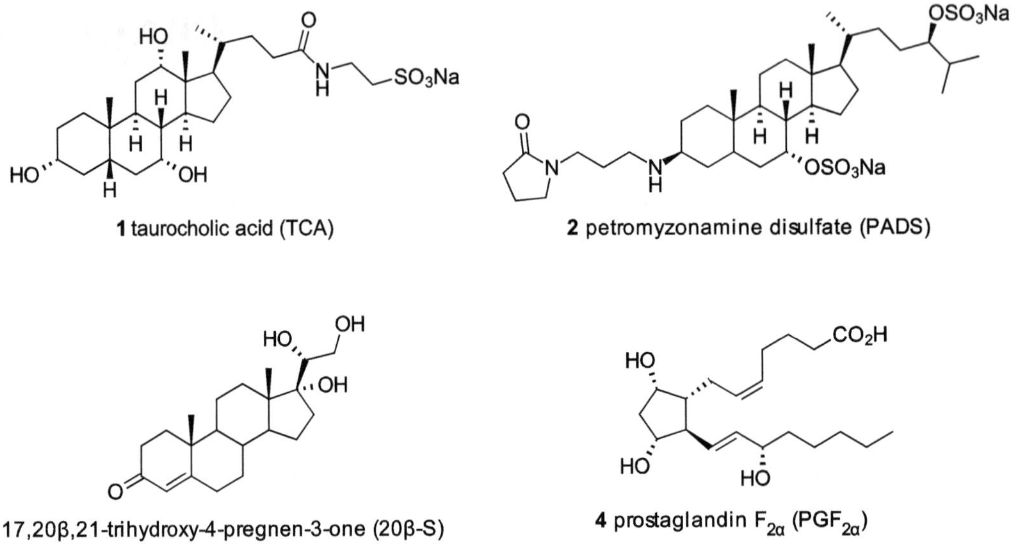

Fig. 1 Four important fish odorants with pheromonal function

increasing their water solubility. Four important classes of chemicals used by fishes as odorants, and putative pheromones, are illustrated in Fig. 1. Bile acids (e.g., taurocholic acid (TCA; **1**) have been shown to be potent olfactory stimuli for many fish species [3–5]. Amino sterols (e.g., petromyzonamine disulfate (PADS; **2**) are part of the sea lamprey migratory pheromone mixture, which contains at least three related steroids [6]. 17,20β,21-Trihydroxy-4-pregnen-3-one (20β-S; **3**) has been identified as a maturation-inducing hormone and pheromonal precursor in the percid fish, *Gymnocephalus cernuus* [7]. F series prostaglandins (e.g., $PGF_2\alpha$; **4**) are part of the goldfish, *Carassius auratus*, and common carp, *Cyprinus carpio*, postovulatory hormonal pheromone mixture [2, 8].

Polar functional groups that ensure water solubility—i.e., sulfate, glucuronic acid, carboxylic acid, amide, hydroxyl—are present in many pheromones including the four compounds depicted (Fig. 1). These polar functional groups also reduce volatility of the compound, so the methods used to separate them for analysis are different from those used for airborne pheromones; namely liquid chromatography (LC) instead of gas chromatography (GC).

Detection of very low concentrations of fish pheromones requires highly specific and highly sensitive methods. Two techniques used commonly are mass spectrometry (MS)—including hyphenated derivatives (MS/MS)—and enzyme-linked immunosorbent assay (ELISA). Even with these extremely sensitive techniques, a pre-concentration step is usually also required, and solid phase extraction (SPE) chromatography is the method commonly employed.

1.2 SPE

Solid phase extraction (SPE) is a method used to concentrate chemicals of interest from water for analysis. This may also involve separating them from an interfering matrix, such as urine. SPE methods are generically similar but are application specific, so the chromatographic supplier literature should be consulted for the best conditions for a specific application. This chapter will describe a generic method developed in our laboratories to concentrate lamprey bile acids from water samples using Oasis HLB SPE cartridges (Waters Corporation).

1.3 LC/MS

Liquid chromatography/mass spectrometry (LC/MS) is a rapidly evolving technique that is becoming more widespread in the analysis of fish pheromones, due to its high sensitivity and selectivity [8–11]. The analysis of water soluble chemicals is primarily carried out using reverse phase high pressure liquid chromatography (RP HPLC), which employs columns containing small uniform sorbent particles (3–5 μm) to provide a highly efficient, automated separation procedure that delivers chemicals in a suitable format for analysis. For chemicals with acidic or basic functional groups, buffers (e.g., ammonium acetate or ammonium formate) and/or pH modifiers (e.g., formic acid or triethylamine) are usually added to the HPLC solvent. These buffers and modifiers give reliable retention times, good peak shape and improved detection by mass spectrometry.

1.4 ELISA

ELISA is a type of immunoassay which uses competitive binding to antibodies to measure miniscule quantities of compounds, usually via a fluorescent label (see below). ELISA often represents a reasonable alternative to direct chemical measurement because it can be very quick and inexpensive. It can also analyze many samples at once (typically several dozen through the use of a 96 well assay). Additionally, some structures either do not fragment (e.g., androstenedione) or do not fragment in a distinctive manner (e.g., 13,14-dihydro-15-keto-prostaglandin $F_{2\alpha}$) in mass spectrometry, so ELISA may be a good primary method [8]. ELISAs require both an antibody for the structure of interest as well as synthesized standards, both of which can be difficult (and expensive) to produce and beyond the scope of most workers (and this chapter). Thus, almost all workers, including us, use commercially available kits to measure fish pheromones. ELISA kits have been developed for some pheromones that also happen to be common metabolites in mammals (e.g., prostaglandin $F_{2\alpha}$ and metabolites, androstenedione). Many types of ELISA exist (indirect, sandwich, competitive, etc.) and we will not review them here other than to mention that ELISAs either employ an antibody which binds to the antigen (compound of interest), and then either a standard that is labeled or an enzyme that is linked to standard. Either way, the amount of antigen present in the sample is determined by measuring the presence of label bound to an antibody which in turn reflects some type of competition between the quantity of antigen found in the sample and that added as part of the kit's protocol. These reactions are carried out in plates with wells (usually 96) that are also generally treated with antibodies to capture the product. Not surprisingly, different companies have produced a wide variety of kits. These employ different suites of antibodies, enzymes, labels and reagents which need to deployed in kit-specific manners, and whose instructions must be followed very carefully. Generally, the strategy behind the chemical reaction being exploited is provided by the kit. Here, we provide some tips on issues to consider when using these kits.

2 Materials

2.1 SPE

1. Oasis HLB solid phase extraction cartridges (60 mg).
2. Nanopure water.
3. Analytical grade methanol (*see* **Note 1**).
4. Automatic pipette (10–1,000 μL).
5. Vacuum manifold *or* syringe with luer slip end and adaptor for SPE.
6. Rotary evaporator (if available).
7. Nitrogen gas blow down apparatus.
8. 2 mL amber glass vials with lids and septa.
9. Low volume glass inserts (150 μL).
10. Oxygen free nitrogen.
11. Sonicator (recommended).

2.2 LC/MS

1. Acetonitrile (LC/MS grade).
2. Nanopure water.
3. Triethylamine.
4. Automatic pipette (10–1,000 μL).
5. LC/MS: Thermo Finnigan Surveyor LC with MS pump and autosampler. LCQ Advantage Ion Trap mass spectrometer with ESI probe operating in negative mode.
6. Nitrogen generator: Peak Scientific.
7. Column: Phenomenex Gemini, 5 μm, 150 × 2 mm, 110 Å.
8. Standards (prepared at a concentration of 100 μg/mL).
9. 2 mL amber glass vials with lids and septa.

2.3 ELISA

1. ELISA kit (typically contains 96 well antibody-coated microplate, standards, enzyme conjugate, substrate, EIA buffer, wash buffer, instructions, information on antibody specificity) (*see* **Note 2**).

2. Ultrapure water (*see* **Note 3**).
3. Automatic, calibrated pipette (10–1,000 μL) (*see* **Note 4**).
4. Pipette tips.
5. Pheromone standards (can be prepared at a concentration of 100 μg/mL).
6. Clean test tubes to mix standards.
7. Plastic film to cover plates.
8. ELISA microplate reader with appropriate wavelength as well as associated computer and software.
9. Microplate shaker (recommended).

3 Methods

3.1 SPE

The following method describes a common approach used for concentrating moderately polar chemicals from water samples, using Oasis HLB cartridges [10]. This method was designed for low volumes of water (up to 20 mL) but can be scaled up accordingly by increasing the size of the cartridge to allow for larger water volumes.

1. Wash SPE cartridge (60 mg) with methanol (2 mL) (*see* **Notes 5** and **6**).
2. Wash SPE cartridge with H_2O (2 mL).
3. Elute water samples (20 mL) through the SPE, discarding the eluent (*see* **Note 7**).
4. Remove excess H_2O by vacuum or several passes of air through with positive pressure.
5. Immediately elute the material retained by the SPE cartridge into a 2 mL amber glass vial with methanol (2 mL) (*see* **Note 8**).
6. Remove all methanol under a stream of N_2 gas (*see* **Note 9**).
7. Redissolve in 1:1 methanol:H_2O (100 μL), cap vial and sonicate briefly (*see* **Note 10**).
8. Transfer to a low volume 150 μL glass insert, inside the 2 mL amber glass vial. Seal with a lid and septum.
9. Store vials in a −20 °C freezer.

3.2 LC/MS

An LC/MS method was developed within the Stewart laboratory to analyze fish tissues and water for common bile acids (Fig. 2). This particular method uses an alkaline modifier (2 mM triethylamine; *see* **Note 11**) and an LC column that is stable to basic conditions (Phenomenex Gemini; *see* **Note 12**). The rationale behind this was that as all bile acids analyzed had acidic functional groups, an alkaline mobile phase would enhance formation of negatively charged ions and improve the MS analysis.

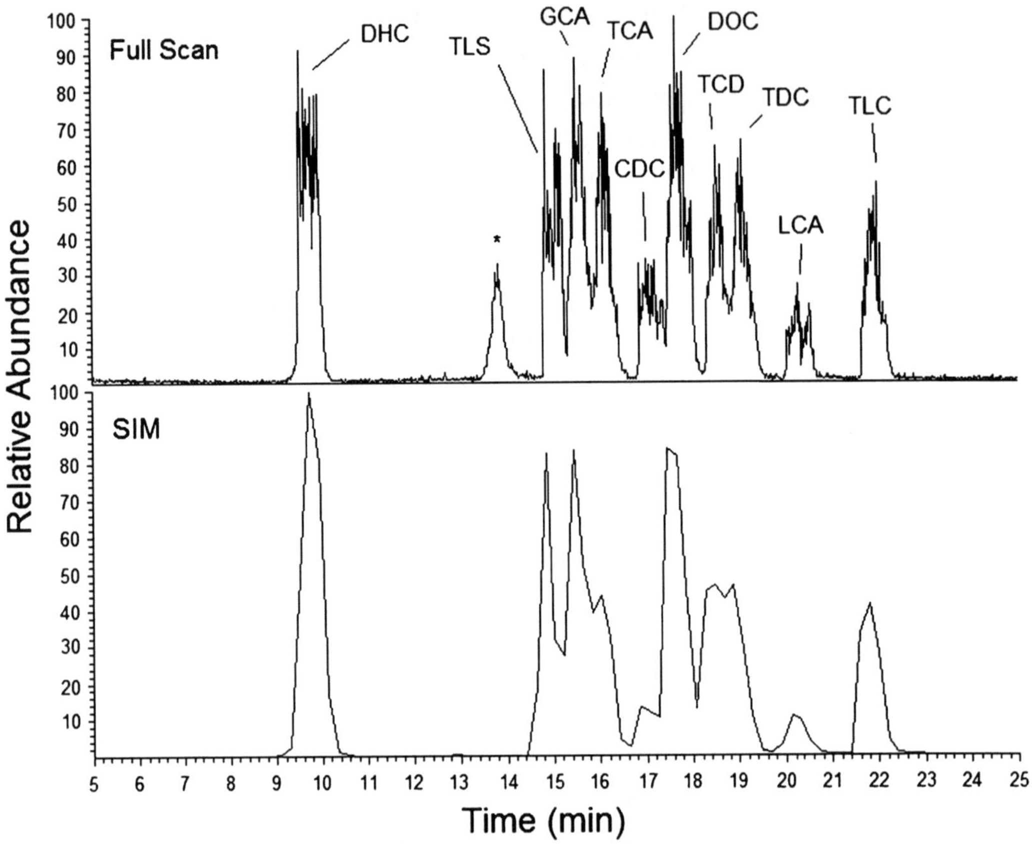

Fig. 2 LC/MS traces of 10 bile acid standards. Top trace is the full scan chromatogram (*m/z* 200–700) and bottom trace is the SIM windows chromatogram. Standards are DHC (Dehydrocholic acid), TLS (Taurolithocholic acid 3-sulfate disodium salt), GCA (Glycocholic acid), TCA (Taurocholic acid), CDC (Chenodeoxycholic acid), DOC (Deoxycholic acid), TCD (Sodium taurochenodeoxycholate), TDC (Sodium taurodeoxycholate hydrate), LCA (Lithocholic acid), and TLC (Sodium taurolithocholate). The peak labelled with an *asterisk* in the full scan chromatogram is an impurity

1. Ensure solvent reservoirs on the LC have sufficient solvent for all the required analyses. Place H_2O in line A and acetonitrile in line C (*see* **Note 13**).
2. Prepare 20 mM triethylamine modifier by pipetting 697 μL of triethylamine into 250 mL of H_2O. Place this in line D (*see* **Note 14**).
3. Set column temperature to 30 °C.
4. Use the following gradient chromatography method: Linear gradient from 80:10:10 to 0:90:10 A:C:D over 60 min. Hold 0:90:10 for 5 min. Linear gradient from 0:90:10 to 80:10:10 ACD over 1 min. Hold 80:10:10 for 9 min. Total run time 75 min. Flow 250 μL/min (*see* **Note 15**).
5. Use the following parameters for MS: Sheath gas 80 arbitrary units (approx. 1.2 L/min), aux/sweep gas 10 arbitrary units

(approx. 0.3 L/min), spray voltage 5.00 V, capillary temperature 280 °C, capillary voltage –17.00 V, Tube lens offset –35.00.

6. Set following analysis windows for MS: negative mode, full scan *m/z* 200–700 (*see* **Note 16**).
7. Equilibrate the LC column with starting solvent (*see* **Note 17**).
8. Run at least one blank analysis to ensure column has been properly cleaned from previous use, is equilibrated for sample analysis and that MS is at correct operating conditions (*see* **Note 18**).
9. Inject 10 μL of each standard using the autosampler (*see* **Note 19**).
10. Set SIM windows for individual compounds being analyzed (*see* **Note 20**).
11. Inject 10 μL of each sample using the autosampler (*see* **Note 21**).
12. Obtain spectrum for each peak observed in full scan analysis (*see* **Note 22**).

Modification of the MS SIM windows (*see* **Note 23**)—but keeping the same LC method—allowed for two subsequent applications in the analysis of fish pheromones.

The first application of this particular method was the analysis of the sea lamprey (*Petromyzon marinus*) migratory pheromones petromyzonol sulfate (PS), petromyzonamine disulfate (PADS; compound 2; Fig. 1), and petromyzosterol disulfate (PSDS) from tank water containing ammocetes of the southern pouched lamprey, *Geotria australis* (unpublished). Migratory pheromones utilized by *G. australis* had not previously been studied. Full scan, SIM (*see* Fig. 3) and SRM (*see* **Note 24**) analyses were undertaken using authentic standards to confirm or deny presence. As Fig. 3 suggests, PS was detected in the lamprey holding water with PADS and PSDS undetected. A subsequent follow-up study was performed using a much more sensitive triple quadrupole MS, which confirmed that PS was the dominant pheromone released, PADS was a very minor component (0.33 %) and PSDS was not detected [12].

A second application involved female redfin perch, *Perca fluviatilis.* Pre-spawning females were injected with a priming hormone (Ova-RH, a salmon GnRH analogue), and the maturation inducing hormone 20β-S [3] and then placed in small (2 L) aerated tanks. Water was replaced every 12 h and released analytes concentrated by SPE. Time series LC/MS analyses were undertaken using both full scan and SIM windows for conjugates and metabolites of 20β-S (Fig. 4). 20-βS was the dominant chemical released after 12 h, but reduced substantially in the 24 and 36 h analyses. Many conjugates were detected in the 24 and 36 h analyses, including a putative sulfate and glucuronides (×2) of 20β-S (no standards are available to confirm the presence of these compounds, however, they had similar retention times to 17,20-βP sulfate and glucuronide and correct pseudo-molecular ions). The LC method

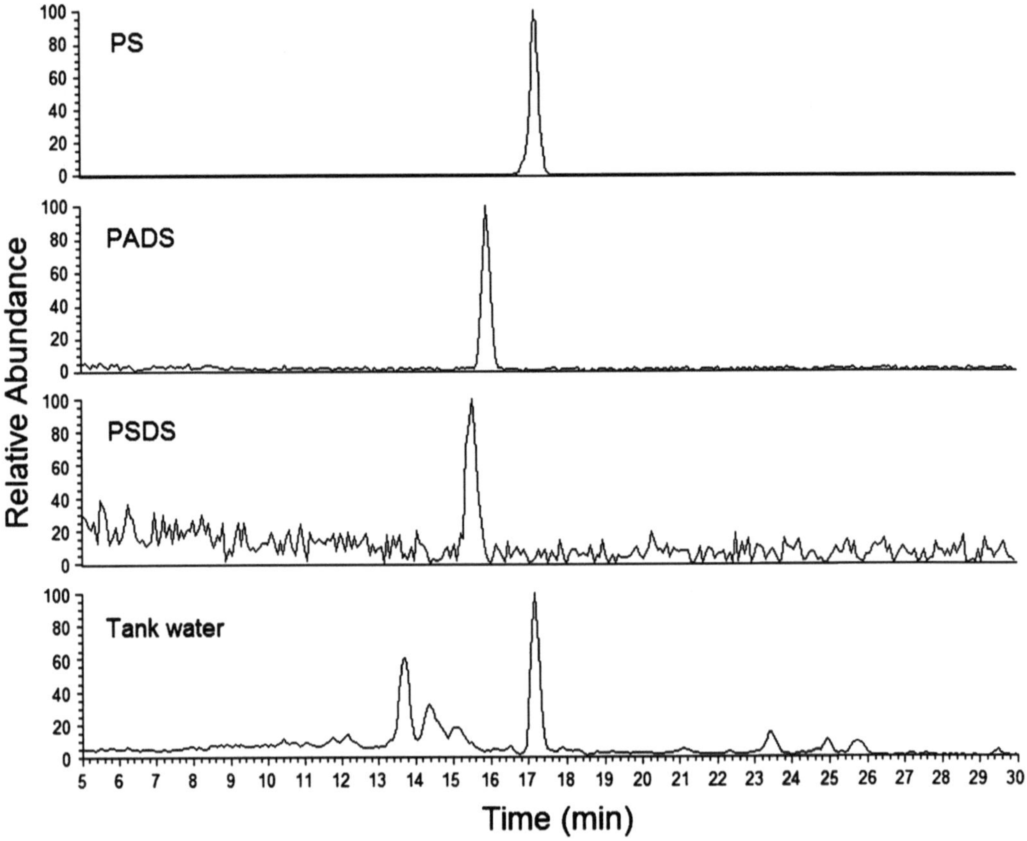

Fig. 3 LC/MS SIM analysis of the southern pouched lamprey (*Geotria australis*) ammocete holding water. The top three traces are the standards: petromyzonol sulfate (PS), petromyzonamine disulfate (PADS), and petromyzosterol disulfate (PSDS). The bottom trace is pre-concentrated holding water which clearly shows presence of PS, but PADS or PSDS were not detected

was subsequently used to fractionate holding water for further chemical analysis, electro-olfactogram (EOG) measurements, and behavior tests (unpublished).

3.3 ELISA

An ELISA method was recently developed by the Sorensen laboratory to analyze water containing ovulated common carp and goldfish for the F prostaglandins that are known to function as sex pheromones in this species [8]. We have successfully used kits produced by both Neogen (Lansing, Michigan; kit #404710) and Cayman Chemical (Ann Arbor, Michigan; kit #5160111) to measure this class of pheromone, as well as sex steroids. These kits employ different approaches, so here we describe commonalities and ask readers to defer to manufacturers for specifics. For ELISA analysis of pheromones, we typically collect 1 L samples of fish holding waters (20–100 g fish are held for 1 h in well water) and extract these waters using SPE cartridges, as described above. In initial tests we also collect three types of controls; at least three samples of well water which lack fish (blank controls to test for

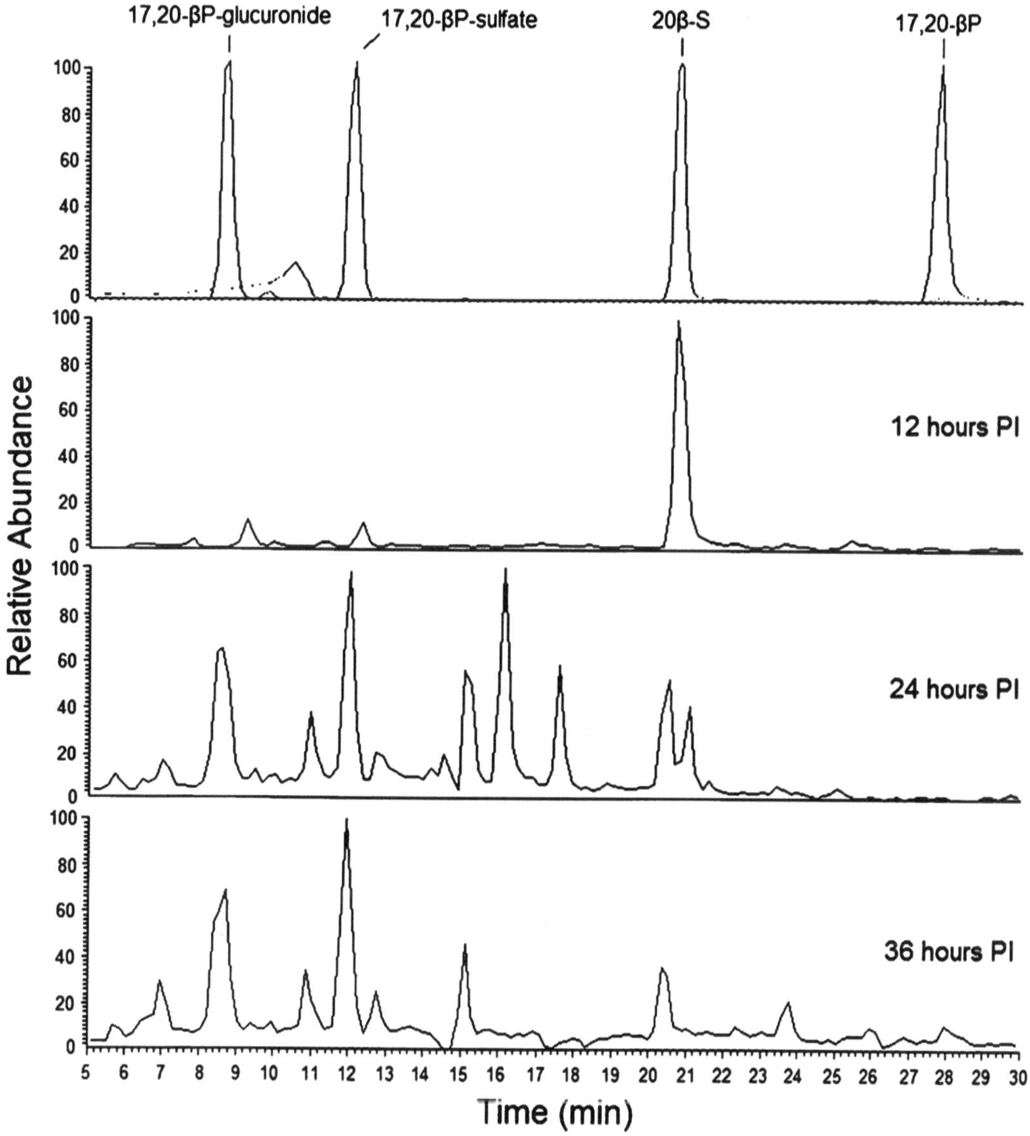

Fig. 4 LC/MS SIM time series analysis of female redfin perch (*Perca fluviatilis*) holding water. The top trace contains four standards including sulfate and glucuronide conjugates of 17,20-βP. No standards were available for 20-βS-sulfate or 20-βS-glucuronide. The bottom three traces contain time series analyses (PI = post injection)

contamination and the possible presence of interfering compounds in water), three samples of water which lack fish but to which we add known amounts of pheromone standards (1 μg) (to confirm antibody specificity and determine extraction efficiency), and three samples which contain fish which we know not to be releasing any pheromone (e.g., another species, or the same species but immature, to confirm antibody specificity, which is never 100 %). Depending on the published specificities of the antibody used in a kit, we may also test compounds closely related to the compound of interest. If in doubt, results may also be confirmed by LC/MS, as described elsewhere [8]. Although all ELISA kits differ, many of the steps they use are same.

The following description is based on that used for acetylcholinesterase competitive ELISAs with some notes for other assay types added:

1. Read kit instructions very carefully. Note that different kits use different combinations of reagents, tracers and standards. Very likely, there will be differences from the protocols we describe here. Expired kits, or kits not kept under specified conditions, should not be used.
2. Plan your experiment to use all 96 wells in an efficient manner. Typically samples (unknowns) are run in duplicate and may include a few blanks and positive controls. Also, at least eight pairs of wells are also reserved for standards to create an eight-point calibration curve. Depending upon the assay type, wells may also be needed for blanks (reagent interference), total activity, non-specific binding (NSB), maximum binding (B_0), B_0/B_{max} (*see* **Note 25**).
3. Prepare EIA buffer and wash buffers as specified by kit directions using supplied materials.
4. Prepare standards as specified by kit instructions using supplied materials. Usually this involves serial dilution.
5. If requested by the kit, prepare enzyme conjugate (tracer), as specified by kit instructions. Store at 4 °C.
6. Prepare antiserum, as specified by the kit. Store at 4 °C.
7. Transfer EIA buffer to wells, as described by kit instructions (*see* **Notes 26** and **27**).
8. Transfer standards to wells by automatic pipette (*see* **Note 28**).
9. Prepare samples (unknowns) and add to the appropriate wells that are not controls or standards. This is accomplished by transferring predetermined aliquots of SPE extracted fish water samples into clean glass vials, drying down under a stream of N_2, and then adding 1 mL of EIA buffer to these vials. Gently shake (or vortex) and allow to sit for 5–10 min to assure complete dissolution. As specified by the kit instructions, then add these (diluted) samples to the wells. 50 μL is a typical volume.
10. If specified by kit instructions, use an automatic pipette to add specified quantities of tracer to all wells, except those that might be designated to measure total activity and blanks.
11. If specified by kit instructions, add specified quantities of antibody to each well except those used for total activity, non-specific binding, and the kit blank. Then very gently move the plate in a circle (or use microplate shaker) to promote mixing,

cover with plastic film, and then incubate per kit instructions (times and conditions will vary greatly by kit) (*see* **Note 29**).

12. Wash and develop (e.g., incubate) the plate as specified by kit instructions. This involves several steps and differs between kits. Usually several washes are suggested using supplied wash buffer which is added by pipette and then the plate is turned upside down onto a paper towel. After washing the plate, assays that measure acetylcholinesterase activity may use Ellman's reagent, after which a tracer is added to the total activity well. The plate is then typically covered and placed on an orbital shaker (or simply shaken by hand). Allow the plate to develop, following kit instructions, which in some cases may involve adding a stop reagent after a set period of time (*see* **Note 30**).
13. After developing the plate, gently shake the plate to ensure even mixing. Clean its bottom with a Kimwipe, remove its plastic film, and place into the plate reader. Operate the reader per kit instructions. Some readers will automatically plot data and determine the necessary relationships.
14. Analyze data, following kit instructions. These typically provide great detail. Briefly, all duplicates should be averaged. Average NSB (substrate background) is then subtracted from the NSB average to yield corrected maximum binding (B_0). Calculate B/B_0 (standard bound/ maximum bound) for the remaining wells after averaging. Plot $\%B/B_0$. The quantity of unknown may then be derived by the logistic equation used to describe the plot. Data that fall outside that of the linear portion of the standard curve should not be used and re-run after dilution or concentration so they fall within the specified relationship. Great attention should be taken to ensure that this relationship is credible; each kit should have its own criteria for fit as well as acceptable intra- and inter-assay variation (*see* **Note 31**).

The first application of this particular method was the analysis of prostaglandin $F_{2\alpha}$ released by ovulated goldfish (unpublished) but later we applied it to the common carp [8]. Analyses were undertaken using a variety of authentic standards in urine and whole waters to confirm or deny presence. A subsequent follow-up study was performed using triple quadrupole MS, which confirmed all results [8].

4 Notes

1. All solvents need to be of analytical grade or higher.
2. Ideally antibody specificity to compounds of likely importance has been predetermined by the kit manufacturer.
3. Ultrapure water generally comes with the kit but you may need extra. Water quality is extremely important for ELISA

because small amounts of organics may interfere with the assay. Even HPLC quality water sometimes is not adequate so this is best purchased from the company.

4. A repeat pipette can be useful, especially if many samples are measured.
5. Keep the SPE covered by solvent at all times, to prevent it drying out. Although polymeric phases like Oasis are not as susceptible to problems with drying as reverse phase cartridges, this is considered best laboratory practice.
6. Apply vacuum (or positive pressure if using a syringe) to ensure the solvent is eluting through the cartridge at a steady drip, not a constant stream, which is too fast. This also ensures adequate time for equilibration of analyte between liquid and solid phases, allowing retention of the analyte.
7. If a pheromone is known to be highly polar, extraction efficiency should be confirmed using standards first. If fish holding water contains a lot debris, it should be passed through filter paper first to prevent clogging of the SPE cartridge.
8. Amber glass vials ensure any compounds that are unstable to light are preserved. This is especially important when samples are exposed to light during work-up and analysis.
9. Oxygen free nitrogen is essential to avoid potential issues with oxidation of your sample. Take care to use a gentle flow of nitrogen gas to avoid splashing and spilling. The process is expedited by concurrent heating of the vial, however, beware of temperature instability and use a maximum temperature of 40 °C.
10. Do not place any part of the hand in the sonicator water while in operation. Observe the inside of the vial to assess when all material is dissolved.
11. Triethylamine has a pK_b of 11.0 and a working pH range of between 10 and 12. For LC/MS it is imperative that volatile buffers are used to avoid build-up of unwanted buffer on the probe and potential degradation (or loss) of signal.
12. Traditional silica based LC columns are susceptible to hydrolysis under alkaline conditions (pH >8), so it is important to use a specially designed column that is stable under these conditions (up to pH 12).
13. All solvents should be filtered through 0.45 mm nylon filters prior to each analysis.
14. Water should be filtered prior to addition of the triethylamine as it is volatile.
15. The final step of 9 min ensures the LC column has been equilibrated before the next injection. Flow rate should be increased

slowly from zero to avoid pressure shock to the LC column. Pressure will change throughout the run and should be monitored over time to assess potential blockages to the system or column.

16. For negative mode ESI under alkaline conditions (2 mM triethylamine) most chemicals produce the pseudo-molecular ions $[M-H]^-$ or $[M-Na]^-$.
17. HPLC columns are usually shipped in the solvent they were tested with. Carry out small incremental (ca. 10 %) changes to the solvent proportions to reach initial conditions. End caps should always be in place when the column is not in use to avoid drying.
18. This also allows the user to assess baseline data and potential contamination on the column or in the MS prior to analysis.
19. Prior to analyzing samples, parameters need to be set for standards. Normally a full scan is used to identify pseudo-molecular ions (*see* **Note 16**), retention time and any fragment ions.
20. SIM windows are user definable, and for this method, were set at *m/z* pseudo-molecular ion ±1. The method can be set to run concurrent full scan and SIM windows. Full scan gives more information (pseudo-molecular ion, adducts, fragment ions) but is less sensitive. SIM analyses are more sensitive but give limited information.
21. The autosampler usually has different modes of injection. If quantitation is required a loop overfill procedure will ensure an exact volume is injected each time. If sample is precious then a partial loop, or no waste, procedure will minimize the amount used.
22. The full scan spectrum gives much higher confidence that the analyte is detected than SIM analysis, as it can be compared with authentic standards. Isotope ratios, fragments and adduct formation can be compared. This information is not available in SIM analysis so it is imperative that both full scan and SIM analyses are run concurrently.
23. Adjusting SIM windows to accommodate new analytes requires method development as each chemical analyzed will exhibit different behavior in the MS (i.e., adduct formation, fragmentation, formation of doubly charged species, and formation of clusters). This cannot necessarily be predicted so it is advisable to run full scan to ascertain the most intense ion formed for each analyte under the elution conditions.
24. SRM analyses can be undertaken with Ion Trap MS. These involve capturing ions of choice and fragmenting to daughter ions, which gives a highly selective analysis. This selectivity however is compromised by lack of sensitivity and (in this case)

was an inferior technique to SIM. Tandem mass spectrometers (MS/MS) use a similar technique (MRM) which are usually highly selective *and* sensitive.

25. Depending on results from pilot studies you may choose to run samples in triplicate instead of duplicate. Some assays will not need all of these controls.
26. Pipetting technique is absolutely critical. Make sure pipettes are calibrated and practice repeatability if you lack experience. Use different pipette tips for each reagent. Equilibrate pipette tips in reagent and then withdraw slowly. Do not insert pipette tips into fluids found in the wells.
27. Pilot studies should be performed using different dilutions of fish extract along with blank and spiked standards to establish the concentrations that are both present and can best be run using the kit. Most assays are designed to measure between 0.001 and 1.0 ng of standard per mL (determined by antibody binding). How much sample and fish water this represents will have to be determined by pilot studies that test various dilutions. For common carp, we diluted samples between 10 (for control samples) and 1,000 times (for fully ovulated fish samples releasing large quantities of prostaglandin $F_{2\alpha}$).
28. Aliquot low concentrations of standards first.
29. Some kits do not require adding antibody because it will already be coating the wells.
30. Increasing development time can sometimes increase assay sensitivity. Developed plates may be kept in the refrigerator and reread at a later date, if in question.
31. Kits usually provide excellent advice on trouble shooting. Here are a few pointers. High variance between duplicates may be a sign of poor pipetting techniques or contamination. A high NSB indicates poor washing. A low B_0 value could indicate contaminants and/or inadequate development time. Low sensitivity could mean the standard is degraded. If only the total activity well develops, then the water source is probably contaminated.

References

1. Stacey NE, Sorensen PW (2011) Hormonal pheromones. In: Farrell AP (eds) Encyclopedia of fish physiology: from genome to environment, vol 2. Academic: San Diego
2. Sorensen PW, Hoye TR (2010) Pheromones in Vertebrates. In: Lew M, Hung-Wen L (eds) Comprehensive Natural Products II. Elsevier, Oxford, pp 225–262
3. Zhang C, Brown S, Hara T (2001) Biochemical and physiological evidence that bile acids produced and released by lake char (*Salvelinus namaycush*) function as chemical signals. J Comp Physiol B 171(2):161–171
4. Michel WC, Lubomudrov LM (1995) Specificity and sensitivity of the olfactory organ of the zebrafish, *Danio rerio*. J Comp Physiol A Neuroethol Sens Neural Behav Physiol 177(2):191–199
5. Baker CF, Carton AG, Fine JM, Sorensen PW (2006) Can bile acids function as a migratory pheromone in banded kokopu, *Galaxias fasciatus* (Gray)? Ecol Freshw Fish 15(3):275–283

6. Sorensen PW, Fine JM, Dvornikovs V, Jeffrey CS, Shao F, Wang JZ, Vrieze LA, Anderson KR, Hoye TR (2005) Mixture of new sulfated steroids functions as a migratory pheromone in the sea lamprey. Nat Chem Biol 1(6):324–328
7. Sorensen PW, Murphy CA, Loomis K, Maniak P, Thomas P (2004) Evidence that 4-pregnen-17,20β,21-triol-3-one functions as a maturation-inducing hormone and pheromonal precursor in the percid fish, *Gymnocephalus cernuus*. Gen Comp Endocrinol 139(1):1–11
8. Lim H, Sorensen P (2011) Polar metabolites synergize the activity of prostaglandin F2α in a species-specific hormonal sex pheromone released by ovulated common carp. J Chem Ecol 37(7):695–704
9. Fine JM, Sorensen PW (2008) Isolation and biological activity of the multi-component sea lamprey migratory pheromone. J Chem Ecol 34(10):1259–1267
10. Stewart M, Baker C, Cooney T (2011) A rapid, sensitive, and selective method for quantitation of lamprey migratory pheromones in river water. J Chem Ecol 37(11):1203–1207
11. Xi X, Johnson NS, Brant CO, Yun S-S, Chambers KL, Jones AD, Li W (2011) Quantification of a male sea lamprey pheromone in tributaries of Laurentian Great Lakes by Liquid Chromatography – Tandem Mass Spectrometry. Environ Sci Technol 45(11):6437–6443
12. Stewart M, Baker C (2012) A sensitive analytical method for quantifying petromyzonol sulfate in water as a potential tool for population monitoring of the Southern Pouched Lamprey, *Geotria Australis*, in New Zealand streams. J Chem Ecol 38(2):135–144

Chapter 6

Analysis of Ascarosides from *Caenorhabditis elegans* Using Mass Spectrometry and NMR Spectroscopy

Xinxing Zhang, Jaime H. Noguez, Yue Zhou, and Rebecca A. Butcher

Abstract

The nematode *Caenorhabditis elegans* secretes a family of water-soluble small molecules, known as the ascarosides, into its environment and uses these ascarosides in chemical communication. The ascarosides are derivatives of the 3,6-dideoxysugar ascarylose, modified with different fatty acid-derived side chains. *C. elegans* uses specific ascarosides, which are together known as the dauer pheromone, to trigger entry into the stress-resistant dauer larval stage. In addition, *C. elegans* uses specific ascarosides to control certain behaviors, including mating attraction, aggregation, and avoidance. Although in general the concentration of the ascarosides in the environment increases with population density, *C. elegans* can vary the types and amounts of ascarosides that it secretes depending on the culture conditions under which it has been grown and its developmental history. Here, we describe how to grow high-density worm cultures and the bacterial food for those cultures, as well as how to extract the culture medium to generate a crude pheromone extract. Then, we discuss how to analyze the types and amounts of ascarosides in that extract using mass spectrometry and NMR spectroscopy.

Key words Dauer, Ascarosides, Pheromone, *C. elegans*, Mass spectrometry, NMR spectroscopy, dqf-COSY

1 Introduction

C. elegans secretes ascarosides as chemical signals to coordinate its development and behavior. All ascarosides are derivatives of the 3,6-dideoxysugar ascarylose, with different fatty acid-derived side chains [1–9]. The ascarosides that are found in *C. elegans* culture medium can be divided into several structural classes (Table 1). In this chapter, we describe a general method for growing high-density worm cultures and extracting the culture medium in order to generate a crude pheromone extract that contains secreted ascarosides (Subheadings 3.1 and 3.2). This crude pheromone extract can be directly used in biological assays, such as the dauer formation assay (*see* Chapter 20), it can be fractionated chromatographically for bioactivity-guided purification of ascarosides [1–4, 6], or the ascarosides in the extract can be analyzed using analytical

Kazushige Touhara (ed.), *Pheromone Signaling: Methods and Protocols*, Methods in Molecular Biology, vol. 1068,
DOI 10.1007/978-1-62703-619-1_6,

Table 1
Summary of ascarosides found in wild-type (N2) *C. elegans* culture medium that have had their structures confirmed through chemical synthesis[a]

Structure-based name	Other name(s)	Orig. ref.	Mol. formula	[M+Na]+	[M–H]–	*n*	R	Structure
Ascarosides with a saturated fatty acid side chain that is attached to the sugar at the penultimate (ω–1) carbon								
asc-C4	ascr#11	[7]	$C_{10}H_{18}O_6$	257.1001	233.1025	1	H	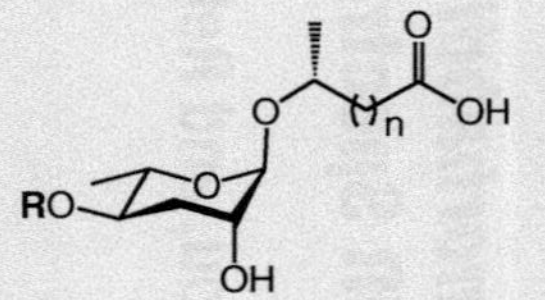
asc-C5	ascr#9	[7, 8]	$C_{11}H_{20}O_6$	271.1158	247.1182	2	H	
IC-asc-C5	indolecarboxyl-ascaroside C5; ascaroside C5; icas#9	[6]	$C_{20}H_{25}NO_7$	414.1529	390.1553	2	IC	
OS-asc-C5	osas#9	[9]	$C_{23}H_{33}NO_{10}$	506.2002	482.2026	2	OS	
asc-C7	daumone 1; ascr#1; ascaroside C7	[1]	$C_{13}H_{24}O_6$	299.1471	275.1495	4	H	
IC-asc-C7	icas#1	[7, 8]	$C_{22}H_{29}NO_7$	442.1842	418.1866	4	IC	
asc-C9	ascr#10	[7, 8]	$C_{15}H_{28}O_6$	327.1784	303.1808	6	H	
Ascarosides with an α,β-unsaturated (Δ) fatty acid side chain that is attached to the sugar at the penultimate (ω–1) carbon								
asc-ΔC7	ascr#7	[5]	$C_{13}H_{22}O_6$	297.1314	273.1339	2	H	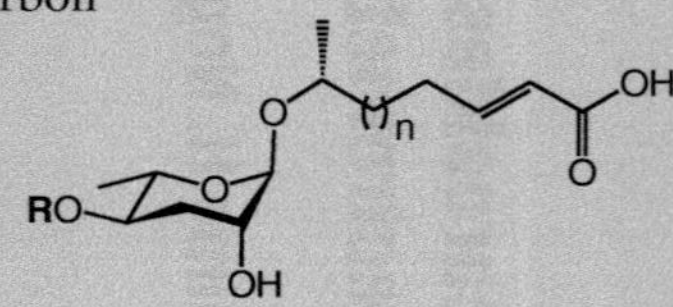
IC-asc-ΔC7	icas#7	[7, 8]	$C_{24}H_{33}NO_7$	440.1685	416.1709	2	IC	
asc-ΔC9	ascaroside C9; ascr#3; daumone 3	[2]	$C_{15}H_{26}O_6$	325.1627	301.1651	4	H	
IC-asc-ΔC9	icas#3	[7, 8]	$C_{24}H_{31}NO_7$	468.1998	444.2022	4	IC	
HB-asc-ΔC9	hbas#3	[7]	$C_{22}H_{30}O_8$	445.1838	421.1862	4	HB	
MB-asc-ΔC9	mbas#3	[7]	$C_{20}H_{32}O_7$	407.2046	383.2070	4	MB	
Ascarosides with a side chain that is attached to the sugar at the penultimate (ω–1) carbon and terminates in a methyl ketone (MK)[b]								
asc-C6-MK	ascaroside C6; ascr#2; daumone 2	[2]	$C_{12}H_{22}O_5$	269.1365	n.a.	2	H	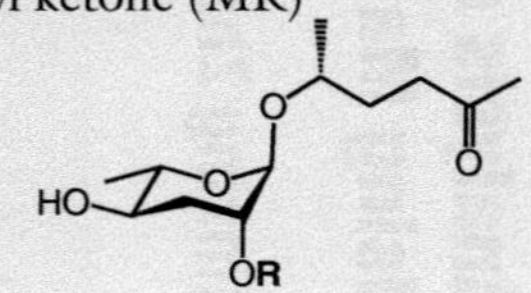
Glu-asc-C6-MK	ascr#4; daumone 4	[4]	$C_{18}H_{32}O_{10}$	431.1893	n.a.	2	Glu	

Ascarosides with a saturated fatty acid side chain that is attached to the sugar at the terminal (ω) carbon								
asc-ωC3	ascaroside C3; ascr#5; daumone 5	[3]	$C_9H_{16}O_6$	243.0845	219.0869	1	–	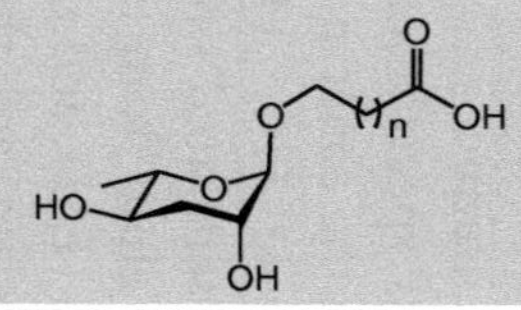
asc-ωC5	oscr#9	[7]	$C_{11}H_{20}O_6$	271.1158	247.1182	3	–	
asc-ωC9	oscr#10	[7]	$C_{15}H_{28}O_6$	327.1784	303.1808	7	–	
Ascarosides with an α,β-unsaturated (Δ) side chain attached at the penultimate (ω–1) carbon and linked to *p*-aminobenzoic acid (PABA)								
asc-ΔC7-PABA	ascr#8	[5]	$C_{20}H_{27}NO_7$	n.a.	392.1709	2	–	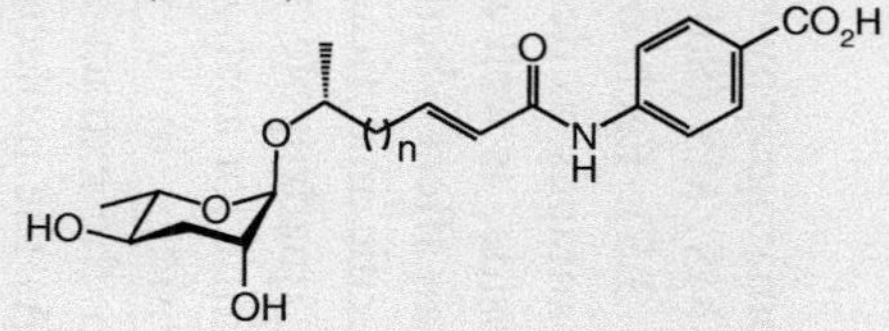

R group definitions:

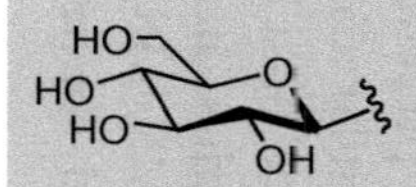

Glu, *glu*cosyl

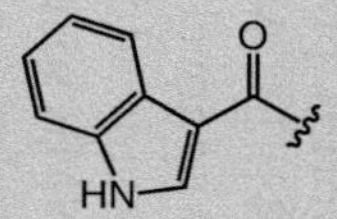

IC, *i*ndole-3-*c*arbonyl

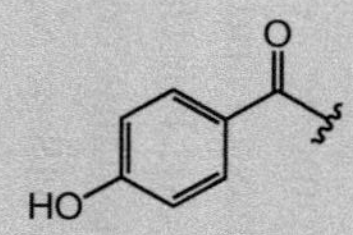

HB, 4-*h*ydroxy*b*enzoyl

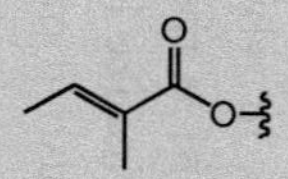

MB, 2-(*E*)-*m*ethyl-2-*b*utenoyl

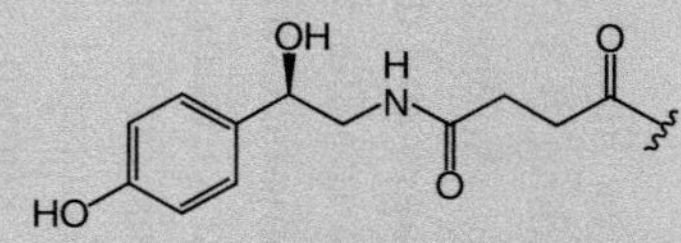

OS, *o*ctopamine *s*uccinyl

The structure-based name given in the first column assumes that the canonical ascaroside structure is an ascaroside with a saturated fatty acid side chain that is attached to the sugar at the penultimate (ω–1) carbon. Deviations from the canonical structure are indicated such that each name conforms to the following:
(modifications to the sugar)-asc-(deviations in the fatty acid)C#-(modifications at the end of the fatty acid)
Modifications to the sugar and their abbreviations are listed at the bottom of the table. Deviations in the fatty acid include attachment of the fatty acid to the sugar at the terminal (ω) carbon, instead of penultimate (ω–1) carbon, as well as α-β (Δ) unsaturation of the fatty acid. Modifications at the end of the fatty acid include the methyl ketone (MK) moiety and addition of *p*-aminobenzoic acid (PABA)

[a]Many other ascarosides, such as asc-C6 (ascr#12), asc-C8 (ascr#14), IC-asc-C9 (icas#10), asc-C11 (ascr#18), asc-ωC11 (oscr#18), asc-C12 (ascr#20), OS-asc-C9 (osas#10), and OS-asc-C6-MK (osas#2), have also been detected in wild-type *C. elegans* culture medium, but their structures have not yet been fully confirmed through chemical synthesis [7–9]

[b]Reduced versions of asc-C6-MK (ascr#6.1 and ascr#6.2) with a hydroxyl instead of a ketone have also been reported [5]

techniques, such as liquid chromatography-mass spectrometry (LC-MS) (Subheading 3.3), LC-MS/MS (Subheading 3.4), and NMR spectroscopy (Subheading 3.5). Each of these techniques has its own advantages and disadvantages, as described below. The structural class of a specific ascaroside, as well as its concentration in the crude pheromone extract, can affect whether or not it can be detected in the extract using a specific analytical technique.

LC-MS analysis on a single quadrupole mass spectrometer (Subheading 3.3) is a rapid technique for analyzing the concentrations of specific known ascarosides in a sample [10, 11]. In LC-ESI-MS, the molecules in the extract are separated on a column based on their polarity. As the molecules elute from the column, they are ionized by electrospray ionization (ESI), and the mass-to-charge (*m/z*) ratios of the ionized species are then detected. Most ascarosides will ionize most strongly as positively charged, sodium adducts. Thus, they can be detected by positive ion mode LC-MS, in which the ionized ascaroside is detected as $[M+Na]^+$, where M is the mass of the ascaroside (*see* Table 1 for which ascarosides ionize in positive ion mode). Many ascarosides, but not all, will also ionize as negatively charged, deprotonated species. Thus, they can also be detected by negative ion mode LC-MS, in which the ionized ascaroside is detected as $[M-H]^-$ (*see* Table 1 for which ascarosides ionize in negative ion mode). In order to use LC-MS to identify the presence of a specific ascaroside and to determine the concentration of that ascaroside in the extract, a synthetic standard is needed for the ascaroside of interest. The crude pheromone extract can contain non-ascaroside compounds with masses and retention times that are similar to the ascarosides of interest. For this reason, LC-MS-based identification of ascarosides is usually confirmed with additional techniques, including LC-MS/MS or NMR spectroscopy. LC-MS can be used to analyze the ascarosides in (1) the crude pheromone extract from the culture medium of a non-synchronized culture (generated by method given in Subheading 3.2), (2) the crude pheromone extract from the culture medium of a synchronized culture [11], (3) the crude extract from the worms themselves [10], or (4) the secretions from worms produced over a short period of time ("worm water") [4]. LC-MS is not useful for characterizing the chemical structures of novel ascarosides, as this must be done using techniques that provide additional structural information, such as LC-MS/MS and NMR spectroscopy.

LC-MS/MS analysis on a triple quadrupole mass spectrometer (Subheading 3.4) allows the simultaneous identification of many of the ascarosides present in the crude pheromone extract [7]. LC-MS/MS is similar to LC-MS in that the molecules in the extract are separated on a column based on their polarity and ionized. The *m/z* ratios of the ionized species are then detected. However, in LC-MS/MS, the molecules are not only ionized to detect the ionized species (called the precursor ion), but also broken apart, resulting in product ions that provide additional structural information about the molecules. When LC-MS/MS is performed on a triple quadrupole mass spectrometer, a precursor ion scan can be used to detect only molecules that give rise to a specific product ion and report back the precursor ions that produced the specific product ion. This technique is particularly useful for the detection of classes of molecules with similar structures, such as the ascarosides, that all produce a common product ion. Those ascarosides that ionize in negative ion mode can be fragmented in LC-MS/MS to produce a common product ion of *m/z* 73 [$C_3H_5O_2$]. Thus, this technique, which was developed by von Reuss and coworkers [7], can be used to detect all ascarosides in a crude pheromone extract that ionize as negative ions and report back the precursor ions (that is, $[M-H]^-$) corresponding to those ascarosides (Subheading 3.4). Those ascarosides that ionize only in positive ion mode can be monitored by looking for a neutral loss of 130 amu [$C_6H_{10}O_3$] that corresponds to the loss of the ascarylose sugar [7]. However, fragmentation of the ascarosides in positive ion mode is not as efficient as that in negative ion mode, leading to lower and less reliable signal, and will not be described in this review.

NMR spectroscopy can be used to analyze the ascarosides in a crude pheromone extract or chromatography fractions derived from that extract (Subheading 3.5). One type of NMR experiment that is particularly informative is the double quantum-filtered correlation spectroscopy (dqf-COSY) experiment [5, 10]. Like a standard COSY experiment, this experiment detects correlations between protons on neighboring carbons in a molecule, and these correlations appear as peaks off the diagonal of the spectrum. However, the dqf-COSY experiment is generally run in a phase-sensitive mode and gives a much cleaner spectrum near the diagonal than COSY, which is important for analysis of mixtures containing multiple components, such as crude pheromone extract, which may have many peaks near the diagonal. Each dqf-COSY correlation provides additional information about the protons involved in the correlation, including the coupling constant between those two protons (active coupling), as well as the coupling constants of those protons with other neighboring protons (passive coupling). The dqf-COSY spectrum of the crude pheromone extract contains peaks that are diagnostic for the ascarosides. Furthermore, dqf-COSY spectra have been acquired for many of the known ascarosides, and thus, the presence of a particular ascaroside in a crude pheromone extract or chromatography fraction can be substantiated by dqf-COSY. It should be noted, however, that the dqf-COSY spectrum of ascarosides from the same structural class, but with different side chain lengths, will be similar. One drawback of using NMR spectroscopy to analyze the ascarosides in the crude pheromone extract is that less abundant ascarosides cannot be detected because NMR spectroscopy is an inherently less sensitive technique than mass spectrometry. Further fractionation of the crude pheromone extract to give a less complex sample containing the specific ascaroside of interest is needed. However, chromatographic purification of individual ascarosides is dependent on the polarity of those ascarosides and done on an ad hoc basis, and thus, it is beyond the scope of this review.

2 Materials

2.1 25× Bacterial Stock

1. *E. coli* strain OP50 or HB101, grown on Luria-Bertani (LB)-agar.
2. LB broth.
3. Falcon tube (50 mL) for bacterial mini-culture.
4. Laminar flow hood.
5. 2.8 L baffled flask for large bacterial culture.
6. Shaker for shaking bacterial culture at 37 °C.
7. Refrigerated table-top centrifuge.
8. Potassium phosphate (1 M, pH 6): In 0.8 L water, dissolve 136.1 g KH_2PO_4. Add potassium hydroxide until the pH reaches 6. Then add water to a final volume of 1 L. Autoclave.
9. S basal: 0.1 M NaCl, 0.05 M potassium phosphate (pH 6). Autoclave.
10. Microfuge and 1.5 mL Eppendorf tubes.

2.2 Worm Culturing

1. *C. elegans*, wild-type (N2) or mutant strain.
2. Stereomicroscope and worm pick for manipulating worms.
3. Cholesterol (5 mg/mL in ethanol): Dissolve 50 mg of cholesterol in 10 mL of ethanol (200 proof). Sonicate for approximately 10 min or shake until the cholesterol dissolves.
4. 25× bacterial stock from Subheading 3.1 for feeding worm cultures. 25 mL of 25× stock (corresponding to 625 mL-worth of bacterial culture) is needed for every 150 mL of worm culture that will be grown.
5. 10 cm NGM-agar plates with bacterial lawn: In 500 mL water, add 1.5 g NaCl, 8.5 g granulated agar, and 1.25 g peptone. Autoclave, and then add: 0.5 mL 1 M $CaCl_2$, 0.5 mL 1 M $MgSO_4$, 12.5 mL potassium phosphate (pH 6), and 0.5 mL 5 mg/mL cholesterol. Pour into 10 cm plates, and let them dry on the benchtop overnight. Pipet 0.75 mL 25× bacterial stock onto each plate, spread on plate (but do not allow lawn to touch plate edge), and dry in laminar flow hood or on benchtop. Approximately one plate is needed for every 150 mL of worm culture that will be grown.

6. Potassium citrate (1 M, pH 6): In 0.8 L water, dissolve 210.14 g citric acid monohydrate. Add potassium hydroxide until the pH reaches 6. Then add water to a final volume of 1 L. Autoclave.
7. Trace metals solution: In 1 L water, dissolve 1.86 g disodium EDTA (5 mM), 0.69 g $FeSO_4 \cdot 7H_2O$ (2.5 mM), 0.20 g $MnCl_2 \cdot 4H_2O$ (1 mM), 0.29 g $ZnSO_4 \cdot 7H_2O$ (1 mM), 0.025 g $CuSO_4 \cdot 5H_2O$ (0.1 mM). Autoclave. Wrap bottle in foil, and store in the dark.
8. S medium: Autoclave 150 mL of S basal in a 500 mL Erlenmeyer flask, covering the top with foil. Just before use, add 1.5 mL 1 M potassium citrate (pH 6), 1.5 mL trace metals solution, 0.45 mL 1 M $CaCl_2$, 0.45 mL 1 M $MgSO_4$, and 150 μL cholesterol (5 mg/mL).
9. Optional: To help prevent contamination of the cultures, add the antifungal nystatin (100 U/mL final concentration) and the antibiotics penicillin–streptomycin (50 U/mL final concentration for penicillin and 50 μg/mL final concentration for streptomycin).
10. Refrigerated shaker for culturing worms in shaker flasks at 22.5 °C.
11. Graduated cylinder (250 mL) and ice bath.
12. Refrigerated table-top centrifuge and polypropylene centrifuge bottles with screw caps.
13. Lyophilizer and lyophilization flask (600 mL).
14. Mortar and pestle.
15. Erlenmeyer flask (500 mL) and funnel.
16. Ethanol (190 proof).
17. Tabletop shaker for performing extractions.
18. Filter paper (Whatman, No. 41), Buchner funnel, neoprene adaptor, and Erlenmeyer filtering flask (500 mL).
19. Rotary evaporator and round bottom flask (250 mL).
20. Short (5.75 in.) and long (9 in.) Pasteur pipets and bulbs.
21. Clean ball of cotton.
22. Bendable spatula.
23. Speedvac, rotor for 8 mL glass vials, and 8 mL glass vials (e.g., Fisher, cat. 03-340-60B) (*see* **Note 1**).

2.3 LC-MS

1. Methanol.
2. Sonicator.
3. Speedvac and rotor for 8 mL glass vials (*see* **Note 1**).
4. Microfuge and methanol-safe 1.5 mL Eppendorf tubes (e.g., Fisher, cat. 05-408-129).

5. Mass spectrometer: Agilent 6130 single quad mass spectrometer with API-ES source and operating in dual negative/positive scan mode.
6. LC: Agilent 1260 infinity binary pump with autosampler.
7. Column: Phenomenex 5 μm Luna C_{18} 2, 100 Å (100 × 4.6 mm).
8. LCMS-grade water with 0.1 % formic acid (solvent A).
9. LCMS-grade acetonitrile with 0.1 % formic acid (solvent B).
10. Autosampler vials and glass inserts with springs.

2.4 LC-MS/MS

1. Methanol.
2. Sonicator.
3. Speedvac and rotor for 8 mL glass vials (*see* **Note 1**).
4. Microfuge and methanol-safe 1.5 mL Eppendorf tubes (e.g., Fisher, cat. 05-408-129).
5. Mass spectrometer: Thermo TSQ quantum ultra triple quadrupole mass spectrometer using argon collision gas, operating in negative ion, heated (H)-ESI, precursor scanning mode, selecting for a product ion of 73.0.
6. LC: Accela UHPLC (max pressure of 1,000 bar) with autosampler.
7. Column: Kinetex 2.6 μm C18 100 Å (100 × 2.10 mm).
8. LCMS-grade water (solvent A).
9. LCMS-grade acetonitrile with 0.1 % acetic acid (solvent B).
10. Autosampler vials and glass inserts with springs.

2.5 NMR Spectroscopy

1. Lyophilizer.
2. Methanol-d_4.
3. Sonicator.
4. Short (5.75 in.) and long (9 in.) Pasteur pipets and bulbs.
5. Clean ball of cotton.
6. 5 mm methanol-matched Shigemi NMR tube (Shigemi, cat. MMS-005v) (*see* **Note 2**).
7. Parafilm.
8. NMR spectrometer: Varian 500 MHz NMR spectrometer.
9. NMR probe: 5 mm Varian indirect-detection, triple resonance probe ($^{1}H/^{13}C/(^{31}P$-$^{15}N)$PFG).
10. Pasteur pipets with extra long tips for removing samples from NMR tubes (Fisher, cat. 13-678-6).
11. Speedvac, rotor for 8 mL glass vials, and 8 mL glass vials (e.g., Fisher, cat. 03-340-60B).

3 Methods

3.1 Preparing 25× Bacterial Stock

This method prepares the bacterial food (25× bacterial stock) needed for growing the worm cultures in Subheading 3.2. Sterile technique is important in the preparation of the 25× bacterial stock. If the 25× bacterial stock is contaminated with bacteria or fungus, this contamination may overwhelm the worm cultures that are grown in Subheading 3.2, regardless of whether antibiotics and antifungals are added to the worm culture medium.

1. Inoculate 5 mL LB with a colony of OP50 or HB101 bacteria. Let the mini-culture grow for 6–8 h in a shaker at 250 rpm at 37 °C (*see* **Note 3**).
2. Use 1 mL of this mini-culture to inoculate 1 L of LB broth in a 2.8 L baffled flask. Grow culture overnight in a shaker at 250 rpm at 37 °C.
3. Collect bacteria by centrifuging at 2,700 × *g* for 10 min. Discard the supernatant, and resuspend in enough S basal to make the final volume equal to 40 mL to give a 25× bacterial stock. Transfer to a 50 mL falcon tube, and store at 4 °C (*see* **Notes 4–6**).

3.2 Unsynchronized Worm Cultures and Generating Crude Pheromone Extract

This method describes how to grow unsynchronized worm cultures and how to extract the culture medium to generate crude pheromone extract. Production of many of the ascarosides is maximized by feeding the worm cultures regularly with sufficient bacterial food to allow reproduction and increase worm density. It is important not to feed the worms too much on any 1 day because they will suffocate, but it is also important not to deprive the worms of food. The exact culture conditions used will influence the amounts of the ascarosides produced.

1. Passage 20 N2 worms onto a 10 cm NGM-agar plate with a bacterial lawn.
2. Let worms grow for ~4 days at room temperature until just before the bacterial food supply runs out.
3. Wash the plate with 10 mL of S medium, and put in a 15 mL tube. Count the number of worms in 3 × 10 μL drops under the stereomicroscope to gauge the density of worms. Make sure that you invert the tube several times before taking out each 10 μL sample (*see* **Note 7**).
4. Add ~90,000 worms per 150 mL culture.
5. Feed the worms 3 mL of 25× bacterial stock (110–125 mg/mL). For our feeding schedule, this day is considered day 0.
6. Shake at 22.5 °C at 225 rpm for 9 days. Feed worms 1 mL 25× bacterial stock on day 1, and 3 mL 25× bacterial stock on days 2–8.

7. On day 9, harvest the culture by transferring the culture to a tall container, such as a 250 mL graduated cylinder, and place the container in an ice-bath for 30 min to 1 h in order to settle the worms. Transfer the supernatant to a centrifuge bottle until you cannot pipet any more of the supernatant without disturbing the settled worms. Transfer the remaining worms and culture medium from the bottle to a 50 mL falcon tube, centrifuge at 200 × *g* for 5 min. Pipet the remaining supernatant from the falcon tube into the centrifuge bottle together with the previously pipetted supernatant. Centrifuge the centrifuge bottle at 2,700 × *g* for 10 min to remove the bacteria from the supernatant. Then carefully transfer the supernatant to a 600 mL lyophilization flask, and freeze by placing it at –20 °C overnight.
8. Lyophilize the frozen culture medium to dryness (2–3 days) (*see* **Note 8**).
9. Scrape out solid material into a mortar, and grind with a pestle to a fine powder.
10. Place the powder in a 500 mL Erlenmeyer flask using a funnel, and extract by adding 150 mL ethanol (190 proof) and shaking in a table-top shaker at 250 rpm for 2 h.
11. Filter the extraction mixture through filter paper in a Buchner funnel.
12. Place the retentate back in the Erlenmeyer flask from **step 10**, and extract a second time with 150 mL of ethanol (190 proof), shaking overnight (*see* **Note 9**).
13. Filter the extraction mixture through filter paper in a Buchner funnel, and combine the filtrate with the filtrate from **step 11**.
14. Rotovap the combined filtrates in a round bottom flask to a small volume (~3 mL).
15. Place a small amount of cotton in a short Pasteur pipet, and pack it down inside the Pasteur pipet using a long Pasteur pipet, such that the packed cotton plug is about 3 mm in length.
16. Wash the cotton in the Pasteur pipet with a small amount of ethanol to remove any impurities from the cotton. Use a bulb to push the ethanol through the Pasteur pipet and dry the cotton.
17. Filter the sample through the cotton plug in the Pasteur pipet into a 8 mL glass vial. Use a bulb to push the sample through the Pasteur pipet.
18. Wash the round bottom with ~3 mL of ethanol (190 proof). Scrape the sides of the round bottom with a metal spatula to make sure that as much of the solid is exposed to the ethanol as possible.

19. Filter this ethanol wash through the cotton plug in the Pasteur pipet into the 8 mL glass vial from **step 17**. Use a small amount of ethanol to wash the plug, and add this last wash to the glass vial as well.
20. Speedvac the sample in the glass vial to dryness. Store the sample at −20 °C until use.

3.3 LC-MS Quantitation of Ascarosides

This method can be used to quantify the concentrations of the ascarosides in a particular sample, such as in the crude pheromone extract (Fig. 1) or in worm secretions produced over a short period of time ("worm water") [4]. The mass spectrometer must be well maintained with low background signal. It should be noted that this method can underestimate the amount of an ascaroside in a sample, especially if the sample is complex, if ion suppression occurs.

1. Resuspend the sample from Subheading 3.2, **step 20**, by adding 1 mL methanol. Sonicate for 20 min, vortex briefly to dislodge any remaining solids on the sides of the vial, and sonicate again for 20 min. Centrifuge the vial (with the cap closed) in the speedvac for 2 min to pellet most of the particulates.
2. Pipet 100 μL of the supernatant into a 1.5 mL Eppendorf tube, and centrifuge at 17,600 × *g* for 1 min to pellet any remaining particulates. Pipet 50 μL from the top of the sample to an autosampler vial. Pipet the remaining 50 μL of the sample from the tube back into vial (*see* **Note 10**).
3. Make 0.2, 0.4, 2, 4, 8, 20, 40, 60, 80, and 100 μM stocks of each synthetic standard in methanol. Transfer 20 μL of each stock to an autosampler vial (*see* **Note 11**).
4. For an Agilent 6130 single quadrupole mass spectrometer, use the following parameters for the spray chamber of the ion source: Drying gas flow of 12 L/min, nebulizer pressure of 35 psig, capillary voltage of 3,000 V, and drying gas temperature of 350 °C. Acquire signals in dual positive–negative scan mode with a 50 % cycle time between the two modes, peak width of 0.1 min, and cycle time of 1.08 s/cycle. Use a mass range of 100–1,000, fragmentor voltage of 125 V, gain of 1.00, threshold of 150, and stepsize of 0.10 (*see* **Note 12**).
5. Set the column temperature to 25 °C.
6. Set up the following chromatography method, where the solvent A is water with 0.1 % formic acid and solvent B is acetonitrile with 0.1 % formic acid: (1) ramp from 5 % B to 100 % B over 20 min, (2) hold at 100 % B for 2 min, (3) ramp from 100 % B to 5 % B over 1 min, and (4) hold at 5 % B for 2 min. Set the flow rate at 0.7 mL/min (*see* **Note 13**).
7. In order to clean the column, wash the column with 100 % acetonitrile for 10 min at 0.7 mL/min.

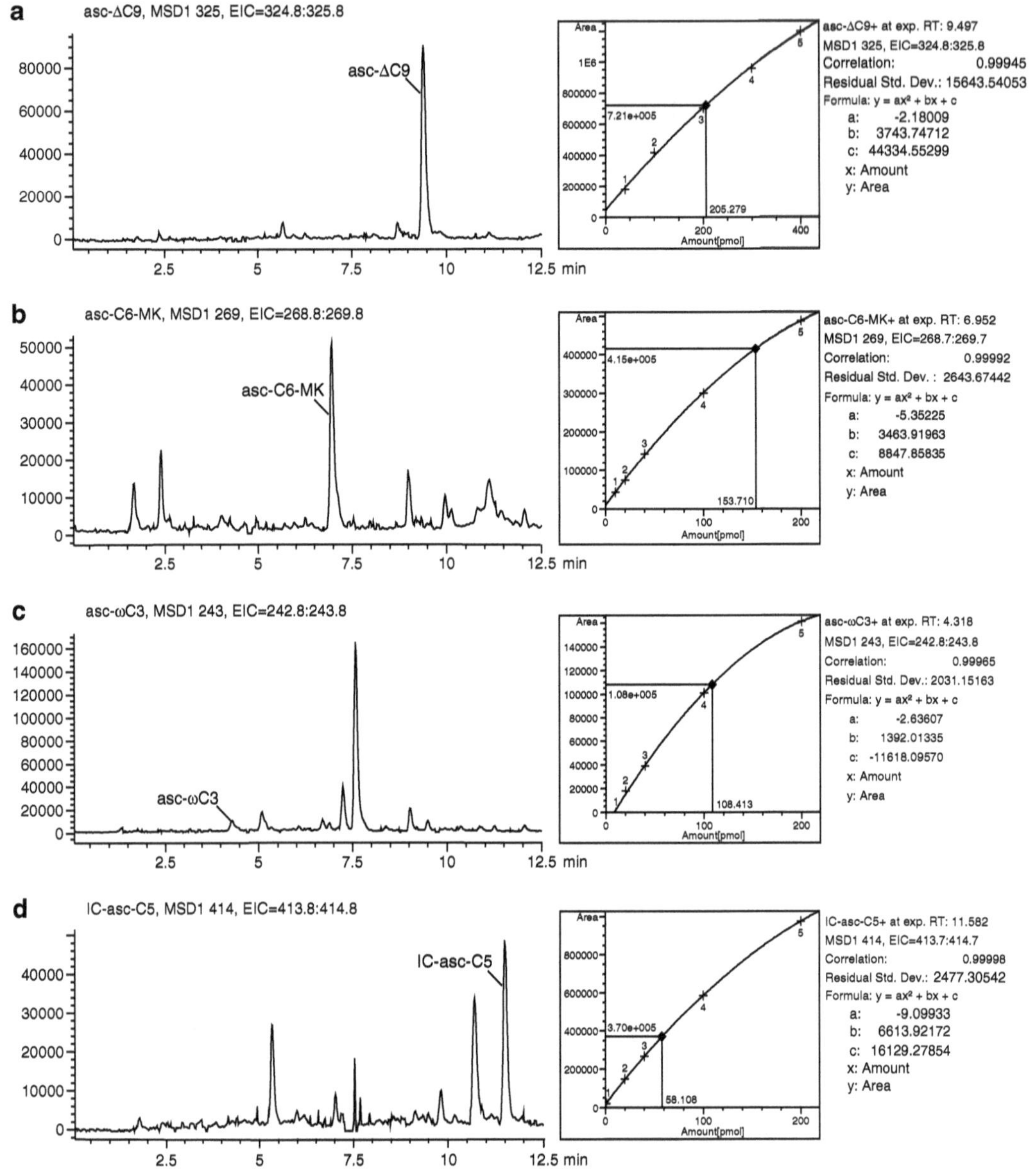

Fig. 1 LC-MS-based determination of the concentrations of asc-ΔC9, asc-C6-MK, asc-ωC3, and IC-asc-C5 in crude pheromone extract. A calibration curve of synthetic ascarosides was generated by injecting 5 μL of several concentrations of the ascarosides and fitting a quadratic curve to the data. 5 μL of crude pheromone extract was also injected, and the relevant ion for each ascaroside was extracted from the positive total ion count trace. The area under the peak for each ascaroside was used to estimate the number of pmol in 5 μL of crude pheromone extract. This number then was used to estimate the concentration of each ascaroside in the conditioned medium that was used to generate the crude pheromone extract. (**a**) The concentration of asc-ΔC9 in the conditioned medium is ~273.7 nM. (**b**) The concentration of asc-C6-MK in the conditioned medium is ~204.9 nM. (**c**) The concentration of asc-ωC3 in the conditioned medium is ~144.5 nM. (**d**) The concentration of IC-asc-C5 in the conditioned medium is ~77.4 nM

8. In order to equilibrate the column, wash the column with 5 % B for 10 min at 0.7 mL/min.
9. Use the autosampler to inject 5 μL of methanol into the LC-MS. This blank run will allow you to assess the background level of signal (*see* **Note 14**).
10. Use the autosampler to inject 5 μL of sample into the LC-MS.
11. Extract the relevant ion for the ascaroside of interest. That is, extract $[M+Na]^+$ for the ascaroside of interest from the positive total ion count trace, or extract $[M-H]^-$ for the ascaroside of interest from the negative total ion count trace. Obtain the area under the peak for the extracted ion.
12. Use the autosampler to inject 5 μL of each concentration of the synthetic standards in turn into the LC-MS.
13. For each concentration of synthetic standard injected into the LC-MS, extract the relevant ion from the total ion count trace, and use the area under the peak to generate a standard curve.
14. Use the standard curve generated to estimate the concentration of the ascaroside of interest in the sample (*see* Fig. 1).
15. In order to clean and store the column, wash the column with 100 % acetonitrile for 10 min.

3.4 LC-MS/MS Analysis of Ascarosides

This method can be used to analyze the ascarosides in the crude pheromone extract that ionize in negative ion mode (Fig. 2). This method assumes that the user has a basic knowledge of LC-MS/MS and is assisted by an experienced mass spectrometry facility manager. The mass spectrometer must be well maintained with low background signal.

1. Resuspend the sample from Subheading 3.2, **step 20**, by adding 1 mL methanol. Sonicate for 20 min, vortex briefly to dislodge any remaining solids on the sides of the vial, and sonicate again for 20 min. Centrifuge the vial (with the cap closed) in the speedvac for 2 min to pellet most of the particulates.
2. Pipet 100 μL of the supernatant into a 1.5 mL Eppendorf tube, and centrifuge at 13,700 rpm for 1 min to pellet any remaining particulates. Pipet 50 μL from the top of the sample to an autosampler vial with glass insert with spring. Pipet the remaining 50 μL of the sample from the tube back into vial (*see* **Note 10**).
3. Set the mass spectrometer to operate in negative ion, heated (H)-ESI, precursor scanning mode, selecting for a product ion of 73.0. If you are using a Thermo TSQ quantum ultra mass spectrometer, use the following parameters for the mass spectrometer: spray voltage of 3,000 V, vaporizer temperature of 300 °C, sheath gas pressure of 40 arb, auxillary gas pressure of 20 arb, capillary temperature of 350 °C, skimmer offset of

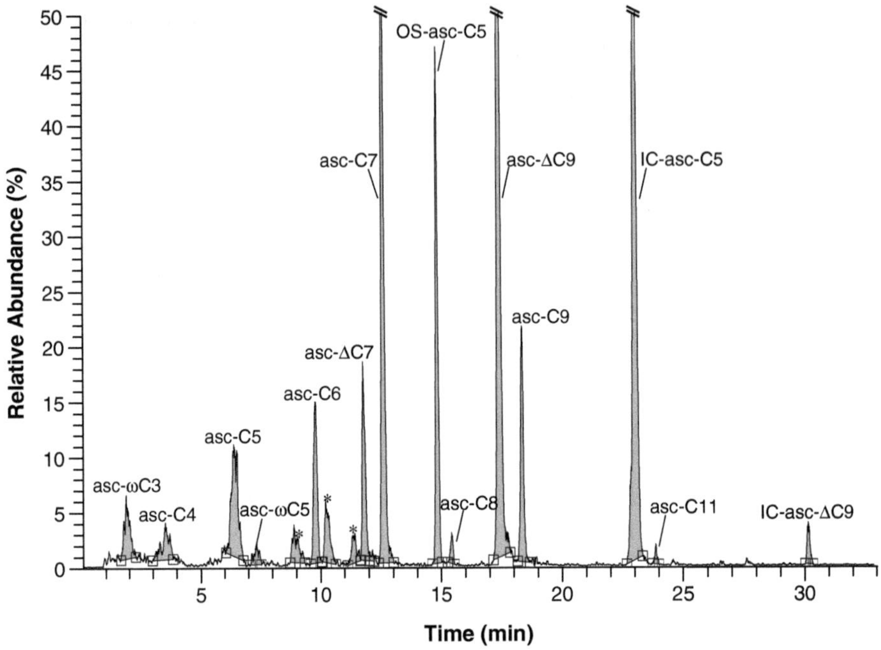

Fig. 2 LC-MS/MS trace in precursor ion scanning mode (selecting for *m/z* 73.0 product ion) for crude pheromone extract. *Asterisks* indicate signals from unidentified compounds or non-ascarosides

5 V, collision pressure of 0.8 mTorr, collision energy of 25 eV, and scan time 0.4 s. We usually set the mass range to *m/v* 200–500 (*see* **Notes 15** and **16**).

4. Set the column temperature to 30 °C.
5. Set up the following chromatography method, where solvent A is water and solvent B is acetonitrile with 0.1 % acetic acid: (1) hold at 2 % B for 1 min, (2) ramp from 2 % B to 50 % B over 34 min, (3) ramp from 50 % B to 100 % B over 1 min, (4) hold at 100 % B for 4 min, (5) ramp from 100 % B to 2 % B over 1 min, (6) hold at 2 % B for 4 min. Set the flow rate at 0.25 mL/min (*see* **Note 17**).
6. In order to clean the column before use, wash the column with 100 % acetonitrile for 10 min at 0.25 mL/min.
7. In order to equilibrate the column, wash the column with 2 % solvent B for 10 min at 0.25 mL/min.
8. Use the autosampler to inject 5 μL of methanol into the LC-MS/MS. This blank run will allow you to assess the background level of signal (*see* **Note 14**).
9. Use the autosampler to inject 5 μL of sample into the LC-MS/MS.
10. After the run is complete, smooth the trace using a Boxcar or Gaussian function, with 5–7 smoothing points (*see* **Note 18**).

11. Obtain the areas under the peaks for each ascaroside by extracting each specific precursor ion from the total ion count trace (*see* **Note 19**).
12. In order to validate that a given peak is a given ascaroside, you will need to acquire a synthetic standard. Dissolve the standard(s) in ethanol (200 proof) at 270 μM. Transfer some of the sample to an autosampler vial, and use the autosampler to inject 5 μL from the autosampler vial into the LC-MS/MS.
13. After the run is complete, verify that the synthetic standard has the same retention time as the tentatively assigned peak from **step 11**.
14. In order to clean and store the column, wash the column with 100 % acetonitrile for 10 min.

3.5 NMR Spectroscopy-Based Analysis of Ascarosides Using the dqf-COSY Experiment

This method can be used to analyze the ascarosides in the crude pheromone extract (Fig. 3) or chromatography fractions generated from the crude pheromone extract. The ascarosides must be present at sufficient concentrations in the crude pheromone extract for detection by NMR spectroscopy. Subheading 3.5 assumes that the user has a basic knowledge of NMR and is assisted by an experienced NMR facility manager. The NMR spectrometer must be well maintained with an updated shim set. Although we use a Varian NMR spectrometer in our own experiments, we employ terminology in this method that should be interpretable by users familiar with either Agilent (Varian) or Bruker NMR spectrometers. In order to have the necessary chemical shift separation in the complex spectrum obtained from crude pheromone extract, an NMR spectrometer with a magnetic field of 500 MHz or higher should be used. The number of points in both dimensions should provide for a resolution of 1–2 Hz per point. The concentrations of the ascarosides in the crude pheromone extract are relatively low compared to the concentrations of other more abundant metabolites. Therefore, it is critical to maximize the signal-to-noise ratio for the sample, and thus, an indirect detection probe or a cold probe is recommended.

1. Resuspend the sample from Subheading 3.2, **step 20**, by adding 1 mL methanol-d_4. Sonicate for 20 min, vortex briefly to dislodge any remaining solids on the sides of the vial, and sonicate again for 20 min. Centrifuge in the speedvac (with the cap closed) for 2 min to pellet any particulates that do not dissolve.
2. Transfer the supernatant to a new vial, and speedvac or rotovap the sample to dryness (*see* **Note 20**).
3. Place the vial of dry sample, a short Pasteur pipet, and an empty 5 mm methanol-matched Shigemi tube on the lyophilizer for 1–2 h. This drying step will remove any trace amounts

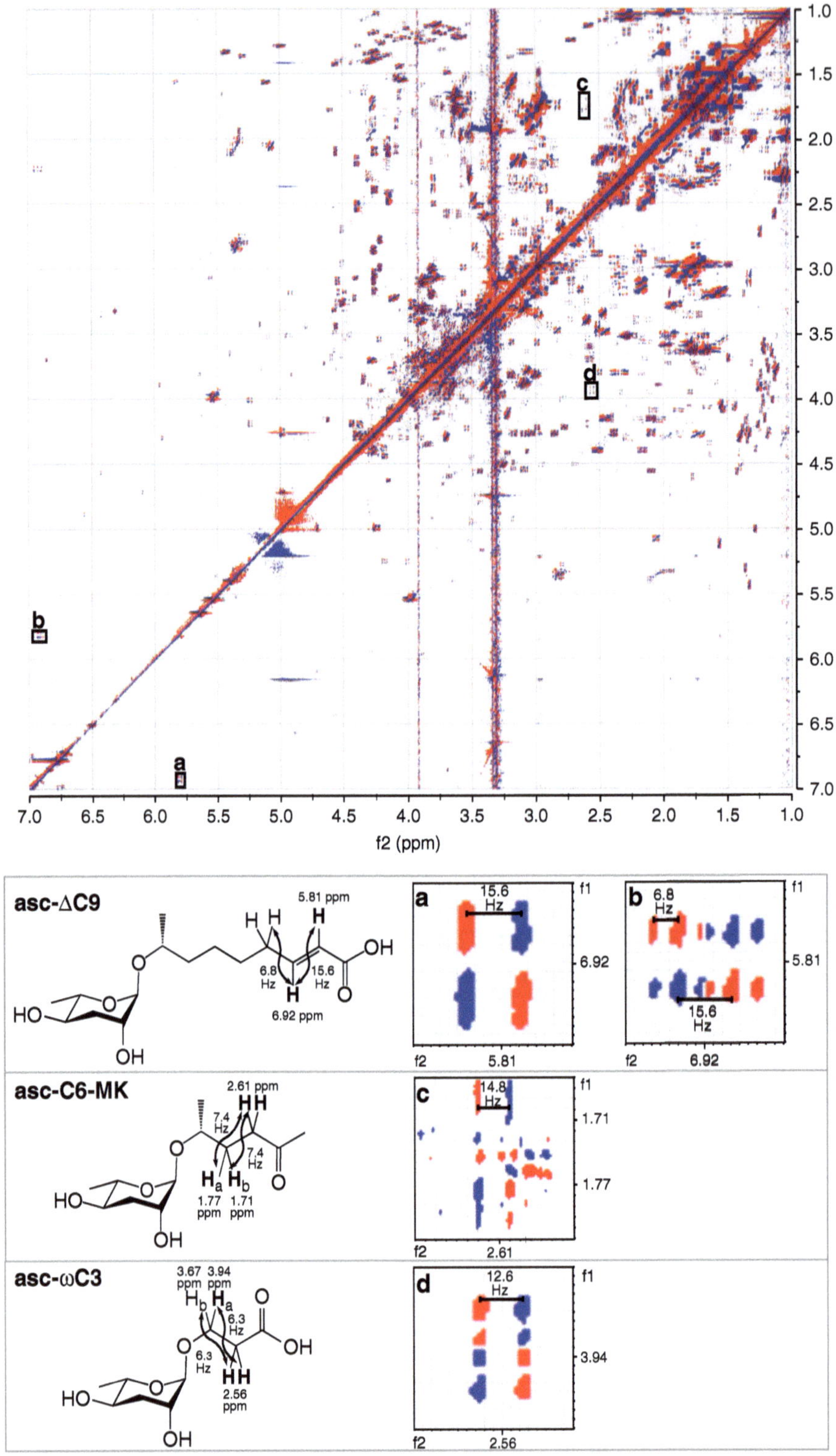

Fig. 3 dqf-COSY spectrum of crude pheromone extract, highlighting specific correlations for *bolded protons* for specific ascarosides (asc-ΔC9, asc-C6-MK, and asc-ωC3) that are visible in the spectrum. (**a** and **b**) Highlight

of methanol and water in the sample that would otherwise interfere with the signal of the sample in the NMR spectrum.

4. Resuspend your sample by adding 300 μL of methanol-d_4 and sonicating for 5 min.
5. Place a small amount of cotton in the short Pasteur pipet, and pack it down inside the Pasteur pipet using a long Pasteur pipet, such that the packed cotton plug is about 3 mm in length.
6. Wash the cotton in the Pasteur pipet with a small amount of methanol-d_4 to remove any impurities from the cotton. Use a bulb to push the methanol-d_4 through the Pasteur pipet and dry the cotton.
7. Filter the sample through the cotton plug in the Pasteur pipet, directly into the Shigemi tube. Use a bulb to push the sample through the Pasteur pipet. Use a few drops of methanol-d_4 to wash the cotton plug, allowing the wash to drip directly into the Shigemi tube (*see* **Note 21**).
8. Insert the solvent-matched Shigemi tube insert into the Shigemi tube, taking care to avoid any bubbles between the sample and the insert. Use a small strip of parafilm (about 0.5 × 2 cm in length) to wrap around the top of the tube to prevent any evaporation of the sample while it is in the NMR spectrometer (*see* **Notes 22** and **23**).
9. Insert your sample into the spectrometer. Setup parameters for a ^1H spectrum in methanol (*see* **Note 24**).
10. Lock the sample, and optimize the lock phase.
11. Gradient shim the sample.
12. Tune the ^1H channel of the probe.
13. Turn off the spinner (*see* **Note 25**).
14. Acquire a ^1H spectrum with 16 scans (and with the spinner off) (*see* **Notes 26** and **27**).

Fig. 3 (continued) the below diagonal and above diagonal correlations between the *bolded proton* at 5.81 ppm and the *bolded proton* at 6.92 ppm in asc-ΔC9. The active coupling can be measured in (**a**) and (**b**) to give the coupling constant between the two *bolded protons* (15.6 Hz). The passive coupling can be measured in (**b**) to give the coupling constant between the *bolded proton* at 6.92 ppm and the other protons to which it is coupled (6.8 Hz). (**c**) Highlights the above diagonal correlations between the *bolded protons* at 2.61 ppm and the two nonequivalent *bolded protons* at 1.71 ppm and 1.77 ppm in asc-C6-MK. The active coupling between the *bolded protons* can be measured in (**c**). The measured value of 14.8 Hz must be divided in two (to give coupling constants of approximately 7.4 Hz) because the *bolded proton* at 2.61 ppm is a triplet and oppositely phased peaks in the center of the triplet cancel each other. The below diagonal correlation is not highlighted as it is too faint. (**d**) Highlights the below diagonal correlation between the two *bolded protons* at 2.56 ppm and the *bolded proton* at 3.94 ppm in asc-ωC3. The above diagonal correlation is not highlighted as it is obscured by a vertical stripe

15. Set the sweep width to the region of the ^{1}H spectrum that you are interested in for the 2D experiment. Usually you will select the region where there are peaks, plus 0.5 ppm on either side. In general, for the crude pheromone extract, the sweep width will extend from 0.5 to 9.5 ppm.
16. Determine the 90° pulse width. Make sure that the pulse width is equal to this new optimized value (*see* **Note 28**).
17. Acquire another ^{1}H spectrum (*see* **Note 29**).
18. Move the ^{1}H spectrum into a new experiment in order to move all optimized parameters, including sweep width, 90° pulse width, and gain, into the new experiment. Setup up a dqf-COSY experiment with the default parameters.
19. Change the default parameters in the following manner: the number of scans to 16, the number of iterations to 1,024, the number of points to 4,096, and the relaxation delay to 1 (*see* **Notes 30** and **31**).
20. Start the experiment, and once finished, process the data (*see* **Note 32**).

4 Notes

1. If a speedvac with a rotor for 8 mL glass vials is not available, the material can be dried on a rotovap with a 24/40 glass adaptor (e.g., Chemglass cat. CG-1318-10) and connector (e.g., Chemglass cat. CG-1318-21) for the 8 mL glass vials.
2. This tube is magnetic susceptibility-matched to methanol. The catalog number given is for a tube that is designed for use with a 5 mm Varian NMR probe.
3. Because sterile technique is important, we perform as many of the steps as possible in a laminar flow hood.
4. To gauge the sterility of the 25× bacterial stock, streak it out on an LB-agar plate and allow colonies to grow. There should be only one colony type.
5. The 25× bacterial stock can be stored at 4 °C for up to 1 month.
6. Typically, if HB101 or OP50 is grown in LB broth as described, 110–125 mg/mL of bacteria will be obtained for the 25× stock. Calculate the stock concentration by weighing an empty 1.5 mL Eppendorf tube and then adding 1 mL of 25× stock to it. Centrifuge the tube in the microfuge at 13,700 rpm for 1 min, and remove the supernatant by pipetting. Centrifuge again, and remove any remaining supernatant by pipetting. Reweigh the tube with the pellet in it. The weight of the pellet gives the stock concentration in mg/mL.

7. Generally, for N2, ~90,000 worms are obtained from a single 10 cm plate, but for some mutants, especially for those that like to dig into the agar, this number can be lower.
8. It is important that there are no chunks of frozen material remaining, since it will interfere with the extraction process. After 2 days, it is helpful to use a spatula to break apart any remaining material, and place the flask back on the lyophilizer.
9. The second extraction significantly increases the pheromone yield.
10. If the sample will be used again within a week, store at −20 °C. If the sample will not be used again within a week, speedvac it to dryness, and store at −20 °C.
11. If a standard curve will be generated for more than one synthetic ascaroside, it is convenient to put all synthetic standards together at a given concentration for simultaneous injection.
12. These parameters are instrument dependent. The fragmentor voltage was optimized on our instrument using a synthetic ascaroside standard (20 μM asc-ΔC9 in ethanol). If a mass spectrometer other than an Agilent 6130 single quadrupole mass spectrometer will be used, these parameters may need to be re-optimized.
13. This chromatography protocol was optimized for monitoring the concentrations of specific ascarosides (asc-ωC3, asc-C6-MK, asc-ΔC9, IC-asc-C5, as shown in Fig. 1). It should be noted that this gradient is a short one, and so two or more ascarosides in the crude pheromone extract may elute at the same time. For example, two ascarosides with similar polarities and the same mass (e.g., asc-C5 and asc-ωC5) may elute from the column at the same time, which would be problematic if the objective was to analyze their concentrations. To prevent co-elution of two or more ascarosides, a longer gradient may be required such as the gradient used in Subheading 3.4.
14. If there is a high background level of signal, it most likely indicates that the column is dirty and must be cleaned more thoroughly before use.
15. These parameters are instrument dependent. They were optimized on our instrument based on direct infusion of a synthetic ascaroside standard (4 μM asc-ΔC9 in ethanol). If a mass spectrometer other than a Thermo TSQ quantum mass spectrometer will be used, these parameters will need to be re-optimized.
16. This mass range should be shifted if you are interested in analyzing ascarosides with larger masses, such as the long-chain ascarosides.

17. This chromatography protocol was optimized for the separation and analysis of more polar, short-chain ascarosides, but not for the separation and analysis of less polar, long-chain ascarosides. If you are interested in separating and analyzing long-chain ascarosides, you will need to extend the ramp from 50 % B to 100 % B over a longer period of time.
18. Using greater than seven smoothing points may create artificial peaks.
19. Different ascarosides ionize and fragment with different efficiencies, so the area of a given peak is not a very good indication of its relative abundance.
20. This drying step will remove trace amounts of methanol and water in your sample that would otherwise interfere with the signal of the sample the NMR spectrum.
21. The sample should extend exactly 2 cm in the Shigemi tube. This volume of sample will allow the sample to extend the full length of the receiver coil when inserted into the NMR spectrometer. When washing the cotton filter with a few drops of methanol-d_4, take great care not to add too much volume, since the sample will then extend longer than 2 cm in the Shigemi tube and will be more dilute. A more dilute sample will give lower signal in the NMR spectrum.
22. If bubbles appear, they must be removed since they may interfere with optimal shimming. They can be removed by slightly pushing the insert 1–2 mm further into the tube with a quick turning motion.
23. Shigemi tubes have a smaller sample window than regular 5 mm NMR tubes because both the bottom of the tube and the insert that fits into the tube are made of solvent-matched glass. Thus, the volume of solvent required to dissolve the sample and fill the tube is smaller, thereby increasing the concentration of the sample and the signal-to-noise ratio in the NMR spectrum.
24. Be sure to align the Shigemi tube in the spinner using the depth gauge so that once the tube is inserted into the NMR spectrometer, the deuterated solvent will be centered within the receiver coil and will extend the full length of the receiver coil. If the deuterated solvent does not extend the full length of the receiver coil, it will cause issues with shimming.
25. If the lock level drops significantly (>10 %) when the spinner is turned off, it indicates poor shimming. Adjust the non-spinning (x, y) shims accordingly.
26. The number of scans determines the signal-to-noise ratio of the spectrum. For concentrated samples such as the crude pheromone extract, 16 scans is generally good, but for more

dilute samples, such as chromatography fractions from the crude pheromone extract, the number of scans should be set higher (e.g., 32). In general, if a sample is more dilute by a factor of 2, then the number of scans must be increased by a factor of 4 in order to obtain a spectrum with a similar signal-to-noise ratio. The number of scans in a dqf-COSY should be a multiple of 8 due to phase-cycling.

27. Make sure that the peak shapes are sharp and that the methanol peak at 3.31 ppm is a fully defined quintet. If the peak shape is not sharp or is skewed, it indicates poor shimming. Spinning the tube improves peak shape, but since spinning will be turned off during the 2D dqf-COSY experiment, it is important to assess peak shape in a ^{1}H spectrum in which the spinner has been turned off.

28. Because the crude pheromone extract has a high concentration of salt, the 90° pulse width can be much higher than for more dilute samples. Optimizing the 90° pulse width will improve the signal-to-noise ratio in the 2D dqf-COSY experiment.

29. Acquiring this second ^{1}H spectrum with an optimized 90° pulse width will give you a lower gain value than that obtained in the original ^{1}H spectrum. Use this new lower gain value in the dqf-COSY experiment.

30. The number of iterations determines the resolution in the indirect (f_1) dimension and should be at least 512 for a dqf-COSY experiment performed on a complex mixture. If the experiment time (which is proportional to the number of iterations) is not an issue, then the number of iterations should be increased to 1,024.

31. The resolution in the direct (f_2) dimension depends on the number of points. Since increasing the number of points does not significantly increase the experiment time, there is no disadvantage to setting the number of points to as high as 4,096 (except that it will increase the file size). The number of points should be equal to a power of two since the Fourier transform uses 2^n points. Given a sweep width of approximately 9 ppm, setting the number of points to 4,096 will give an acquisition time of approximately 0.5.

32. The desktop processing software, MestreNova, is a convenient way to process NMR data. In the f_2 dimension, use a sine-bell window function with a 90° shift (to obtain high resolution) or a squared sine-bell window function with a 55° shift (if high noise is present). In the f_1 dimension, use a sine-bell window function with a 90° shift. For zero-filling, the size of the Fourier transform should be 4,096 in both dimensions. The spectrum will also need to be phased.

Acknowledgments

Funding for this work was provided by the National Institutes of Health (GM87533 to R.A.B.) and the Human Frontiers Science Program (RGY0042/2010 subcontract to R.A.B.). Our work on the triple quadrupole mass spectrometer was supported in part by the National Institutes of Health and National Center for Research Resources CTSA grant 1UL 1RR029890. We would like to thank Dr. Tim Garrett (University of Florida) for helpful suggestions on the LC-MS/MS parameters and Dr. Ion Ghiviriga (University of Florida) for helpful suggestions on the NMR parameters. We would like to thank Prof. Justin R. Ragains (Louisiana State University) for providing the synthetic ascaroside standards that are used in our work.

References

1. Jeong PY, Jung M, Yim YH et al (2005) Chemical structure and biological activity of the *Caenorhabditis elegans* dauer-inducing pheromone. Nature 433:541–545
2. Butcher RA, Fujita M, Schroeder FC et al (2007) Small-molecule pheromones that control dauer development in *Caenorhabditis elegans*. Nat Chem Biol 3:420–422
3. Butcher RA, Ragains JR, Kim E et al (2008) A potent dauer pheromone component in *C. elegans* that acts synergistically with other components. Proc Natl Acad Sci USA 105:14288–14292
4. Srinivasan J, Kaplan F, Ajredini R et al (2008) A blend of small molecules regulates both mating and development in *Caenorhabditis elegans*. Nature 454:1115–1118
5. Pungaliya C, Srinivasan J, Fox BW et al (2009) A shortcut to identifying small molecule signals that regulate behavior and development in *Caenorhabditis elegans*. Proc Natl Acad Sci USA 106(19):7708–7713
6. Butcher RA, Ragains JR, Clardy J (2009) An indole-containing dauer pheromone component with unusual dauer inhibitory activity at higher concentrations. Org Lett 11:3100–3103
7. von Reuss SH, Bose N, Srinivasan J et al (2012) Comparative metabolomics reveals biogenesis of ascarosides, a modular library of small-molecule signals in *C. elegans*. J Am Chem Soc 134:1817–1824
8. Srinivasan J, von Reuss SH, Bose N et al (2012) A modular library of small molecule signals regulates social behaviors in *Caenorhabditis elegans*. PLoS Biol 10:e1001237
9. Artyukhin AB, Yim JJ, Srinivasan J et al (2013) Succinylated octopamine ascarosides and a new pathway of biogenic amine metabolism in *C. elegans*. J Biol Chem 288:18778–18783
10. Butcher RA, Ragains JR, Li W et al (2009) Biosynthesis of the *Caenorhabditis elegans* dauer pheromone. Proc Natl Acad Sci USA 106:1875–1879
11. Kaplan F, Srinivasan J, Mahanti P et al (2011) Ascaroside expression in *Caenorhabditis elegans* is strongly dependent on diet and developmental stage. PLoS One 6: e17804

Part II

Approaches to Characterize Gene and Function of Pheromone Receptors

Chapter 7

Identification of Chemosensory Receptor Genes from Vertebrate Genomes

Yoshihito Niimura

Abstract

Chemical senses are essential for the survival of animals. In vertebrates, mainly three different types of receptors, olfactory receptors (ORs), vomeronasal receptors type 1 (V1Rs), and vomeronasal receptors type 2 (V2Rs), are responsible for the detection of chemicals in the environment. Mouse or rat genomes contain >1,000 OR genes, forming the largest multigene family in vertebrates, and have >100 V1R and V2R genes as well. Recent advancement in genome sequencing enabled us to computationally identify nearly complete repertories of OR, V1R, and V2R genes from various organisms, revealing that the numbers of these genes are highly variable among different organisms depending on each species' living environment. Here I would explain bioinformatic methods to identify the entire repertoires of OR, V1R, and V2R genes from vertebrate genome sequences.

Key words Olfactory receptor, Vomeronasal receptor, Multigene family, Bioinformatics, Vertebrate, G protein-coupled receptor

1 Introduction

Chemical senses are essential for the survival of most animals. In vertebrates, chemical molecules in the environment are mainly detected by three different types chemosensory receptors, named olfactory receptors (ORs), vomeronasal receptors type 1 (V1Rs), and vomeronasal receptors type 2 (V2Rs) (for review, Ref. 1). All of them are G protein-coupled receptors (GPCRs), membrane proteins having seven transmembrane (TM) α-helical regions (Fig. 1). Each type of receptors is encoded by a multigene family. The three gene families share almost no sequence similarity, though their molecular structures are similar to one another.

Among the three gene families, the OR gene family is by far the largest (for review, Ref. 2). OR genes are predominantly expressed in the olfactory epithelium of the nasal cavity and were first identified from rats by Linda Buck and Richard Axel in 1991 [3]. The genomes of many mammalian species harbor ~1,000 or more

Kazushige Touhara (ed.), *Pheromone Signaling: Methods and Protocols*, Methods in Molecular Biology, vol. 1068, DOI 10.1007/978-1-62703-619-1_7, © Springer Science+Business Media, LLC 2013

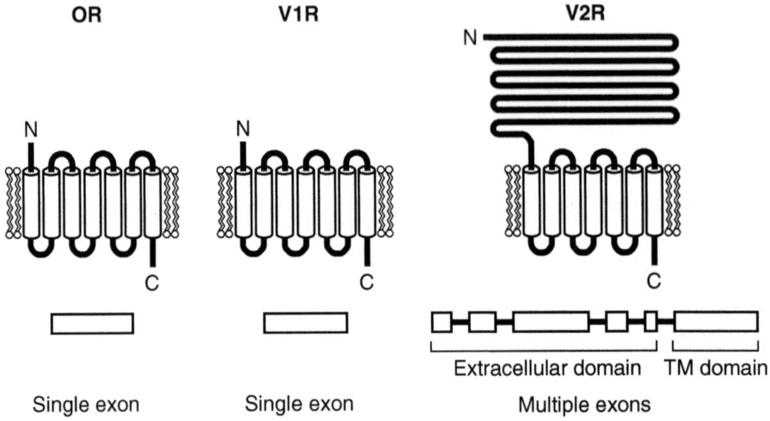

Fig. 1 Structures of ORs, V1Rs, and V2Rs. Membrane topologies and exon–intron structures of genes are shown

OR genes, which comprise 4–5 % of their proteomes. An OR is ~310 amino acids long on average, and typically the gene does not have any introns in its coding region. OR genes are present in all vertebrate species. Fishes have much smaller repertoires of OR genes (~100) than mammals. However, despite smaller OR gene repertoires in fishes, OR gene sequences in fishes are more diverse than those in mammals [4, 5]. Amphibians and fishes have some additional groups of OR genes that are not present in mammals.

V1R and V2R genes were discovered in 1995 [6] and 1997 [7–9], respectively. They are expressed in vomeronasal organs in mammals and are involved in pheromone detection. Fishes do not have a discrete vomeronasal organ, but they do have both V1R and V2R genes, which are expressed in olfactory epithelium in fishes [10, 11]. V1R genes are intronless like OR genes, whereas V2R genes consist of multiple exons (Fig. 1). A V2R gene is characterized by a long N-terminal extracellular domain, which is encoded by several exons, while the TM domain typically corresponds to a single exon.

Recently the whole genome sequences became available from diverse organisms, which enabled us to identify nearly complete repertoires of chemosensory receptor genes in a given species. Comparative genomic studies revealed that the repertoires of chemosensory receptor genes are highly variable among different species (Refs. 4, 5, 12–14 for OR genes, Refs. 10, 15–18 for V1R genes, and Refs. 17, 19, 20 for V2R genes). In this chapter, I describe the methods to identify OR, V1R, and V2R genes from vertebrate genomes. OR and V1R genes do not have any introns; therefore, detection of these genes from genome sequences is relatively easy. However, because the numbers of the genes are huge, development of computation methods is requisite for the identification of the entire gene repertoires. On the other hand, identification of V2R genes is more difficult due to their complex gene structures.

2 Materials

1. The following computer programs need to be installed: BLAST (http://blast.ncbi.nlm.nih.gov/; Ref. 21) for homology searches, MAFFT (http://mafft.cbrc.jp/alignment/software/; Ref. 22) for constructing multiple alignments, and LINTREE (http://www.personal.psu.edu/nxm2/software.htm; Ref. 23) for phylogenetic tree construction. Moreover, the HMMER package (http://hmmer.janelia.org; Ref. 24) and the Wise2 package (http://www.ebi.ac.uk/Tools/Wise2/; Ref. 25) are used for the identification of V2R genes.
2. The genome sequences of various vertebrate species can be downloaded from the Web sites of the University of California Santa Cruz (http://genome.ucsc.edu), Ensembl Genome Browser (http://ensembl.org), the Genome Sequencing Center at Washington University School of Medicine (http://genome.wustl.edu), the Broad Institute (http://www.broadinstitute.org), etc.

3 Methods

Here I would explain the method for identifying OR genes in some detail. Figure 2 illustrates the flowchart of OR gene identification. OR genes are classified into three categories: intact genes, truncated genes, and pseudogenes. An intact gene putatively encodes a functional OR. A pseudogene is defined as a sequence containing interrupting stop codons, frameshifts, and/or gaps within conserved regions. A truncated gene is a partial intact sequence encoding a part of OR and is located at the contig end. It is either a functional gene or a pseudogene. Therefore, the fraction of pseudogenes in a given species is overestimated if only intact genes are considered to be functional. This effect is critical for the species with low-coverage genome data, which contain many short contigs [5, 12, 14]. For this reason, some caution is necessary to estimate the fraction of pseudogenes.

The methods to identify V1R and V2R genes were described in Refs. 15, 16 and Refs. 19, 20, 26, respectively. The methods explained here are based on these articles with some modification.

3.1 Identification of OR Genes

1. Conduct TBLASTN searches [21] with a cutoff *E*-value of 1e−5 against a given genome sequence using known OR genes as queries (*see* **Note 1**). Use the option "-F F" (filtering low-complexity regions is not used) (*see* **Note 2**).
2. Because multiple sequences are used as queries, a number of different queries may hit the same genomic region. In such a case, choose a BLAST hit with the lowest *E*-value (called a

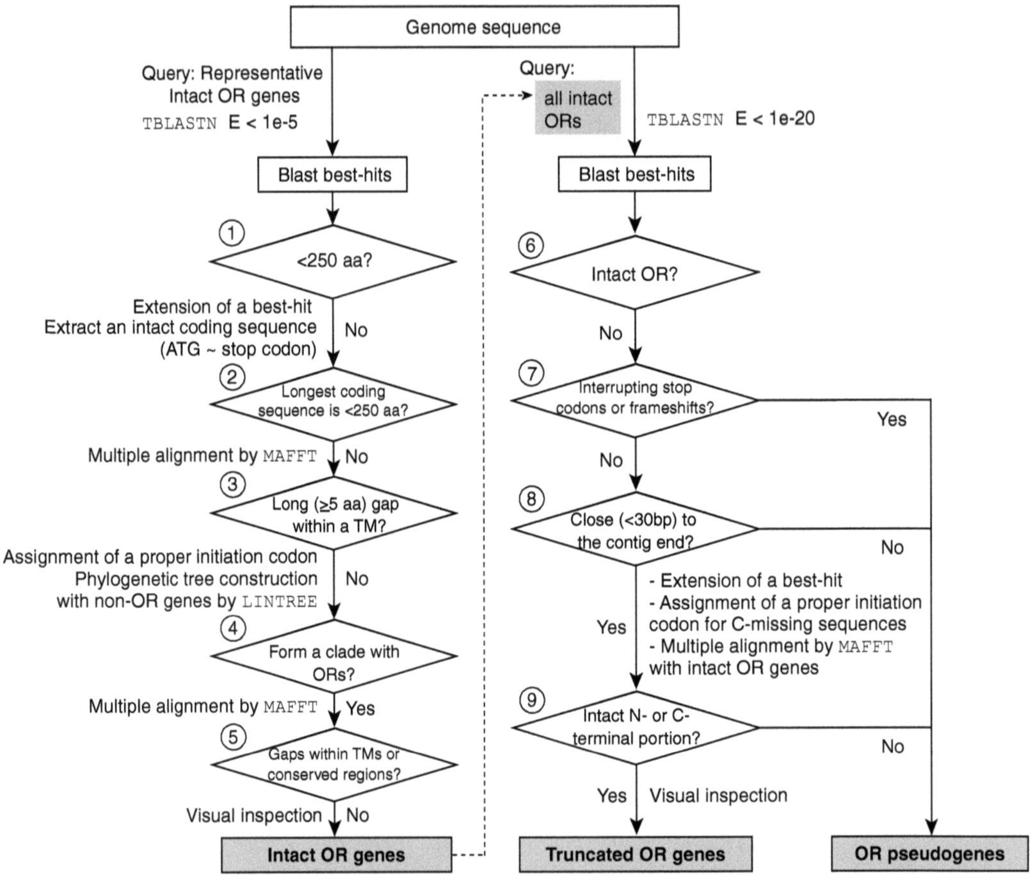

Fig. 2 Flowchart for the identification of intact OR genes, truncated OR genes, and OR pseudogenes from vertebrate genome sequences

"best-hit") among all BLAST hits to a given genomic region. Extract all best-hits from the genome sequence.

3. To identify intact OR genes from the best-hit sequences, several criteria are used. First, discard the best-hit sequences shorter than 250 amino acids (criterion 1 in Fig. 2; *see* **Note 3**).

4. For each of the best-hit sequences remaining after **step 3**, extend it to both directions along the genome sequence. Then, extract the longest coding sequence from an ATG codon to a stop codon (Fig. 3a). If the length of the sequence after extension is less than 250 amino acids, discard it (criterion 2 in Fig. 2). If the extended sequence contains undetermined nucleotides, the sequence should also be discarded.

5. Construct a multiple alignment from the remaining sequences after **step 4** by using the program MAFFT [22] (*see* **Note 4**), and assign the positions of seven TM regions according to Ref. 27.

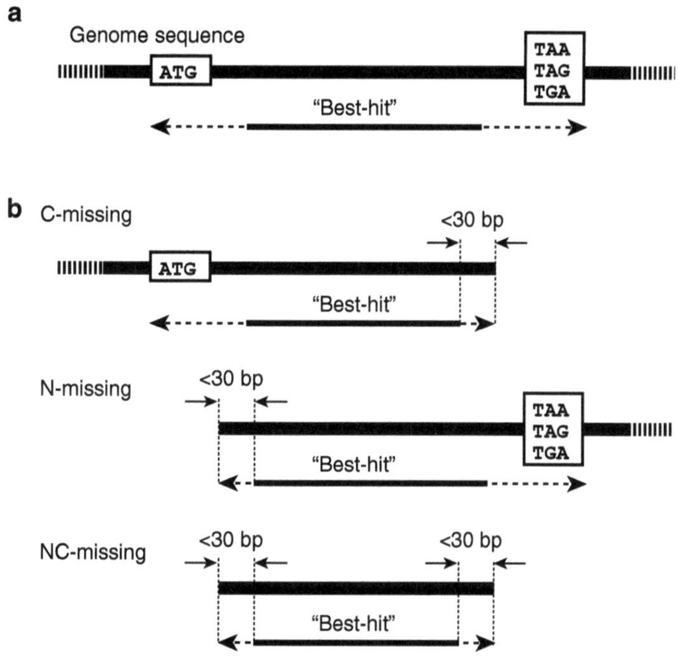

Fig. 3 (**a**) Extension of a best-hit along the genome sequence to take the longest coding sequence. (**b**) Extension of a best-hit that is located near the end of a contig. C-missing, N-missing, and NC-missing sequences are shown separately

6. If a sequence has a gap of five or more amino acids within TM regions, exclude it (criterion 3 in Fig. 2).
7. Construct a multiple alignment once again from the remaining sequences after **step 6** by using MAFFT [22] (*see* **Note 4**). Then choose the most proper ATG codon as the initiation codon in case that a sequence examined contains two or more ATG codons in the N-terminal tail region (the upstream of the first TM region; *see* **Note 5**).
8. Construct a neighbor-joining phylogenetic tree [28] using the remaining sequences after **step 7** together with several non-OR GPCR genes as the outgroup (*see* **Note 6**). Run the program njboot in LINTREE [23] with the option "-d28" (Poisson correction distance) and "-b500" (bootstrap resamplings for 500 times).
9. Eliminate non-OR genes on the basis of the phylogenetic tree constructed in **step 8** (criterion 4 in Fig. 2; *see* **Note 7**). When a given sequence is located out of the OR gene clade in the phylogenetic tree, it should be discarded (*see* **Note 8**).
10. Construct a multiple alignment from the remaining sequences after **step 9** by MAFFT [22] (*see* **Note 4**), and eliminate the sequences having gaps within TM regions or at other conserva-

tive amino acid sites by visual inspection (criterion 5 in Fig. 2). The remaining sequences are regarded as intact OR genes.

11. To identify non-intact OR genes, perform TBLASTN searches [21] against the genome sequence using all intact OR genes identified in **step 10** as queries with the *E*-value below 1e–20 (*see* **Note 9**).
12. Extract all best-hit sequences in the same way as **step 2**.
13. Exclude all of the intact OR genes obtained in **step 10** (criterion 6 in Fig. 2). All remaining sequences are regarded to be truncated genes or pseudogenes.
14. Extract the best-hit sequences that meet both of the following conditions. (1) There are no interrupting stop codons and frameshifts (criterion 7 in Fig. 2). (2) The distance between the end of the sequence and the end of the contig containing the sequence is less than 30 bp (criterion 8 in Fig. 2).
15. Classify the remaining sequences after **step 14** into three category, C-missing, N-missing, and NC-missing (Fig. 3b). For a C-missing sequence, the upstream of the best-hit is present in the contig examined, whereas its downstream is missing. Conversely, for an N-missing sequence, its downstream is present in the contig, while its upstream is missing. An NC-missing sequence lacks both upstream and downstream portions (*see* **Note 10**).
16. For a C-missing sequence, extend it along the genome sequence and extract a sequence from the most upstream ATG codon to the most downstream in-frame codon without any interrupting stop codons (Fig. 3b, top).
17. Construct a multiple alignment using the extended C-missing sequences obtained in **step 16** together with some representative intact OR genes by MAFFT [22], and choose the most proper ATG codon as the initiation codon for each sequence in the same manner as **step 7** and **Note 5**.
18. For an N-missing sequence, extend it and extract a sequence from the most upstream in-frame codon to the stop codon (Fig. 3b, middle).
19. For an NC-missing sequence, extend it and extract the longest sequence from the most upstream in-frame codon to the most downstream one (Fig. 3b, bottom).
20. Construct a multiple alignment using all of the extended C-missing, N-missing, and NC-missing sequences obtained in **steps 17–19** together with some representative intact OR genes by MAFFT [22].
21. Exclude the sequences that contain gaps within TM regions or at other conservative amino acid sites by visual inspection

(criterion 9 in Fig. 2). The remaining sequences are regarded to be truncated genes.

22. Exclude all truncated genes from the best-hit sequences obtained in **step 13**. All remaining sequences are regarded as OR pseudogenes.

3.2 Identification of V1R Genes

1. Conduct TBLASTN searches [21] with a cutoff *E*-value of 1e−5 against a given genome sequence using known V1R genes as queries. The following options should be used: "-F F" for that filtering low-complexity regions is not used and "-v 1000 -b 1000" for the number of hits reported.
2. From the results obtained in **step 1**, extract all best-hit sequences in the same manner as Subheading 3.1, **step 2**.
3. For each of the best-hit sequences, conduct BLASTP searches [21] against the nr database of GenBank to ensure that the best hit is a V1R gene. Discard the sequences showing higher similarity to non-V1R genes (*see* **Note 11**).
4. Among the remaining sequences after **step 3**, extract sequences that do not contain any interrupting stop codons or frameshifts. For each of these sequences, extend it to both directions along the genome sequence and extract the longest coding sequence from an ATG codon to a stop codon.
5. Construct a multiple alignment from the sequences obtained in **step 4** by MAFFT [22] and choose the most proper initiation codon from each sequence.
6. Assign the location of TM regions in the multiple alignment. If a sequence contains gaps within TM regions or other highly conserved regions, it should be regarded as a pseudogene. The remaining sequences are regarded to be intact V1R genes.
7. Exclude all intact V1R genes (**step 6**) from the sequences obtained after **step 3**. The remaining sequences are regarded as V1R pseudogenes.

3.3 Identification of V2R Genes

1. The first step is to perform TBLASTN searches [21] against a given genome sequence using known V2R genes as queries. To this end, construct a multiple alignment by MAFFT [22] from the known V2R query sequences (*see* **Note 12**).
2. Trim the multiple alignment to extract the TM domain (from the first TM region to the C-terminal end) according to Ref. 8. The TM domain in each V2R gene is used as a query sequence of TBLASTN searches (*see* **Note 13**).
3. Conduct TBLASTN searches [21] with a cutoff *E*-value of 1e−5 using the TM domain of known V2R genes as queries. The following options should be used: "-F F" for that filtering low-complexity regions is not used and "-v 1000 -b 1000" for the number of hits reported.

4. From the results obtained in **step 3**, extract all best-hit sequences in the same manner as Subheading 3.1, **step 2**.
5. For each of the best-hit sequences, extract the genomic sequence together with 200 kb upstream and 1 kb downstream regions (*see* **Note 14**).
6. Construct a profile Hidden Markov Model (HMM) from the alignment generated in **step 1**. (Here use the alignment including the entire coding region rather than a TM domain). Run the program hmmbuild in the HMMER package, version 2.3.2 [24] for the MAFFT output file with options "--fast --gapmax 0.95 -s" [20]. Then run the hmmcalibrate program subsequently.
7. Align each of the elongated best-hit sequences obtained in **step 5** with a profile HMM created in **step 6** by the program genewisedb in the Wise2 package [25]. Use the following options: "-splice flat -cut 20 -alg 623 -aalg 623 -pretty -para -pseudo -genes -sum -cdna -trans -gff -gener" [20].
8. When two neighboring elongated best-hit sequences (from **step 5**) are overlapped, the same genomic region may be detected as a result of the genewisedb searches. In such cases, extract only one genewisedb hit showing the highest score among all overlapping hits.
9. For each of the genewisedb hits obtained in **step 8**, conduct BLASTP searches [21] against the nr database of GenBank to ensure that the best hit is a V2R gene. Discard the sequences showing higher similarity to non-V2R genes (*see* **Note 15**). All remaining sequences are regarded to be V2R genes.
10. Discard the sequences (1) that contain interrupting stop codons or frameshifts, (2) that have gaps within TM regions or other highly conserved regions, and (3) that are shorter than 750 amino acids. The remaining sequences are regarded to be intact V2R genes.
11. Exclude all intact V2R genes (**step 10**) from the sequences obtained after **step 9**. The remaining sequences are regarded as V2R pseudogenes.

4 Notes

1. As for query sequences, an OR gene set with sufficient sequence diversity should be used to retrieve all putative OR genes from the genome. However, currently thousands of OR genes are available in the databases; therefore, to reduce a computational time, highly similar sequences should be eliminated from the queries. To this aim, classify candidate query genes into groups using a given sequence similarity cutoff, e.g., 50 % amino acid

identity, and choose one representative sequence from each group. To examine amniote (mammalian, avian, and reptilian) genomes, human and/or mouse OR genes can be used as queries. On the other hand, to investigate amphibian or fish genomes, OR genes from fishes (e.g., zebrafish) and/or frogs should be used as queries, because amphibian and fish OR genes are more diverse than amniote OR genes [5]. OR gene sequences in mammals and other vertebrates can be obtained from Refs. 12, 14, 29 and Ref. 5, respectively.

2. If the number of BLAST hits is expected to large, e.g., when mammalian genome sequences are examined, "-v" and "-b" options should also be used to change the number of hits and alignments shown.
3. The cutoff length of 250 amino acids is sufficiently shorter than that of any known functional OR genes.
4. When the number of sequences is large (e.g., >400), it is better to separate them into several groups to reduce a computational time.
5. For most of the known functional OR genes in mammals, the length of the N-terminal tail is between 21 and 34 amino acids. Therefore, to choose the initiation codon, the following criteria can be used. If an ATG codon is present in the region (named "region A") between the positions -34 and -21 (here the amino acid position is indicated as the relative position to the boundary between the N-terminal tail and the first TM region), choose the one as the initiation codon. In case that two or more ATG codons are present within the region A, choose the most downstream one among them. If ATG codons are not present in the region A, choose the closest one to the region A.
6. The following genes can be used as the outgroup sequences: alpha-1A-adrenergic receptor isoform 1 (GenBank protein id, NP_000671), beta-1-adrenergic receptor (NP_000675), adenosine A2b receptor (NP_000667), histamine receptor H2 (NP_071640), 5-hydroxytryptamine (serotonin) receptor 1B (NP_000854), 5-hydroxytryptamine (serotonin) receptor 1F (NP_000857), 5-hydroxytryptamine (serotonin) receptor 6 (NP_000862), galanin receptor 1 (NP_001471), and somatostatin receptor 4 (NP_001043). These genes are relatively close to OR genes among the genes belonging to the rhodopsin-like GPCR superfamily [30].
7. When the number of sequences is large (e.g., >200), it is better to separate them into several groups to reduce a computational time.
8. In a phylogenetic tree, OR genes form a monophyletic clade with a high bootstrap support [5]. Therefore, non-OR genes are easily distinguishable from OR genes.

9. The reason for using the cutoff *E*-value of 1e–20 is as follows. First, the *E*-value of a best-hit to the genomic region corresponding to a non-OR gene is 1e–18 or larger. Second, all best-hit sequences with the *E*-value below 1e–20 obtained by OR gene queries are more similar to OR genes than to any known non-OR genes. Therefore, non-intact OR genes can be distinguished from non-OR genes by using the cutoff *E*-value of 1e–20.
10. Note that NC-missing sequences are found only on a contig shorter than the length of an OR gene (~930 bp).
11. This step is necessary to exclude some other receptors (e.g., T2R taste receptors) that are homologous to V1Rs.
12. A V2R gene set with sufficient sequence diversity should be used as query sequences. These sequences are also used to construct a profile HMM (*see* **step 6**). To search for mammalian V2R genes, for example, 75 intact V2R genes in mice with the prefix "mouseMay04V2R" in Ref. 20 can be used as queries. Fish V2R gene sequences are available in Ref. 19.
13. The N-terminal extracellular domain shows a higher extent of sequence diversity [26]; therefore, if N-terminal extracellular domain is used as a query, homology searches give a large number of non-V2R hits. For this reason, it is better to use only a TM domain rather than the entire sequence of a V2R gene as a query.
14. For extracting the entire coding exons of a V2R gene, it is necessary to examine a long upstream genomic sequence, because the N-terminal extracellular domain is encoded by multiple exons (Fig. 1). (Note that a query sequence for homology searches contains only a TM domain.) The reason for using the 200 kb limit is that it is longer than the genomic extent of all previously described V2R genes [20].
15. This step is necessary, because some other receptors (e.g., calcium-sensing receptors and T1R taste receptors) are known to be homologous to V2Rs.

Acknowledgments

This work was supported by grant (20770192 and 23770271) from the Ministry of Education, Culture, Sports, Science, and Technology, Japan.

References

1. Nei M, Niimura Y, Nozawa M (2008) The evolution of animal chemosensory receptor gene repertoires: roles of chance and necessity. Nat Rev Genet 9:951–963
2. Niimura Y (2012) Olfactory receptor multigene family in vertebrates: from the viewpoint of evolutionary genomics. Curr Genomics 13:103–114
3. Buck L, Axel R (1991) A novel multigene family may encode odorant receptors: a molecular basis for odor recognition. Cell 65:175–187
4. Niimura Y, Nei M (2005) Evolutionary dynamics of olfactory receptor genes in fishes and tetrapods. Proc Natl Acad Sci USA 102:6039–6044
5. Niimura Y (2009) On the origin and evolution of vertebrate olfactory receptor genes: Comparative genome analysis among 23 chordate species. Genome Biol Evol 1:34–44
6. Dulac C, Axel R (1995) A novel family of genes encoding putative pheromone receptors in mammals. Cell 83:195–206
7. Herrada G, Dulac C (1997) A novel family of putative pheromone receptors in mammals with a topographically organized and sexually dimorphic distribution. Cell 90:763–773
8. Matsunami H, Buck LB (1997) A multigene family encoding a diverse array of putative pheromone receptors in mammals. Cell 90:775–784
9. Ryba NJ, Tirindelli R (1997) A new multigene family of putative pheromone receptors. Neuron 19:371–379
10. Saraiva LR, Korsching SI (2007) A novel olfactory receptor gene family in teleost fish. Genome Res 17:1448–1457
11. Hashiguchi Y, Nishida M (2005) Evolution of vomeronasal-type odorant receptor genes in the zebrafish genome. Gene 362:19–28
12. Niimura Y, Nei M (2007) Extensive gains and losses of olfactory receptor genes in mammalian evolution. PLoS One 2:e708
13. Hayden S, Bekaert M, Crider TA et al (2010) Ecological adaptation determines functional mammalian olfactory subgenomes. Genome Res 20:1–9
14. Matsui A, Go Y, Niimura Y (2010) Degeneration of olfactory receptor gene repertories in primates: No direct link to full trichromatic vision. Mol Biol Evol 27:1192–1200
15. Grus WE, Shi P, Zhang YP et al (2005) Dramatic variation of the vomeronasal pheromone receptor gene repertoire among five orders of placental and marsupial mammals. Proc Natl Acad Sci USA 102:5767–5772
16. Young JM, Kambere M, Trask BJ et al (2005) Divergent V1R repertoires in five species: Amplification in rodents, decimation in primates, and a surprisingly small repertoire in dogs. Genome Res 15:231–240
17. Shi P, Zhang J (2007) Comparative genomic analysis identifies an evolutionary shift of vomeronasal receptor gene repertoires in the vertebrate transition from water to land. Genome Res 17:166–174
18. Young JM, Massa HF, Hsu L et al (2010) Extreme variability among mammalian V1R gene families. Genome Res 20:10–18
19. Hashiguchi Y, Nishida M (2006) Evolution and origin of vomeronasal-type odorant receptor gene repertoire in fishes. BMC Evol Biol 6:76
20. Young JM, Trask BJ (2007) V2R gene families degenerated in primates, dog and cow, but expanded in opossum. Trends Genet 23:212–215
21. Altschul SF, Madden TL, Schaffer AA et al (1997) Gapped BLAST and PSI-BLAST: a new generation of protein database search programs. Nucleic Acids Res 25:3389–3402
22. Katoh K, Kuma K, Toh H et al (2005) MAFFT version 5: improvement in accuracy of multiple sequence alignment. Nucleic Acids Res 33:511–518
23. Takezaki N, Rzhetsky A, Nei M (1995) Phylogenetic test of molecular clock and linearized trees. Mol Biol Evol 12:823–833
24. Eddy SR (2011) Accelerated Profile HMM Searches. PLoS Comput Biol 7:e1002195
25. Birney E, Clamp M, Durbin R (2004) GeneWise and Genomewise. Genome Res 14:988–995
26. Yang H, Shi P, Zhang YP et al (2005) Composition and evolution of the V2r vomeronasal receptor gene repertoire in mice and rats. Genomics 86:306–315
27. Man O, Gilad Y, Lancet D (2004) Prediction of the odorant binding site of olfactory receptor proteins by human-mouse comparisons. Protein Sci 13:240–254
28. Saitou N, Nei M (1987) The neighbor-joining method: a new method for reconstructing phylogenetic trees. Mol Biol Evol 4: 406–425
29. Go Y, Niimura Y (2008) Similar numbers but different repertoires of olfactory receptor genes in humans and chimpanzees. Mol Biol Evol 25:1897–1907
30. Fredriksson R, Lagerström MC, Lundin LG et al (2003) The G-protein-coupled receptors in the human genome form five main families. Phylogenetic analysis, paralogon groups, and fingerprints. Mol Pharmacol 63: 1256–1272

Chapter 8

Functional Assays for Insect Olfactory Receptors in Xenopus Oocytes

Tatsuro Nakagawa and Kazushige Touhara

Abstract

Reconstitution of olfactory or pheromone receptors in heterologous expression systems greatly facilitates the functional analysis of these receptors. *Xenopus laevis* oocytes can be used to efficiently express insect olfactory or pheromone receptors. In this chapter, we describe how to use *Xenopus laevis* oocytes for functional assays of insect olfactory receptors. The procedure can be subdivided into the four following steps: (1) in vitro complementary RNA (cRNA) synthesis, (2) isolation of oocytes from female *Xenopus laevis*, (3) cRNA microinjection into oocytes, and (4) two-electrode voltage-clamp recording. This system can be used to identify odor or pheromone ligands and to analyze structure–function relationships involving receptor proteins of interest.

Key words *Xenopus laevis* oocyte, In vitro cRNA transcription, Heterologous expression, Electrophysiology

1 Introduction

To understand the function of olfactory or pheromone receptors in insects, it is necessary to identify cognate ligands, analyze signal transduction mechanisms, and examine structure–function relationships within the receptor proteins. Currently, there are two types of expression systems—in vivo and in vitro—for recording the responses of olfactory or pheromone receptors in insects. The in vivo recording systems, such as electroantennograms, single sensillum recordings, and antennal lobe imaging techniques (*see* Chapters 11 and 12), have been used to accumulate vast amounts of knowledge as to physiological roles of some pheromone compounds and some olfactory or pheromone receptors [1–3]. However, these in vivo systems are not suitable for exhaustive, efficient, and high-throughput analysis of many insect olfactory or pheromone receptors because it is difficult to generate a variety of mutants quickly. Alternative in vitro recording systems would be invaluable for overcoming such situations because in vitro systems

Kazushige Touhara (ed.), *Pheromone Signaling: Methods and Protocols*, Methods in Molecular Biology, vol. 1068,
DOI 10.1007/978-1-62703-619-1_8, © Springer Science+Business Media, LLC 2013

could make it much easier and faster to analyze the function of olfactory or pheromone receptors.

Xenopus laevis oocyte has long been used as an excellent expression system for functional analyses of myriad G-protein coupled receptors (GPCRs) and ion channels. In functional assay using *Xenopus* oocytes, which is called two-electrode voltage-clamp recording (TEVC), the activity of the expressed receptor is detected as changes in membrane current. *Xenopus* oocytes have two major advantages to analyze the function of membrane proteins; first, oocytes have an abundance of ribosomal RNAs and transfer RNAs and can, therefore, express exogenously introduced mRNA as functional receptor proteins; second, oocytes possess many kinds of endogenous signal transduction components, such as G-proteins ($G\alpha_q$, $G\alpha_s$, and $G\alpha_{i/o}$), adenylyl cyclase, phospholipace C, and Ca^{2+}-activated Cl^- channels, and therefore, the responses of GPCRs, as well as those of ion channels, can be detected by co-expressing some ion channels that are activated downstream of the signal transduction cascade of the GPCRs.

Many researchers have put their efforts into reconstituting insect olfactory or pheromone receptors in *Xenopus* oocytes. In 2001, Wetzel and colleagues succeeded, for the first time, in detecting odor responses of *Drosophila* Or43a expressed in *Xenopus* oocytes [4]. After that, Nakagawa and colleagues found that co-expression of BmOrco (the Orco family receptor in *Bombyx mori*) with either BmOr-1 (the bombykol receptor) or BmOr-3 (the bombykal receptor) in oocytes greatly enhances the sensitivities to the respective BmOr-1 or BmOr-3 cognate ligand [5]. This result has been shown to be applicable to other combinations of Or and the Orco family receptors, and to other heterologous expression systems as well [6–11]. These findings provided invaluable methods for efficient analysis of the function of many insect olfactory and pheromone receptors in vitro and lead a rapid increase in our knowledge of the function of olfactory and pheromone receptors in insects. For instance, it was shown that insect Or and Orco proteins form heteromeric receptors of unknown stoichiometry [12]. And more recently, several groups have shown that the insect Or–Orco complexes function as odor- or pheromone-activated nonselective cation channels [8–10]. Furthermore, some site-directed mutagenesis studies using *Xenopus* oocytes provided important insights into the functional domains within insect Or–Orco complexes, specifically the domains that are involved in ligand binding and in ion permeation properties [13, 14].

Method for expressing insect olfactory or pheromone receptors in *Xenopus* oocytes is now firmly established. This method can be subdivided into the following four steps: (1) in vitro transcription of cRNAs of a canonical Or and a Orco protein, (2) isolation of oocytes from female *Xenopus laevis*, (3) microinjection of the synthesized cRNAs into oocytes, (4) electrophysiological recording. In this section we will describe a detailed protocol for performing functional analysis of an Or–Orco complex of interest in *Xenopus* oocytes.

2 Materials

2.1 In Vitro cRNA Synthesis

1. cDNA clones encoding Ors in a *Xenopus* oocyte expression vector (*see* **Note 1**). Store these reagents at −20 °C.
2. RNA synthesis solution (for 1 sample) (*see* **Note 2**). All the reagents should be stored at −20 °C.

   For RNA synthesis reactions using T7 polymerase, please use the following volumes and concentrations for each reagent listed in each reaction mixture: 46.5 μl of DEPC water, 20 μl of 5× transcription optimized buffer (40 mM Tris–HCl, pH 7.9, 6 mM $MgCl_2$, 2 mM spermidine, and 10 mM NaCl), 5 μl of ATP (10 mM), 2.5 μl of GTP (10 mM), 5 μl of CTP (10 mM), 5 μl of UTP (10 mM), 5 μl of DTT (0.1 M), 5 μl of m7G(5′)ppp(5′)G RNA Capping Analog (10 mM), 3 μl of Ribonuclease inhibitor, and 3 μl of T7 polymerase (10–20 units/μl).

   For RNA synthesis reactions using SP6 polymerase, please use the following volumes and concentrations for each reagent listed in each reaction mixture: 41.5 μl of DEPC water, 20 μl of 5× transcription optimized buffer (40 mM Tris–HCl, pH 7.9, 6 mM $MgCl_2$, 2 mM spermidine and 10 mM NaCl), 5 μl of ATP (10 mM), 2.5 μl of GTP (10 mM), 5 μl of CTP (10 mM), 5 μl of UTP (10 mM), 10 μl of DTT (10 mM), 5 μl of m7G(5′)ppp(5′)G RNA Capping Analog (10 mM), 3 μl of Ribonuclease inhibitor, and 3 μl of SP6 polymerase (10–20 units/μl).
3. RNase-Free DNase (1 unit/μl), 10× reaction buffer (400 mM Tris–HCl, pH 8.0, 100 mM $MgSO_4$ and 10 mM $CaCl_2$). Store at −20 °C.
4. RNA electrophoresis buffer: (Percentages of total volume) 77 % DEPC water, 18 % formaldehyde, 5 % 20× MOPS (0.4 M MOPS, 160 mM NaOAc, 20 mM EDTA, pH 7.0). Store at 4 °C.
5. 1 % denaturing agarose gel: (for 100 ml volume) 1 g agarose, 77 ml DEPC water, 5 ml 20× MOPS (0.4 M MOPS, 100 mM CH_3COONa, 20 mM EDTA, pH7.0 with NaOH), 18 ml formaldehyde (*see* **Note 3**).
6. RNase-free phenol–chloroform–isoamyl alcohol. Store at 4 °C.
7. RNase-free chloroform.
8. Ethanol.
9. RNase-free 3 M sodium acetate (NaOAc), pH 5.2
10. RNase-free 4 M LiCl.
11. UltraPure DNase/RNase-free distilled water.

12. RNA denaturing buffer: for 1 sample, use 5 μl formamide, 1.75 μl formaldehyde, 0.5 μl 20× MOPS (0.4 M MOPS, 100 mM CH_3COONa, 20 mM EDTA, pH 7.0 with NaOH), 10× loading buffer, 0.25 μl EtBr (10 mg/ml), and 0.5 μl DEPC water.

2.2 Isolation of Oocytes from Female *Xenopus laevis* (Fig. 1)

As *Xenopus* oocytes are susceptible to bacterial contamination, all the implements, solutions, and buffers—except for the anesthesia—must be sterilized prior to each experiment.

1. Female *Xenopus laevis*. Keep at 16 °C.
2. Anesthesia: (in 200 ml water) 0.3 % 3-aminobenzoic acid ethyl ester, 0.1 % $NaHCO_3$.
3. MBSH buffer: 88 mM NaCl, 1 mM KCl, 0.33 mM $Ca(NO_3)_2 \cdot 4H_2O$, 0.41 mM $CaCl_2 \cdot 2H_2O$, 0.82 mM $MgSO_4 \cdot 7H_2O$, 2.4 mM $NaHCO_3$, 10 mM HEPES, 60 μg/μl penicillin, and 50 μg/μl streptomycin (*see* **Note 4**). Store at 16 °C.

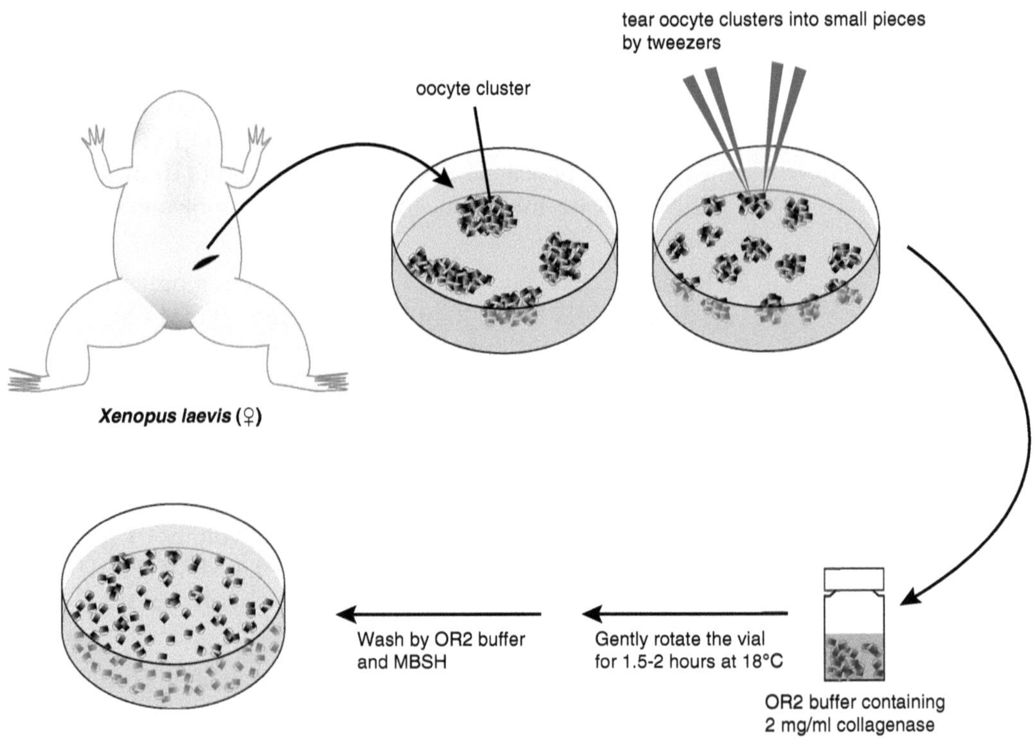

Fig. 1 A flow diagram for the isolation oocytes from female *Xenopus laevis.* Clusters of oocytes were taken from the stomach of each anesthetized frog. Tweezers were used to tear each cluster into small pieces. The follicle cells surrounding the oocytes were digested by treating oocyte clusters with 2 mg/ml collagenase diluted in Ca^{2+}-free buffer (OR2 buffer) for 1–2 h at 16 °C. Finally, the collagenase solution was washed away with OR2 buffer and then with Ca^{2+}-containing MBSH. Isolated oocytes were then collected for injections

4. OR2 buffer: 82.5 mM NaCl, 2 mM KCl, 1 mM $MgCl_2 \cdot 6H_2O$, 5 mM HEPES, pH 7.5 with NaOH. Store at 16 °C.
5. Collagenase B (Roche Applied Science).
6. Surgical implements: scalpel, surgical needles, tweezers, and sutures (*see* **Note 5**).
7. Sterile 10-cm dish.
8. Shaker for glass vials.

2.3 Microinjection of cRNA into Oocytes (Fig. 2)

1. Standard dissecting scope equipped with a cold light source.
2. Injector mounted on a micromanipulator (for example; NANOJECT II Variable Volume (2.3–69 nl) Automatic Injector with Glass Capillaries (110 V), Drummond Scientific Company).
3. Glass capillaries (for example; 3 1/2 inch capillary (8.9 cm), OD = 1.14 mm, ID = 0.53 mm, Drummond Scientific Company).
4. Sterile gauze.
5. Needle puller (for example; P-97 Flaming/Brown type micropipette puller, Sutter Instruments).
6. Mineral oil.
7. Sterile Pasteur pipettes.

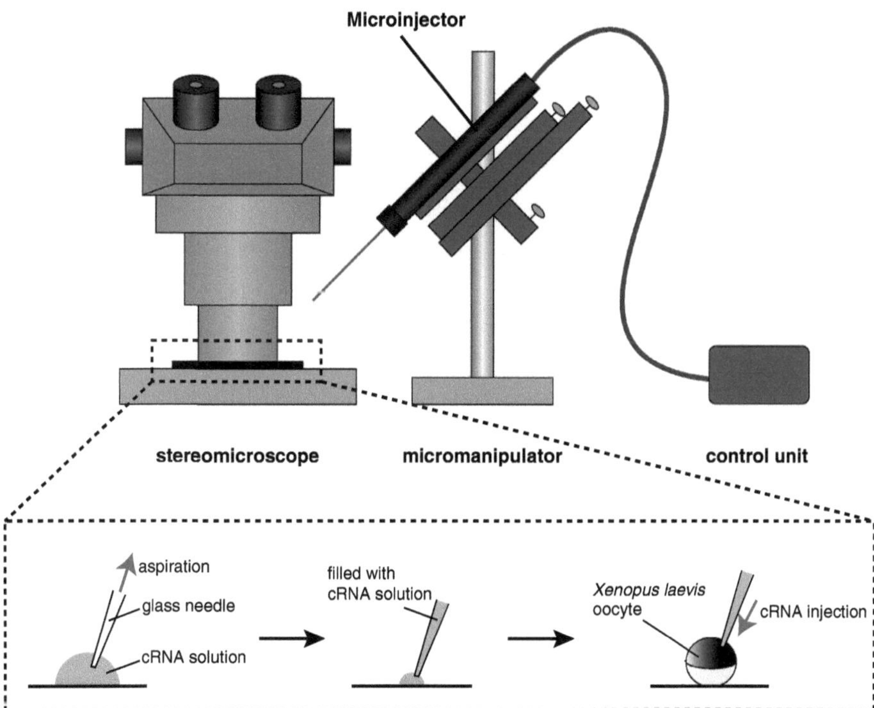

Fig. 2 A schematic drawing of the apparatus used for microinjection of cRNA into oocytes. The micropipette Puller P-97 (SUTTER Ins.) was used to make the glass needles. A microinjector fitted with a glass needle filled with cRNA solution was used to inject cRNA solution (50 nl) into each oocyte

2.4 Two-Electrode Voltage Clamp Method (Fig. 3)

1. Standard dissecting scope equipped with a cold light source.
2. Standard bath solution (*see* **Note 6**): 115 mM NaCl, 2.5 mM KCl, 1.8 mM $BaCl_2$, 10 mM HEPES, pH 7.2 with NaOH.
3. Oocyte Clamp Amplifier Model OC-725C (WARNER instruments).

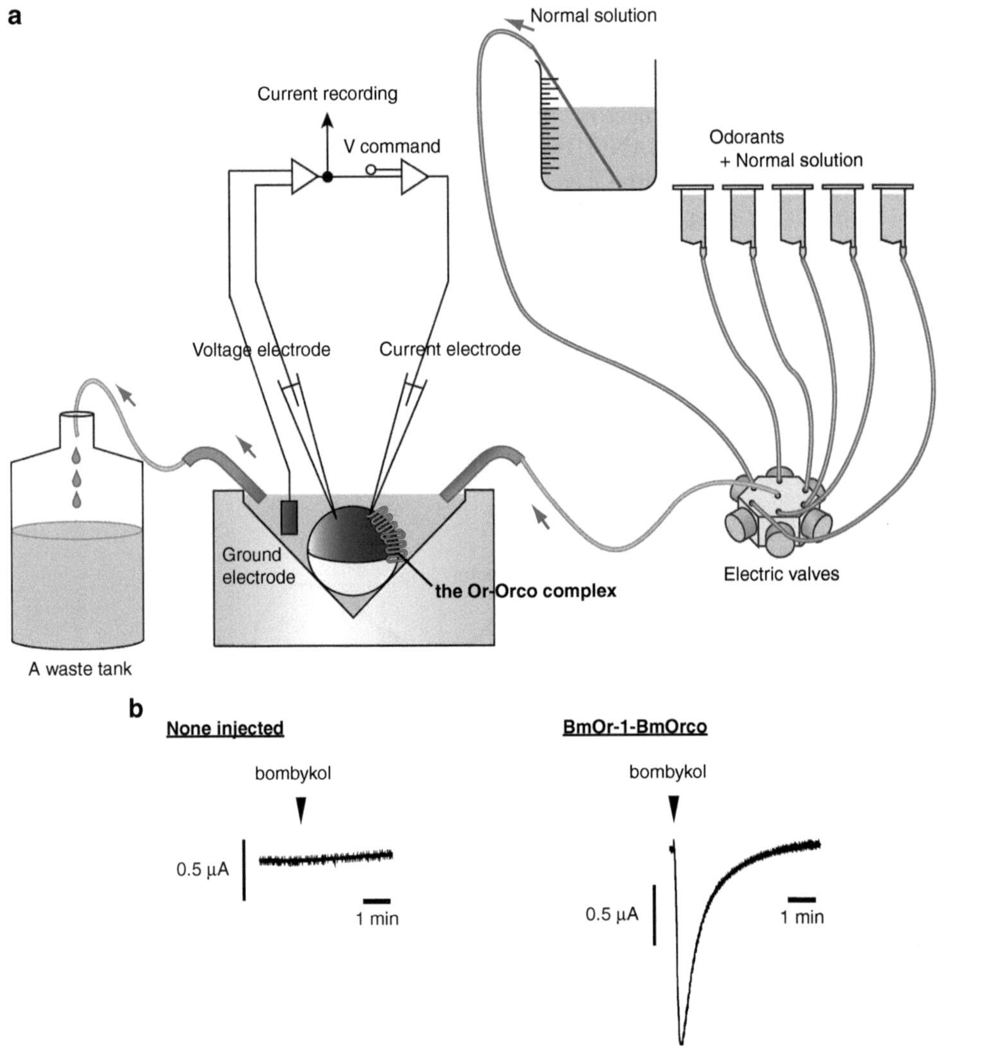

Fig. 3 Electrophysiological recordings from oocytes that express an insect Or–Orco complex (two-electrode voltage clamp method). (**a**) A schematic diagram of the apparatus used for two-electrode voltage clamp recording. Ligands are applied via a superfusing bath solution, which is supplied via a silicon tube that is connected to a computer-driven solenoid valve; alternatively, peristaltic pumps can be used to apply the ligand. Membrane voltage is measured via the voltage electrode, and the recorded signal is compared to the command voltage and amplified through a feedback amplifier. The difference between the two voltages is then converted into current and applied to the cell via the current electrode. (**b**) Representative current traces of none-injected oocyte (*left*) and BmOr-1-BmOrco expressing oocyte (*right*) in response to bombykol. The responses are measured as the changes of membrane current

4. Glass capillaries (For example; Capillary glass tubing with flame polished ends Borosilicate-thin wall with filament, OD = 1.50 mm, ID = 1.17 mm, Length = 10 cm, WARNER instruments).
5. Needle puller (for example; P-97 Flaming/Brown type micropipette puller, Sutter Instruments).
6. Peristaltic pump (*see* **Note 7**).
7. Perfusion chamber (for example; Oocyte Recording Chamber, WARNER instruments or a hand-made chamber.).
8. 3 M KCl solution.
9. A needle and a syringe for loading the KCl solution into the glass capillaries.
10. Software for recording and analyzing electrophysiological data.

3 Methods

Carry out all procedures at room temperature unless otherwise specified.

3.1 In Vitro cRNA Synthesis

Use gloves throughout this procedure.

1. Linearize the plasmid DNA (5 μg) by digesting it for at least 2 h with a restriction enzyme specific for a site located on the 3′ end of polyadenylation region of the plasmid vector (*see* **Note 8**). The reaction volume should be 100–200 μl. Check that the template is completely digested by electrophoresis using 1 μl of the digestion mix.
2. Purify the digested DNA either by phenol/chloroform/isoamyl alcohol extraction or by using a DNA purification kit (elute the DNA with 100 μl of RNase-free water).
3. Ethanol precipitate the DNA for 25 min at −80 °C using 10 μl of RNase-free 3 M NaOAc and 250 μl of 100 % ethanol. Centrifuge at 20,000 × *g* for 25 min at 4 °C, wash once with 200 μl of 70 % ethanol, and remove supernatant. Spin the tube briefly and collect the remaining ethanol from the bottom with a pipette, and thus completely remove all ethanol from the tube (*see* **Note 9**).
4. Resuspend the pellet with 100 μl of RNA synthesis solution by repeated pipetting. Incubate the mixture for 2 h at 37 °C.
5. To digest template DNA, add 5 μl of RNase-free DNase and 11 μl of 10× reaction buffer, and incubate the mixture for 5–10 min at room temperature.
6. Purify synthesized RNA either by phenol/chloroform/isoamyl alcohol extraction or using a RNA purification kit (elute with 100 μl of RNase-free water).

7. Ethanol precipitate the RNA for 25 min at −80 °C in 10 μl of RNase-free 4 M LiCl and 250 μl of 100 % ethanol. Centrifuge at 20,000×*g* for 25 min at 4 °C, wash once with 200 μl of 70 % ethanol, and decant supernatant completely. Ethanol should be removed completely by spinning the tube briefly and collecting all the ethanol at the bottom with a pipette (*see* **Note 10**). Resuspend the pellet in 21 μl of ultrapure-distilled RNase/DNase free water.
8. Use 1 μl of the RNA solution to check the amount and size of the synthesized RNA via electrophoresis through a denaturing agarose gel. Briefly, denature RNA by mixing 1 μl of RNA solution and 9 μl of RNA denaturing buffer, incubate it at 70 °C for 10 min, and put it on ice for 2 min. Load the mixture onto a denaturing agarose gel, and perform electrophoresis (*see* **Note 11**). Estimate RNA concentration by spectrophotometry, and make the final concentration of RNA 1 μg/μl.

3.2 Isolation of Oocytes from Female *Xenopus laevis*

1. Anesthetize an adult female *Xenopus laevis* by immersing it in iced-anesthesia for 20–30 min prior to surgery. After the frog becomes unconscious, place the frog on its back on top of a bed of crushed ice.
2. Make a 1–2 cm incision in the lower right (or left) abdominal wall near right (or left) leg to expose an ovary.
3. Use tweezers to carefully remove about one third of one ovary; place this tissue sample in OR2 buffer. Use tweezers or surgical scissors to tear the oocyte clusters into small pieces (~ 0.5 cm in diameter).
4. Suture the skin and the abdominal wall of the stomach of the frog separately using thin silk sutures; let the frog recover from anesthesia in shallow water (*see* **Note 12**).
5. Treat the oocytes with collagenase to remove follicular cells. Add 5 ml of collagenase B-diluted OR2 solution (2 mg/ml (*see* **Note 13**)) to 20 ml glass vials, and add approximately an equal volume of ovary suspension that contains ovaries torn into small pieces. Gently rotate the vial for 1–2 h at 16 °C (*see* **Note 14**). Remove the collagenase from the solution by washing four times in fresh OR2 buffer, and then by washing in MBSH four times. Transfer the defolliculated oocytes into an MBSH-filled 10-cm dish. Prior to microinjection, incubate the dish at 16 °C for a few hours or overnight.

3.3 Microinjection of cRNA into Oocytes

Use gloves throughout the procedure

1. Use a needle-puller (e.g., the P-97 puller) to pull microinjection needles (*see* **Note 15**).
2. Use tweezers to snap off the tips of the needles and to adjust the size of the tips to 10–30 μm (*see* **Note 16**).

3. "Backfill" the needles with mineral oil before attaching the needle to the injector by using 30 g × 2″ needle and a syringe.
4. Eject air out of the needle by pushing the "empty" button on the injector pedal because air bubbles often affect the accuracy of the injection volumes.
5. Place equal volumes (0.5–1.0 μl) of canonical Or cRNA (1 μg/μl) and of the Orco family cRNA (1 μg/μl) on a small piece of Parafilm, and mix them by pipetting. Immerse the tip of the needle into the drop of RNA solution and aspirate the solution by pushing the "fill" button on the injector pedal (*see* **Note 17**).
6. Fill a 10-cm dish with MBSH, and immerse a small piece (2 × 2 cm) of sterile gauze in the MBSH forcing it to the bottom of the dish (*see* **Note 18**). Use a Pasteur pipette to transfer 20–30 oocytes onto the gauze by.
7. For each injection, select a large oocyte (approximately 1 mm diameter, the size of stage V or VI oocyte) that has lost the follicle cell layer (*see* **Notes 19** and **20**). Gently inject 50 nl of the RNA solution into each oocyte by piercing the membrane; an oocyte will "plump-up" if injected correctly.
8. Incubate the injected oocytes in MBSH at 16 °C for 3–4 days (*see* **Note 21**).

3.4 Two-Electrode Voltage Clamp Method

Before performing TEVC, precise setting-up of the measurement system is necessary. This system is well established and has been widely used to analyze various types of ion channels; therefore, we will describe this method only very briefly here. Importantly, the information provided here is insufficient for a reader who is unfamiliar with the TEVC to successfully perform the technique; more detailed instructions about how to set-up the system can be found at and freely downloaded from a Web site maintained by WARNER instruments (http://www.warneronline.com/).

1. Use a needle puller (e.g., the P-97 puller) to pull two capillaries (*see* **Notes 22** and **23**); use a syringe needle and a syringe to "backfill" each glass needle with 3 M KCl.
2. Install one capillary each onto the current-electrode holder and the voltage-electrode holder.
3. Perfuse the Standard bath solution in the perfusion chamber.
4. Use the micromanipulator to dip the tip of the current-electrode capillary and of the voltage-electrode capillary into the Standard bath solution, and then adjust the voltage of both electrodes to 0 mV. Also, check if the resistances of the electrodes are 1–10 MΩ.
5. Place an oocyte at the center of the perfusion chamber (*see* **Note 24**).

6. Pierce the oocyte with the current electrode capillary and with the voltage electrode capillary (*see* **Note 25**).
7. Hold the membrane potential at –80 mV, and start recording.
8. Apply the odorant or pheromone of interest, and record the responses as changes in membrane current (*see* **Notes 26** and **27**).

4 Notes

1. Currently available *Xenopus* oocyte expression vectors possess specified features that facilitate stable and efficient translation of cRNAs in *Xenopus* oocytes. First, within the upstream region of multicloning site, the vectors contain RNA polymerase promoter sites (T7 or SP6) for in vitro transcription. Second, the vectors possess the 5′ and 3′ UTRs from the *Xenopus laevis* β-globin mRNA, one at each terminus of the multicloning site; these UTRs greatly enhance translation efficiency, [15]. In addition, polyadenylation (poly(A)) sequences are positioned at the 3′ end of multicloning site; these sequences enhance the stability of transcribed RNA [16].
2. GTP inhibits the RNA capping reaction; therefore, the concentration of GTP should be lower than that of the other NTPs.
3. Make the gel in a fume hood because vapor from formaldehyde is very harmful. Formaldehyde should be added to the solution after the temperature of the mixture of other reagents drops below 70 °C.
4. For convenience, make and store a 25× high-salt stock (Stock A) and a 25× divalent cation stock (Stock B) of MBSH. Compositions of the stocks are as follows: Stock A contains 2.2 M NaCl, 25 mM KCl, 60 mM $NaHCO_3$, 375 mM HEPES, titrated to pH 7.6; Stock B contains 8.25 mM $Ca(NO_3)_2 \cdot 4H_2O$, 10.25 mM $CaCl_2 \cdot 2H_2O$, 20.5 mM $MgSO_4 \cdot 7H_2O$. Both solutions should be filter sterilized with a 0.20 μm pore diameter filter. Store both stock solutions at 4 °C. Prior to oocyte manipulation, make the 1× MBSH solution by diluting Stock A (4 % of total volume), Stock B (4 % of total volume), 60 mg/ml penicillin (0.1 % of total volume), and 50 mg/ml streptomycin (0.1 % of total volume) into sterilized water.
5. All the implements should be sterilized in 70 % EtOH before use.
6. To avoid activation of endogenous Ca^{2+}-activated Cl^- channels, Ca^{2+} should not be included in the extracellular solution. For example, Ba^{2+} is included instead of Ca^{2+} in the Standard bath solution described here.

7. Choose a peristaltic pump that has small flow pulsation that causes a low S/N ratio.
8. If the template remains a circular plasmid, RNA polymerases generate long and heterogeneous RNA transcripts. Linearization of the template is necessary to avoid this problem. Also, check that the plasmid has only one site for the restriction enzyme that is used to linearize the plasmid.
9. Drying the pelleted DNA is not recommended because drying the pellet may reduce the DNA solubility and the success of the subsequent transcription. Make sure that almost all the ethanol is removed from the DNA because ethanol will inhibit the transcription reaction.
10. Drying the pelleted DNA completely is not recommended because drying may drastically reduce the solubility of the DNA. Make sure that all the ethanol is removed since it is toxic for oocytes.
11. Use a scalpel to remove the upper region of the denaturing agarose gel before EtBr flows out of the gel; you can see the EtBr as orange-colored bands.
12. If the frog has already been used to harvest oocytes two or three times, it should be euthanized and discarded; repeated surgery degrades the condition of frogs and oocytes.
13. As collagenase is toxic to oocytes in the presence of Ca^{2+}, you need to use an OR2 buffer that does not contain Ca^{2+}.
14. As efficiency of the collagenase treatment is greatly dependent on collagenase product lot, you must adjust the incubation time according to the actual follicle dissolution rate. If the activity of the collagenase is low, you should perform one additional collagenase treatment. To perform an additional collagenase treatment, remove collagenase from the glass vial by washing the oocytes and the vial with OR2 solution two to three times, and then add fresh collagenase solution to the vial and gently rotate it for another hour.
15. We use the following program to pull the appropriate needles; Step 1: HEAT = 557, PULL = 30, VEL = 30, TIME = 200, Step 2: HEAT = 530, PULL = 85, VEL = 60, TIME = 100.
16. The size of the needle tip is critical for successful microinjection. If a tip is too thick, it will damage oocytes, which will then die before the day of TEVC recording. Conversely, if a tip is too thin, aspirating the RNA solution will be difficult.
17. To avoid aspirating air bubbles, adjust the location of the tip with micromanipulator to keep the tip of needle in the drop of RNA solution drop.

18. Fibers of the gauze prevent the oocytes from rolling over during the microinjection.
19. For each injection, choose a completely defolliculated oocyte because the follicle cell layer makes it difficult to pierce the oocyte; alternatively, you can use tweezers to remove the follicle cell layer, although careful manipulation of the oocytes is needed with this method. Also, you must choose good (healthy) oocytes that have resilient membranes and clearly separated animal (black side) and vegetal (white side) poles. Do not choose bad oocytes such as those that are too small or too large or those for which some or all of animal pole is white.
20. As a result, 25 ng of canonical Or cRNA and 25 ng of the Orco family cRNA are injected to each oocyte.
21. Exchange the MBSH in the dish at least once a day, and remove the weakened or dead oocytes from the dishes; the dead and damaged oocytes degrade the condition of the culture and can make the condition of other, healthy oocytes worse.
22. For instance, we use the following program to pull the appropriate needles; Step 1: HEAT = 570, PULL = Blank, VEL = 25, TIME = 200, Step 2: HEAT = 565, PULL = 30, VEL = 25, TIME = 200, Step 3: HEAT = 580, PULL = 75, VEL = 75, TIME = 200).
23. The size of the tips of capillaries should be thin because thick tips cause KCl solution to flow into the oocyte.
24. Adjust the flow rate of the solution so that the oocyte in the chamber does not move.
25. Here you can see the resting membrane potential (usually −40 mV to −20 mV) of the oocyte.
26. For hydrophobic odorants or pheromones that are difficult to dissolve in the Standard bath solution, make high-concentration stock solution of the ligand in DMSO, and then dissolve this solution in the Standard bath solution prior to each experiment. The final DMSO concentration should be less than 1 %. Alternatively, the solubility of the ligands would be enhanced by adding odor- or pheromone-binding proteins into the solution [17].
27. If you want to measure current–voltage relationship of an Or-Orco receptor, use voltage-steps or voltage ramps to change the membrane current from −80 mV to different currents between +50 and +100 mV before and after the onset of odorant or pheromone exposure. You can obtain the current–voltage relationship by subtracting the traces before the onset of ligand exposure from the trace after the onset.

References

1. Larsson MC, Domingos AI, Jones WD, Chiappe ME, Amrein H, Vosshall LB (2004) *Or83b* encodes a broadly expressed odorant receptor essential for *Drosophila* olfaction. Neuron 43:703–714
2. de Bruyne M, Foster K, Carlson JR (2001) Odor coding in the *Drosophila* antenna. Neuron 30:537–552
3. Wang JW, Wong AM, Flores J, Vosshall LB, Axel R (2003) Two-photon calcium imaging reveals an odor-evoked map of activity in the fly brain. Cell 112:271–282
4. Wetzel CH, Behrendt HJ, Gisselmann G, Stortkuhl KF, Hovemann B, Hatt H (2001) Functional expression and characterization of a *Drosophila* odorant receptor in a heterologous cell system. Proc Natl Acad Sci USA 98:9377–9380
5. Nakagawa T, Sakurai T, Nishioka T, Touhara K (2005) Insect sex-pheromone signals mediated by specific combinations of olfactory receptors. Science 307:1638–1642
6. Wanner KW, Nichols AS, Walden KK, Brockmann A, Luetje CW, Robertson HM (2007) A honey bee odorant receptor for the queen substance 9-oxo-2-decenoic acid. Proc Natl Acad Sci USA 104:14383–14388
7. Mitsuno H, Sakurai T, Murai M, Yasuda T, Kugimiya S, Ozawa R, Toyohara H, Takabayashi J, Miyoshi H, Nishioka T (2008) Identification of receptors of main sex-pheromone components of three Lepidopteran species. Eur J Neurosci 28:893–902
8. Sato K, Pellegrino M, Nakagawa T, Vosshall LB, Touhara K (2008) Insect olfactory receptors are heteromeric ligand-gated ion channels. Nature 452:1002–1006
9. Smart R, Kiely A, Beale M, Vargas E, Carraher C, Kralicek AV, Christie DL, Chen C, Newcomb RD, Warr CG (2008) *Drosophila* odorant receptors are novel seven transmembrane domain proteins that can signal independently of heterotrimeric G proteins. Insect Biochem Mol Biol 38:770–780
10. Wicher D, Schafer R, Bauernfeind R, Stensmyr MC, Heller R, Heinemann SH, Hansson BS (2008) *Drosophila* odorant receptors are both ligand-gated and cyclic-nucleotide-activated cation channels. Nature 452:1007–1011
11. Wang G, Carey AF, Carlson JR, Zwiebel LJ (2010) Molecular basis of odor coding in the malaria vector mosquito *Anopheles gambiae*. Proc Natl Acad Sci USA 107:4418–4423
12. Neuhaus EM, Gisselmann G, Zhang W, Dooley R, Stortkuhl K, Hatt H (2005) Odorant receptor heterodimerization in the olfactory system of *Drosophila melanogaster*. Nat Neurosci 8:15–17
13. Nichols AS, Luetje CW (2010) Transmembrane segment 3 of *Drosophila melanogaster* odorant receptor subunit 85b contributes to ligand-receptor interactions. J Biol Chem 285:11854–11862
14. Nakagawa T, Pellegrino M, Sato K, Vosshall LB, Touhara K (2012) Amino acid residues contributing to function of the heteromeric insect olfactory receptor complex. PLoS One 7:e32372
15. Krieg PA, Melton DA (1984) Functional messenger RNAs are produced by SP6 in vitro transcription of cloned cDNAs. Nucleic Acids Res 12:7057–7070
16. Huez G, Marbaix G, Hubert E, Leclercq M, Nudel U, Soreq H, Salomon R, Lebleu B, Revel M, Littauer UZ (1974) Proceedings: Role of the polyadenylic segment in the translation of globin messenger RNA. Hoppe Seylers Z Physiol Chem 355:1211
17. Grosse-Wilde E, Svatos A, Krieger J (2006) A pheromone-binding protein mediates the bombykol-induced activation of a pheromone receptor in vitro. Chem Senses 31:547–555

Chapter 9

A Protocol for Heterologous Expression and Functional Assay for Mouse Pheromone Receptors

Sandeepa Dey, Senmiao Zhan, and Hiroaki Matsunami

Abstract

Innate social behaviors like intermale aggression, fear, and mating rituals are important for survival and propagation of a species. In mice, these behaviors have been implicated to be mediated by peptide pheromones that are sensed by a class of G protein-coupled receptors, vomeronasal receptor type 2 (V2Rs), expressed in the pheromone-detecting vomeronasal organ (VNO) (Chamero et al., Nature 450:899–902, 2007; Haga et al., Nature 466:118–122, 2010; Kimoto et al., Curr Biol 17:1879–1884, 2007; Leinders-Zufall et al., Nat Neurosci 12:1551–1558, 2009; Papes et al., Cell 141:692–703, 2010). Matching V2Rs with their cognate ligands is required to understand what receptors the biologically relevant pheromones are acting on. However, this goal has been greatly limited by the unavailability of appropriate heterologous tools commonly used to carry out receptor deorphanization, due to the fact that this family of receptors fails to traffic to the surface of heterologous cells. We have demonstrated that calreticulin, a housekeeping chaperone commonly expressed in most eukaryotic cells, is sparsely expressed in the vomeronasal sensory neurons (VSNs). Stable knock down of calreticulin in a HEK293T derived cell line (R24 cells) allows us to functionally express V2Rs on the surface of heterologous cells. In this chapter we describe protocols for maintenance and expansion of the R24 cell line and functional assays for V2Rs using these cells.

Key words Heterologous, Calcium, V2Rs, R24, Cells

1 Introduction

In vertebrates, pheromones are largely detected by the accessory olfactory organ, the vomeronasal organ (VNO). In mice, the VNO has been implicated in detecting a large number of inter and intra specific chemosignals that mediate innate behavior like fear, aggression, mating rituals, pregnancy block, mate recognition, etc. [1–5]. The VNO contains closely packed sensory neurons that express at least 250 putative pheromone receptors spanning three unrelated families of G protein-coupled receptors (GPCRs): vomeronasal receptors classes I and II (V1Rs and V2Rs respectively) and formyl peptide receptors (FPRs) [6–11]. Each sensory neuron expresses either one V1R or an FPR or a combination of two V2Rs in spatially distinct regions. A key question in studying pheromonal

Kazushige Touhara (ed.), *Pheromone Signaling: Methods and Protocols*, Methods in Molecular Biology, vol. 1068,
DOI 10.1007/978-1-62703-619-1_9, © Springer Science+Business Media, LLC 2013

olfaction is what biological ligands activate a receptor or related receptors; however, the complex expression pattern of such a large receptor repertoire makes it challenging to adequately address this question. Heterologous cells provide a convenient platform to carry out large scale functional and deorphanization studies for GPCRs; however, a major road block in studying the pheromone receptors using this tool was the inability to express the receptors functionally on the surface of heterologous cells. We have recently demonstrated that vomeronasal sensory neurons expressing V2Rs show low expression of an otherwise ubiquitous cellular chaperone, calreticulin. Depletion of calreticulin in heterologous cells (HEK293T) enabled export of V2Rs to the surface. Using this knowledge, we established the cell line R24, from HEK293T, with constitutive knock down of calreticulin. We then used this line to carry out V2R ligand binding assays with calcium imaging [12]. In this chapter we describe the protocol to maintain and expand the cells and carry out the functional assays.

Heterologous cells have been extensively exploited to study G protein-coupled receptors. The advantages of deorphanizing receptors heterologously in comparison to in vivo protocols are many. One can selectively transfect a receptor-expressing plasmid into cells, and examine the response of the transfected cells easily to a ligand or set of ligands by applying a series of ligand pulses. To test the specificity of a particular receptor-cognate ligand pair, one can set up control experiments easily by testing either other ligands with the receptor of choice or other receptors with the ligand of choice. The heterologous system allows for identification of receptors for a set of biologically relevant ligands without performing genetic manipulations and/or in vivo assays. The heterologous cells may be further used to study the signal transduction pathway by addition or subtraction of cofactors in the transfection step and comparing how the activity of the cells is affected. One can also create chimeric receptors, receptors with point mutations, or domain switched receptor constructs, transfect into cells, and test their response to a ligand or set of ligands. It allows us to identify motifs or domains and particular amino acids that are important for either ligand binding or signaling or both. Additionally, the heterologous cells provide a means to study various other cell biological areas, such as mechanisms of endoplasmic reticulum targeting of receptors, surface export and interactions with chaperones [13].

To assay for ligand binding, a V2R receptor of interest is co-transfected with H2M-10.4 and B2m as well as Gα15, a Gq class of G protein widely used in functional assays of GPCRs. Activation of Gα15, which couples to many but not all of the GPCRs, leads to a transient intracellular calcium release [14, 15]. Changes in intracellular calcium can be easily detected by chemical dyes with high affinity for calcium which change their spectral properties upon binding calcium ions. Using this principle, we perform calcium dye based imaging because this provides an extremely sensitive and efficient method. We use a combination of two calcium dye indicators, Fluo-4 and Fura Red, which are both visible light excitable (488 nm). The green-fluorescent emission (~525 nm) of Fluo-4 increases on binding calcium while the red fluorescence (~660 nm) of the Fura Red indicator decreases once the indicator binds Ca^{2+}. Using two dyes in combination enables one to carry out ratiometric quantification for changes in cytosolic calcium concentration.

R24 cells, like most other heterologous cell lines pose their own set of disadvantages, which should be closely considered in parallel with the advantages of deorphanizing V2Rs. In an attempt to mimic the vomeronasal sensory neurons, this heterologous cell line was established by knocking down calreticulin, a very essential endoplasmic reticulum chaperone that plays multiple roles in sustenance of cell health and calcium homeostasis. As a result, the cells are slower growing than the parent cell line (HEK293T), an important parameter to bear in mind while planning experiments. The cell must be handled with extreme caution—over-trypsinization should be avoided, cells should be triturated slowly, and all media changes must be done gently and in a timely way. Each batch of cells may be used for a limited number of passages.

Keeping these limitations in mind, one can design experiments using the R24 heterologous cells and calcium imaging to study the ligand specificities and functions of mouse V2Rs.

2 Materials

Use cell culture grade commercially available reagents. Prepare all solutions in a sterile environment, preferably in sterile laminar flow chambers. Diligently follow all waste disposal rules when disposing waste materials.

2.1 R24 Cell Culture and Maintenance

1. Minimum essential medium (MEM) containing Earle's salts and L-glutamine.
2. Fetal bovine serum (FBS): heat inactivated.
3. 10 % supplemented medium (M10): Mix 45 ml MEM, with 5 ml FBS to obtain 10 % supplemented medium (M10). Store M10 in 4 °C.
4. Penicillin-Streptomycin-Amphotericin (PSF) culturing medium: Prepare M10 as above and add Penicillin-Streptomycin and Amphotericin to final concentrations 100 μg/ml and 1.25 μg/ml respectively.
5. Puromycin containing PSF, R24 maintenance medium: Prepare PSF culturing medium as described. Prepare two R24

maintenance media by adding puromycin to final concentrations 5 μg/ml and 20 μg/ml respectively.

6. 100 mm coated sterile cell culture dishes.
7. Centrifuge tubes.

2.2 Cell Transfer and Plating for Calcium Imaging

1. 1× Phosphate Buffer Saline, PBS.
2. Trypsin (0.05 %).
3. 35 mm sterile cell culture dishes with 10 mm optical glass well.

2.3 Transfection

1. Transfection reagent.
2. Plasmid DNA: V2R pheromone receptor, major histocompatibility factor 10.4, β2 microglobulin and mouse Gα15 complete open reading frames cloned in mammalian expression vector pCI.

2.4 Calcium Imaging Assay

1. Fura Red solution: Dissolve 50 μg Fura Red in 12.5 μl DMSO to obtain 4 μg/μl Fura Red.
2. Fluo-4: Dissolve 50 μg Fluo-4 in 12.5 μl DMSO to obtain 4 μg/μl Fluo-4.
3. Pluronic acid: Weigh out 20 mg pluronic acid (amorphous solid) in an eppendorf tube, add 100 μl cell culture grade DMSO and incubate at 37 °C for 10–15 min, intermittently tapping the mixture to dissolve the pluronic acid (*see* **Note 1**).
4. Bovine serum albumin (BSA), 7.5 %.
5. Glucose, 45 %.
6. Hanks Balanced Salt Solution (HBSS) 1×, with glucose: anhydrous calcium chloride 140 mg/L, magnesium chloride-6H2O 100 mg/L, magnesium sulfate-7H_2O 100 mg/L, potassium chloride 400 mg/L, potassium phosphate monobasic 60 mg/L, sodium chloride 8,000 mg/L, sodium phosphate dibasic-7H_2O 48 mg/L, Dextrose 1,000 mg/L; pH = 7.4.
7. 4-(2-hydroxyethyl)-1-piperazineethanesulfonic acid (HEPES), 1 M.
8. Imaging buffer: HBSS, 10 mM HEPES, 0.45 % glucose. Mix 5 ml 1 M HEPES and 5 ml 45 % glucose in 500 ml HBSS to obtain imaging buffer.
9. Loading buffer: Add 13.3 μl, 7.5 % BSA to 1 ml imaging buffer to obtain 0.1 % BSA in imaging buffer.
10. Dye mixture: Mix well 0.5 μl 20 % pluronic acid, 0.5 μl Fluo-4, and 1 μl Fura Red with vigorous pipeting, and add 500 μl loading buffer to obtain dye mixture (*see* **Notes 2** and **3**).

2.5 Equipments

1. 37 °C cell culture incubator with 5 % carbon dioxide.
2. Certified class II biological safety cabinet with laminar flow.
3. Centrifuge with swinging bucket rotor for 15 ml conical tubes.
4. Phase contrast microscope with 10×, 20× objectives.
5. Fluorescence microscope with Fluo 4 and Fura Red filters (excitation, 488 nm; emission, 500–560 nm for Fluo-4 and 605–700 nm for Fura red), 40× objective.
6. Peristalitic pump.

3 Methods

All cell culture procedures must be carried out in a sterile way, in class II biological safety cabinet with laminar flow, at room temperature unless otherwise mentioned.

3.1 R24 Cell Culture and Maintenance

1. Quickly thaw a frozen vial of R24 cells in a 37 °C water bath.
2. Immediately after thawing, transfer cells from vial to sterile 15 ml tube containing 10 ml M10.
3. Centrifuge at 200 × *g* for 5 min.
4. Aspirate supernatant and resuspend cells in 8 ml M10 with gentle trituration.
5. Plate cells in a 100 mm sterile cell culture dish.
6. Observe the cells closely in a phase contrast microscope, they should appear like round particles in suspension in the media.
7. Incubate overnight in a cell culture incubator at 37 °C, 5 % CO_2.
8. Observe the cells again after overnight incubation, they should have settled to the bottom of the cell culture dish and attached to it.
9. If the cells appear to be firmly attached, aspirate M10 plating medium and replace with 8 ml PSF culture medium (*see* **Note 4**).
10. 24 hours later, change medium was to PSF containing 5 μg/ml puromycin (*see* **Note 5**).
11. 24 hours after addition of 5 μg/ml puromycin, change the medium to 20 μg/ml puromycin containing PSF (*see* **Note 6**).
12. Thereafter maintain the cells in 20 μg/ml puromycin and PSF containing medium for experiments.

3.2 Transfer R24 Cells for Transfection

1. Allow cells to grow till confluent or up to desired confluency, checking cell density periodically (*see* **Note 7**).
2. Prior to cell transfer, aspirate media and gently add 8 ml sterile PBS for washing out media (*see* **Note 8**).

3. Incubate cells in 3 ml trypsin (0.05 %) at room temperature; when all cells are loosened from the sterile cell culture dish, gently triturate them a few times to detach them.
4. Stop trypsinization by adding 5 ml M10; gently triturate the mixture several times to dissociate the cells from each other and make a homogeneous cell suspension (*see* **Note 9**).
5. At this point, the cells need to be plated for assay and maintenance. Depending on anticipated frequency of future uses, transfer a portion of the trypsinized cells into a sterile conical tube; for the immediate assay, depending on the number of dishes to be set up, transfer another portion into a second sterile conical tube.
6. Centrifuge both aliquots at $200\times g$, for 5 min at room temperature.
7. Resuspend the aliquot for maintenance in 8 ml 20 μg/ml puromycin containing PSF media by gentle trituration and transfer to a sterile 100 mm cell culture dish. Maintain the dish in the cell culture incubator at 37 °C, 5 % CO_2 till further use.
8. Resuspend the aliquot of cells for imaging in M10 (antibiotic free), 1 ml per 10 mm dish.
9. Transfer 1 ml of media containing resuspended cells into each glass bottomed dish.
10. Incubate dishes overnight in 37 °C, 5 % CO_2.

3.3 Transfection of Cells for Imaging

1. Observe the cells closely using a phase contrast microscope 24 h after plating. If cells appear to have settled on the glass bottomed dish, they are ready for transfection. R24 cells should be ~40 % confluent at the time of transfection in the 35 mm cell culture dishes with 10 mm optical glass wells.
2. Mix plasmid DNA in the following amounts (per glass bottomed dish): V2R pheromone receptor 1,000 ng, H2-M10.4 1,200 ng, β2-microglobulin 200 ng and Gα15 400 ng (*see* **Note 10**).
3. For transfection, follow instructions from the transfection reagents manufacturers' manual.
4. Incubate the cells for ~36 h in 37 °C, 5 % CO_2.

3.4 Calcium Imaging

1. Approximately 36 h after transfection, slowly aspirate the medium out of the glass bottomed culture dish.
2. Gently add 1 ml imaging buffer to wash remaining media, taking care to avoid pipetting the buffer in the glass bottom.
3. Slowly aspirate the buffer from the culture dish.
4. Pipette 50 μl dye mixture to the glass bottom.

5. Incubate the cells in the dye mixture for 45 min in dark (*see* **Note 11**).
6. Dilute stimulant to desired strength in imaging buffer.
7. Wash the stimulant/buffer delivery tubing thoroughly with distilled water.
8. For imaging, expose the cells to a constant flow of imaging buffer (*see* **Note 12**).
9. Collect the data at wavelengths appropriate for Fluo-4 and Fura Red (excitation 488 nm, emission 500–560 nm for Fluo-4, 605–700 nm for Fura-red) at regular intervals. Apply stimulants for desired length of time punctuating with buffer flow to allow the cells to recover in case they respond to a particular pulse of test ligand (*see* **Notes 13** and **14**).
10. At the end of the ligand pulses, apply a pulse of 10 nM isoproterenol to use for positive control (*see* **Note 15**).

3.5 Preparation of R24 Cell Frozen Stocks

1. Pre cool an isopropyl alcohol bath to 4 °C.
2. To prepare frozen stocks of R24 cells, start culture and grow them in PSF containing medium as discussed in Subheading 3.1.
3. When confluent, split them as discussed in Subheading 3.2. After aspirating out the trypsin medium, add 2 ml freezing medium and gently triturate.
4. Transfer 1 ml of resuspended cells to one freezing vial (two vials of frozen stocks from each confluent 100 mm plate of R24 cells).
5. Freeze cell vials in −80 °C.
6. Store frozen vials of R24 cells for long term at −80 °C.
7. Cells split as discussed and resuspended after spinning in freezing mixture (each 100 ml confluent plate of cells resuspended in 1 ml freezing mixture, frozen in a single vial for usage next time) and frozen in −80 °C in isopropyl alcohol bath.

3.6 Data Analysis

1. Base fluorescence intensities, changes in the Fluo-2 and Fura Red intensities, and the ratio of the changes can be plotted with graphing softwares like Microsoft Excel or ImageJ. Use the ratio to analyze the data for responses (Fig. 1).
2. Determine cells that have responded to a positive control stimulant applied at the end of the experimental run, for example 10 nM Isoproterenol (*see* **Note 16**).
3. Response to ligand pulse should appear within 90 s of application of the stimulant (*see* **Note 17**).
4. Responses that appear before the stimulant front reaches the cells should be counted as non specific.

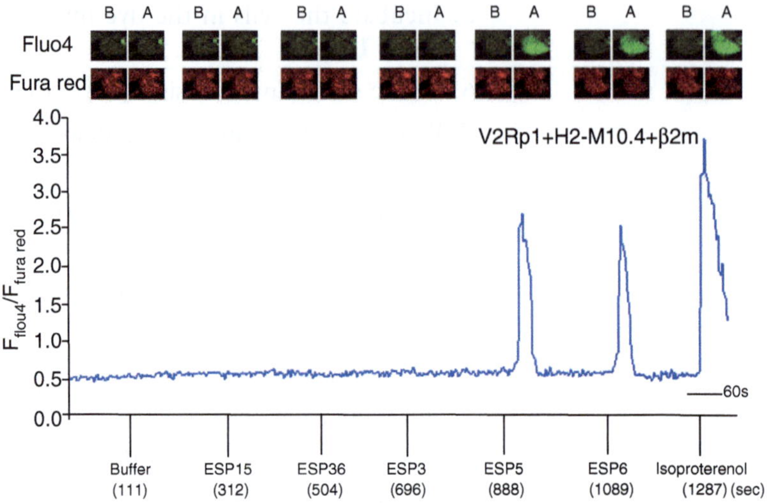

Fig. 1 Response of single R24 cell transfected with V2Rp1 to seriel application of stimulants. *Upper panel*: change of Fluo-4 and Fura Red fluorescence intensities (**b**, before application of stimulant indicated in *X* axis; **a**, after application of stimulant indicated in *X* axis). *Lower panel*: Calcium concentration trace of the cell on application of recombinant peptide stimulants (ESP15, 36, 3, 5, and 6, respectively), 100 nM each and positive control (isoproterenol, 1 nM)

5. A cell that responds nonspecifically outside the range of time interest should be discounted from analysis.
6. A cell must show an increase of Fluo-4 intensity and a concomitant decrease in Fura Red signal intensity; ratio of Fluo-4 to Fura Red signals must exceed 50 % of the base line (*see* **Note 18**).
7. Efficiency of transfection may be calculated by expressing the number of isoproterenol responsive cells as a percentage of the total number of cells in the field of view (*see* **Note 19**). Number of cells responding to a particular ligand should be normalized by expressing as a percentage of the total number of isoproterenol responsive cells. Normalization enables us to compare responses between different dishes, conditions, and days of experiments.

4 Notes

1. For best results, make fresh solutions of 20 % pluronic acid every week.
2. After mixing pluronic acid, Fluo-4, and Fura Red thoroughly, consider incubating in a 37 °C water bath or heating block for 5 min. This increases the total uptake of the dyes by each cell.
3. Once loading buffer is prepared, keep wrapped in aluminum foil in the dark till the end of the experiment, since the fluorescent dyes are light sensitive.

4. Observe the cells closely before replacing the medium. If cells look unsettled, wait until the cells are attached to the dish before replacing the M10 medium.
5. Some cells usually die after addition of lower concentration of puromycin.
6. A larger number of cells usually die after adding media containing higher concentration of puromycin.
7. R24 cells grow and divide approximately every 36 h.
8. It is important to wash out supplemented medium before adding trypsin since any trace of medium will inhibit trypsin activity.
9. R24 cells attach less robustly to cell culture dishes than the parent cell line HEK293T; therefore while carrying out cell transfer, maintain caution to not over-trypsinize. Stop the enzymatic reaction as soon as the cells detach from the culture dish.
10. These amounts of plasmid DNA have been optimized for pCI (mammalian vector) based constructs and transfections carried out with Lipofectamine 2000 (Invitrogen) mediated transfections. For other expression and transfection systems, the relative amounts of the different constructs in the transfection mixture may have to be optimized.
11. For best results, soak transfected cells in dye solution, one dish at a time so that cells can be imaged promptly after loading with dye. Letting the loaded cells wait longer than the required incubation time results in formation of intracellular dye aggregates that appear as tiny fluorescent specks while carrying out the assay and may interfere with the read-out.
12. R24 cells attach lightly to the glass bottomed dish; delivery must be minutely regulated such that buffer or stimulants are delivered at a steady rate and at the same time does not dislodge the cells from the dish. We optimized our assays at delivery rate of ~50 μl/s, regulated by a peristaltic pump (Rainin).
13. We used Leica confocal microscope and the live imaging mode of Leica confocal software for data acquisition. Data was collected at 3 s interval.
14. We pulsed up to 100 nM of ESP family of peptides to stimulate cells transfected with test receptors; ligand pulses were applied for 10 s and buffer bath was applied for 2 min following ligand pulse.
15. Isoproterenol is a cognate ligand for the beta adrenergic receptor, endogenously expressed in the HEK293T cells or cells derived from it. On binding isoproterenol, the receptor couples with transfected Gα15, resulting in an increase in intracellular calcium.

16. Applying the pulse of isoproterenol at the end of the assay enables the experimenter to ensure that cells are healthy and responsive throughout the experiment. Cells not responding to isoproterenol in the end should not be counted.
17. We empirically determined responses of R24 cells transfected with V2Rp receptors to applied concentrations of ESP ligands appeared within 90 s time frame. This may have to be optimized for different ligand-receptor pairs and instrument set up.
18. A concomitant increase in fluorescence intensity of one fluorophore and decrease in the other ensures real signals are selected for analysis over noise. We empirically determined an increase in ratio of Fluo-4 to Fura Red by 50 % from the base line enables us to segregate noise from real signals. This may have to be optimized for different ligand-receptor pairs, calcium dyes and instrument set up.
19. R24 cells are transfection efficient; typically 80–100 % of cells respond to 10 nM isoproterenol within seconds of application. An assay that shows less efficient calcium response to the positive control should not be considered for statistical purposes.

Acknowledgement

This work was supported by NIH grant and Duke University Medical Center, Department of Molecular Genetics and Microbiology.

References

1. Chamero P, Marton TF, Logan DW, Flanagan K, Cruz JR, Saghatelian A, Cravatt BF, Stowers L (2007) Identification of protein pheromones that promote aggressive behaviour. Nature 450:899–902
2. Haga S, Hattori T, Sato T, Sato K, Matsuda S, Kobayakawa R, Sakano H, Yoshihara Y, Kikusui T, Touhara K (2010) The male mouse pheromone ESP1 enhances female sexual receptive behaviour through a specific vomeronasal receptor. Nature 466:118–122
3. Kimoto H, Sato K, Nodari F, Haga S, Holy TE, Touhara K (2007) Sex- and strain-specific expression and vomeronasal activity of mouse ESP family peptides. Curr Biol 17:1879–1884
4. Leinders-Zufall T, Ishii T, Mombaerts P, Zufall F, Boehm T (2009) Structural requirements for the activation of vomeronasal sensory neurons by MHC peptides. Nat Neurosci 12: 1551–1558
5. Papes F, Logan DW, Stowers L (2010) The vomeronasal organ mediates interspecies defensive behaviors through detection of protein pheromone homologs. Cell 141:692–703
6. Dulac C, Axel R (1995) A novel family of genes encoding putative pheromone receptors in mammals. Cell 83:195–206
7. Herrada G, Dulac C (1997) A novel family of putative pheromone receptors in mammals with a topographically organized and sexually dimorphic distribution. Cell 90:763–773
8. Liberles SD, Horowitz LF, Kuang D, Contos JJ, Wilson KL, Siltberg-Liberles J, Liberles DA, Buck LB (2009) Formyl peptide receptors are candidate chemosensory receptors in the vomeronasal organ. Proc Natl Acad Sci USA 106:9842–9847
9. Matsunami H, Buck LB (1997) A multigene family encoding a diverse array of putative pheromone receptors in mammals. Cell 90:775–784

10. Riviere S, Challet L, Fluegge D, Spehr M, Rodriguez I (2009) Formyl peptide receptor-like proteins are a novel family of vomeronasal chemosensors. Nature 459:574–577
11. Ryba NJ, Tirindelli R (1997) A new multigene family of putative pheromone receptors. Neuron 19:371–379
12. Dey S, Matsunami H (2011) Calreticulin chaperones regulate functional expression of vomeronasal type 2 pheromone receptors. Proc Natl Acad Sci USA 108:16651–16656
13. Strader CD, Fong TM, Tota MR, Underwood D, Dixon RA (1994) Structure and function of G protein-coupled receptors. Annu Rev Biochem 63:101–132
14. Kostenis E, Waelbroeck M, Milligan G (2005) Techniques: promiscuous Galpha proteins in basic research and drug discovery. Trends Pharmacol Sci 26:595–602
15. Offermanns S, Simon MI (1995) G alpha 15 and G alpha 16 couple a wide variety of receptors to phospholipase C. J Biol Chem 270:15175–15180

[illegible]

Chapter 10

Genetic Manipulation to Analyze Pheromone Responses: Knockouts of Multiple Receptor Genes

Tomohiro Ishii

Abstract

Gene targeting in the mouse is an essential technique to study gene function in vivo. Multigene families encoding vomeronasal receptor (VR) type 1 and type 2 consist of ~300 intact genes, which are clustered at multiple loci in the mouse genome. To understand the function of VRs and neurons expressing a particular VR in vivo, individual endogenous receptor genes can be manipulated by conventional gene targeting to create loss-of-function mutations or to visualize neurons and their axons expressing the VR. Multiple receptor genes in a cluster can also be deleted simultaneously by chromosome engineering, allowing analysis of function of a particular VR subfamily. Here, we describe protocols for conventional gene targeting and chromosome engineering for deleting a large genomic region in mouse embryonic stem (ES) cells.

Key words Embryonic stem cell, Embryonic fibroblast, Gene targeting, Knock-out, Chromosome engineering, Vomeronasal receptor

1 Introduction

ES cells are pluripotent cells and can be maintained in an undifferentiated state in vitro [1, 2]. To prevent differentiation, ES cells are usually grown on a feeder layer of mitotically inactivated primary mouse embryonic fibroblasts (EFs). Targeting constructs can be introduced into mouse ES cells by electroporation and integrated into ES cell genome by homologous recombination [3]. Gene targeting technology has allowed the generation of mice with desired mutations in the mouse genome. The mutations include gene inactivation, point mutation, insertions, and small deletions. For deletions, removal of 20 kb DNA region has been succeeded in ES cells [4], but larger deletions are difficult to generate by conventional gene targeting. Chromosomal deletions of DNA covering several hundred kb to a few Mb can be achieved by consecutive targeting of *loxP* sites to the end points of the region to be deleted followed by Cre-*loxP* mediated recombination in ES cells [5, 6]. This chromosome engineering technique can also be

Kazushige Touhara (ed.), *Pheromone Signaling: Methods and Protocols*, Methods in Molecular Biology, vol. 1068,
DOI 10.1007/978-1-62703-619-1_10,

applied to create inversions and duplications of a large chromosomal region.

Several mice with targeted mutations in *VR* loci have been reported [7–14]. These mutations include gene knockout by replacement of the coding sequence with a marker gene, replacement of the coding sequence with that of another receptor, insertion of an internal ribosome entry site (*IRES*)-marker cassette into the 3′ untranslated region (UTR) of a *VR* gene to allow for bicistronic expression of the marker gene with the intact *VR* gene, and deletion of a *VR* cluster containing 16 intact receptor genes.

Although the genome-wide knockout programs KOMP, NorCOMM, and EUCOMM [15–17] provide useful resources of mutant ES cell lines, most of *VRs* have not been the subject of gene knockout. Furthermore, mice with targeted mutations of a chemosensory receptor gene using fluorescent marker genes are often useful in the field of the olfactory research, and such experiments need to be done on a customized basis. Whole mouse genome-wide sequence information [18] has enabled to design gene targeting vectors relatively easily and quickly. Databases for bacterial artificial chromosomes (BACs) library from 129 and C57BL/6 strains are useful to obtain genomic DNA fragments for targeting vector constructs [19]. Several convenient gene construct cassettes are available from Addgene: *IRES-taulacZ* [20] and *IRES-tauGFP* [8] are convenient to visualize cell bodies and axons, and *ACNf* cassette [21, 22], which is *neo* auto-excising sequence during germ line transmission of the targeted gene, saves time and labor to remove the *neo*. Furthermore, the MICER project provides a resource with vectors for chromosome engineering [23]. This information and these tools facilitate greatly the generation of mutant mice.

Here we describe our methods of designing targeting vectors, preparation of mouse EF cells, conventional ES cell culture, genetic manipulations for an individual gene in E14 ES cells [20, 24], and deletion of clustered multiple genes in AB2.2 ES cells [5]. It takes approximately 2 months to obtain mutant ES cells from starting a culture of wild-type ES cells by conventional gene targeting. It is possible to perform two or more independent gene targeting procedures simultaneously or in a staggered way, depending on the skill of the investigator. The methods described here have been successful for gene targeting of many genes including chemosensory receptors with high efficiency of homologous recombination rate ranging from 10 to 60 % among drug-resistant clones. Mutant ES cells can be injected into mouse blastocysts to produce chimeric mice. For behavior analysis, it is advisable to use mice in an inbred background by mating the chimeric mouse with mice of 129 strain or backcrossing F1 mice in a heterogeneous background with mice of inbred strain several times.

2 Materials

Tissue culture facility is essential: laminar flow hood, humidified 37 °C, 5 % CO_2 incubator, inverted phase contrast microscope with an objective lens (10× to 20×), water bath, liquid nitrogen storage tanks.

2.1 Conventional Gene Targeting Vectors

1. *neo* auto-excising cassette, *ACNf* (Addgene, Cambridge, MA, USA; ID15510).
2. Mouse BAC clones (*see* Subheading 3).
3. Restriction enzymes and PCR enzymes.

2.2 Preparation of a Stock of Mouse EF Cells for Conventional Gene Targeting

1. Dulbecco's modified Eagle's medium (DMEM) with 4,500 mg/L Glucose and L-Glutamine (Millipore, Benford, MA, USA; cat. no. SLM-121-B).
2. Fetal bovine serum (FBS) (Fisher Scientific, Pittsburgh, PA, USA; cat. no. SH3007103). FBS should be heat-inactivated before use as follows: Remove a bottle of FBS from the −80 °C freezer and place it at 4 °C for >1 day to thaw. Warm the bottle completely in a 37 °C bath. Place the bottle up to its neck in a 56 °C bath for 25 min, with shaking every 5 min. Equilibrate the bottle on ice. Store at 4 °C.
3. 100× penicillin-streptomycin (Millipore, cat. no. TMS-AB2-C). Store 5 mL aliquots at −20 °C.
4. 100× L-glutamine (Millipore, cat. no. TMS-AB-002-C). Store 5 mL aliquots at −20 °C.
5. EF medium: For 500 mL, combine 440 mL of DMEM (item 1) with 50 mL of FBS, 5 mL of 100× penicillin-streptomycin, and 5 mL of 100× L-glutamine. Filter-sterilize using a 0.22 μm filter. Store at 4 °C.
6. Ca^{2+}- and Mg^{2+}-free Dulbecco's Phosphate-buffered saline (PBS).
7. 0.025 % Trypsin and 0.75 mM EDTA without Ca^{2+} and Mg^{2+} (Millipore, cat. no. SM-2004-C).
8. Freezing medium (Millipore, cat. no. S-002-D).
9. 20× Mitomycin C. Dissolve 2 mg Mitomycin C (Sigma, St. Louis, MO, USA; cat. no. M0503) in 10 mL of EF medium. Freshly prepare before use.
10. Embryonic day 13.5 (E13.5) embryos from G418 resistant mouse strains. Pregnant C57BL/6 mice crossed with the TCRδ KO mouse (The Jackson Laboratory, Bar Harbor, ME, USA; stock number 002120, B6.129P2-TcrdtmMom) [25] are used (*see* **Note 1**).
11. Cryovials.
12. Styrofoam boxes.

13. Culture dishes (T75 and T175 flasks).
14. Petri dishes (60 and 100 mm).
15. Dissecting instruments.
16. 70 % Ethanol.

2.3 General ES Cell Culture

1. EmbryoMax ES DMEM (Millipore, cat. no. SLM-220-B) (*see* **Note 2**).
2. FBS (*see* Subheading 2.2, **item 2**).
3. 100× penicillin-streptomycin (*see* Subheading 2.2, **item 3**).
4. 100× L-glutamine (*see* Subheading 2.2, **item 4**).
5. 100× Nonessential Amino Acids (Millipore, cat. no. TMS-001-C). Store 5 mL aliquots at −20 °C.
6. 100× Nucleosides for ES cells (Millipore, cat. no. ES-008-D). Store 5 mL aliquots at −20 °C.
7. 100× β-mercaptoethanol for ES cells (Millipore, cat. no. ES-007-E). Store 5 mL aliquots at −20 °C.
8. Leukemia inhibitory factor (LIF) (Millipore, cat. no. ESG1107) (*see* **Note 3**). Store 50 μL aliquots at −80 °C.
9. ES + LIF medium: For 500 mL, combine 400 mL of EmbryoMax ES DMEM with 75 mL of FBS, 5 mL of 100× penicillin-streptomycin, 5 mL of 100× L-glutamine, 5 mL of 100× Nonessential amino acids, 5 mL of 100× Nucleosides, 5 mL of 100× β-mercaptoethanol, and 50 μL of LIF. Filter-sterilize using a 0.22 μm filter. Store at 4 °C.
10. EF medium (*see* Subheading 2.2, **item 5**).
11. PBS (*see* Subheading 2.2, **item 6**).
12. Trypsin (*see* Subheading 2.2, **item 7**).
13. Freezing medium (*see* Subheading 2.2, **item 8**).
14. E14 ES cell line [24].
15. EF cells (*see* Subheading 3.2).
16. Culture dishes (T12.5 and T25 flasks).
17. Cryovials.
18. Styrofoam boxes.

2.4 Conventional Gene Targeting in ES Cells

In addition to materials in Subheading 2.3, prepare the followings.

1. 100× G418. G418 (Invitrogen, Carlsbad, CA, USA; cat. no. 11811-031). 15 mg/mL active ingredient in PBS. Filter-sterilize using a 0.22 μm filter. Store 5 mL aliquots at −20 °C (*see* **Note 4**).
2. ES + LIF + G418 medium: Add 5 mL of 100× G418 to 500 mL of ES + LIF medium. Filter-sterilize using a 0.22 μm filter. Store at 4 °C.

3. 2× Freezing medium: 80 % FBS, 20 % DMSO (Sigma, cat. no. D2650). Filter-sterilize using a 0.22 μm filter. Freshly prepare before use.
4. PBS with Ca^{2+}, Mg^{2+}.
5. Large scale plasmid preparation kit (*see* **Note 5**).
6. Restriction enzymes.
7. ES lysis buffer: 10 mM Tris–HCl, pH 7.5, 10 mM EDTA, 10 mM NaCl, 0.5 % N-Lauroyl Sarcosine Sodium Salt, 1 mg/mL Protenase K. Protenase K is freshly added before use.
8. Precipitation buffer (for one 96-well plate): Mix 10 mL of 100 % ethanol and 150 μL of 5 M NaCl.
9. TE buffer: 10 mM Tris–HCl, pH 8.0, 1 mM EDTA.
10. Ethanol (70 % and 100 %).
11. Isopropanol.
12. Culture dishes (96-well plate, 48-well plate, 24-well plate, 12-well plate, 6-well plate, 60 mm dish).
13. 96-well round bottom plates.
14. Electroporator Gene Pulser II equipped with a Pulse Controller plus module (Bio-Rad, Richmond, CA, USA).
15. Electroporation cuvettes (Bio-Rad, cat. no. 165-2088).
16. Parafilm.
17. Sealing film for 96-well plates.
18. P20- and P200-μL Multi-channel Pipetmans and tips.
19. Disposable Pippette Basins for Multi-channel Pipetmans.
20. Pipetman and tip for picking colonies (*see* **Note 6**).

2.5 Deletion of a Receptor Gene Cluster by Chromosome Engineering

In addition to all materials above except *ACNf* cassette and E14 ES cell line, prepare the followings.

1. Chromosome engineering cassettes *HprtΔ3′* and *HprtΔ5′* [5] and Cre-expression vector pOG231 [26].
2. E13.5 embryos from puromycin-resistant mouse strains. Pregnant DR4 mice (The Jackson Laboratory, stock number 003208, Tg(DR4)1Jae/J) [27] crossed with the DR4 mouse are used (*see* **Note 7**).
3. 1,000× Puromycin. 1 mg/mL puromycin (Sigma, cat. no. P8833) in tissue culture-grade water. Filter-sterilize using a 0.22 μm filter. Store 500 μL aliquots at −20 °C.
4. 50× Hybri-Max hypoxanthin/aminopterin/thymidine (HAT) Media Supplement (Sigma, cat. no. H0262).
5. ES + LIF + Puromycin medium: Add 500 μL of 1,000× Puromycin to 500 mL of ES + LIF medium (*see* **Note 8**). Filter-sterilize using a 0.22 μm filter. Store at 4 °C.

6. ES + LIF + HAT medium: ES + LIF medium with 1× Hybri-Max HAT Media Supplement. Filter-sterilize using a 0.22 μm filter. Store at 4 °C.
7. AB2.2 ES cell line (*see* **Note 9**).

3 Methods

Before starting gene targeting, visit the International Knockout Mouse Consortium Web site (http://www.knockoutmouse.org/) and ensure that the gene of interest has not yet been mutated in a way that could be useful for the purpose of the project. All tissue culture procedures are carried out under sterile conditions using disposable sterile plasticware and all culture incubations are performed in a humidified 37 °C, 5 % CO_2 incubator. Medium and trypsin should be warmed in a 37 °C water bath before adding them to cells unless otherwise specified. Tissue culture plates are used without any coating procedures in our protocol (*see* **Note 10**). We use the growth surface area of culture plate as a size of cell number for convenience (e.g., 8 cm^2 area of cells) if the area was confluent at the time of freezing, instead of counting cell number.

3.1 Conventional Gene Targeting Vectors

Higher homologous recombination rates in ES cells can be achieved using DNA isogenic to the ES cells. ES cell line E14 and AB2.2 are derived from 129P2/OlaHsd and of 129/SvEv strains, respectively. Therefore, gene targeting vectors should be constructed using DNA of 129 origin.

1. Find the sequence of the gene of interest and research the structure in the Web sites of European Nucleotide Archive at EMBL-EBI (http://www.ebi.ac.uk/ena/). Choose DNA sequences of 129 origin, if available (*see* **Note 11**).
2. Find BAC clones in the Web site of Mus musculus Clone Finder at NCBI (http://www.ncbi.nlm.nih.gov/projects/mapview/mvhome/mvclone.cgi?taxid=10090) (*see* **Note 12**). Choose clones in 129S7/SvEv ES cell BAC library and purchase the clone from Source Bioscience (http://www.lifesciences.sourcebioscience.com/clone-products/sanger-resources/mouse-bmq-bac-library/bmq-mouse-bac-clone---search-page.aspx) (*see* **Note 13**).
3. Design and construct targeting vectors. The basic considerations are the region of homologous sequences to the chromosome integration site (*see* **Note 14**), a positive selectable marker, the region of probes and restriction enzyme digestion sites for screening by Southern blotting, and a linearization restriction enzyme site in the backbone vector (Fig. 1a). Obtain a DNA fragment of homologous sequences by long PCR using

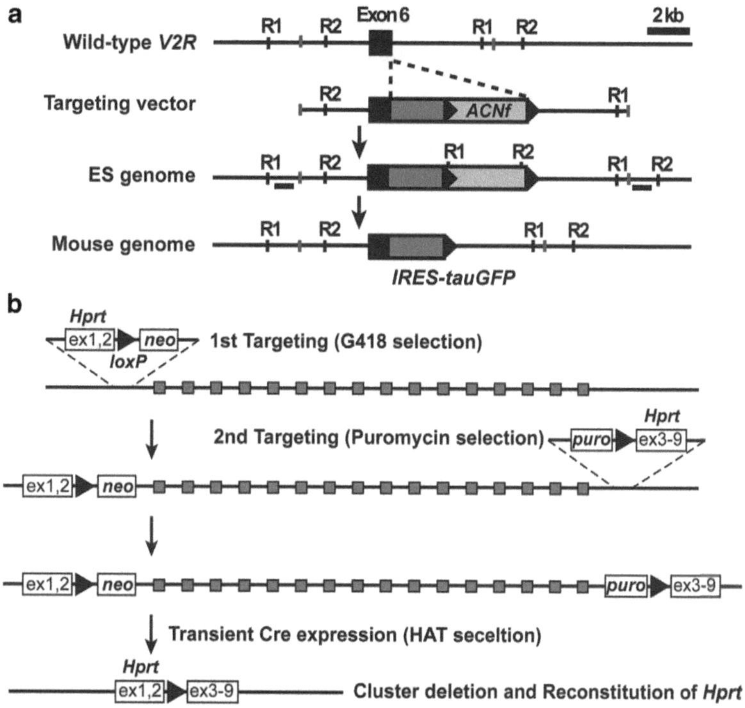

Fig. 1 Gene targeting strategy. (**a**) General strategy for targeted insertion of the *IRES-tauGFP* sequence into *V2R* loci. *Top horizontal line* represents a genomic structure around the last coding exon, mostly exon 6, of the *V2R*. The coding region of exon 6 is shown as a *closed box*. The *IRES-tauGFP-ACNf* cassette is inserted immediately after the stop codon of the *V2R* by homologous recombination in ES cells. The probe for screening clones by Southern blotting for homologous recombination must be located outside the targeting vector. The other probe on the opposite side of the targeting vector, which is also preferably located outside the vector, is used for confirmation of homologous recombination in candidate clones. Restriction enzymes and location of the probes must be chosen so that the fragment sizes detected by Southern blotting can be distinguished between wild type and mutated allele. The *ACNf* cassette is self-excised during transmission through the male germ line, leaving a single *loxP* site behind in the locus. *Filled triangles* represent the *loxP* site. The probes used to detect homologous recombination in Southern blotting are represented as a horizontal bar. R1 and R2 are restriction enzyme sites. The left probes and right probes are used in combination with R1 and R2, respectively. (**b**) General strategy for targeted deletion of a cluster of *VR* genes. *Top horizontal line* represents an unmodified *VR* cluster. Replacement vectors containing the *Hprt* (exons 1, 2)-*loxP-neo* sequence and the *puro-loxP-Hprt* (exons 3–9) sequence are consecutively targeted to the end points of the cluster in ES cells. Transient expression of Cre induces recombination between the *loxP* sites, resulting in a deletion of the VR cluster and the reconstitution of functional *Hprt* gene structure allowing the positive selection in HAT medium. The *grey boxes* and *closed triangles* represent a *VR* gene and the *loxP* site, respectively

BAC clones as a template or excising the fragment from BAC clones with restriction enzymes. For knockout of a *VR* gene by replacement of the coding sequence with *GFP*, for example, replace the coding sequence with *GFP-ACNf* by conventional recombinant PCR and DNA manipulation (*see* **Note 15**). For bicistronic expression of *tauGFP* with a *VR* gene, insert an *IRES-tauGFP-ACNf* cassette into a restriction enzyme site introduced in the 3′ UTR of the *VR* gene (Fig. 1a).

3.2 Preparation of a Stock of Mouse EFs for Conventional Gene Targeting

ES cells are usually cultured on monolayers of mitotically inactivated EF cells. We use E13.5 embryos of the TCRδ KO mouse as a source of G418-resistant EF cells (*see* **Note 16**). Use sterile dissection instruments washed in 70 % ethanol.

3.2.1 Isolation of Embryonic Fibroblasts

1. Kill the 13.5 post-coitus pregnant female by cervical dislocation or another approved method.
2. Swab the abdomen with 70 % ethanol.
3. Open the body wall with a pair of forceps and scissors, and expose the uterine horns.
4. Using a fresh set of forceps and scissors dissect out the uterus and place it in a 100 mm petri dish. Avoid dissection instruments and uterus touching the fur.
5. Dissect the embryos away from the uterus and place them in a 60 mm dish containing PBS.
6. Transfer the dish of embryos to a laminar flow hood. At this point start warming 10 mL of trypsin in a 15 mL conical tube in a 37 °C water bath for **step 13**.
7. Transfer the embryos into the second new 60 mm dish containing PBS.
8. Remove heads using scissors and transfer to the third new 60 mm dish containing PBS.
9. Carefully remove all internal organs using forceps and transfer to the fourth new 60 mm dish containing PBS.
10. Remove as many red blood cells as possible and rinse in the fifth new 60 mm dish containing PBS.
11. Transfer the remaining embryo carcasses to a new 60 mm dish.
12. Mince the carcasses thoroughly with scissors for 5 min.
13. Transfer the minced embryos to pre-warmed trypsin in a 15 mL conical tube.
14. Incubate in a 37 °C water bath for 30 min. Shake the tube every 5 min. Pipette up and down ten times with a 10 mL pipette during the incubation if the minced embryos are not dissociated well.

15. Pipette 20 times with a 10 mL pipette (*see* **Note 17**) and allow large tissue pieces to settle to the bottom for 1 min.
16. Carefully transfer the supernatant to 20 mL of EF medium in a 50 mL conical tube.
17. Spin at 50 × *g* for 1 min to remove large chunks of tissue and transfer the supernatant (~25 mL) to a new 50 mL conical tube.
18. Plate the cell suspension (3–5 mL) on T75 flasks containing 20 mL of EF medium using approximately one embryo per flask and culture overnight (*see* **Note 18**).
19. Change medium for continuous culture, or if confluent wash cells with 10 mL of PBS, add 2 mL of trypsin per flask and incubate in a tissue culture incubator for 5 min.
20. Tap the flasks to detach the cells.
21. Add 18 mL of EF medium and triturate with a pipette to break cell clumps.
22. Plate cells on two T175 flasks from one T75 flask (i.e., dilution of 1:4 to 1:5), in a total volume of 50 mL per flask, and culture for ~2 days (*see* **Note 19**).
23. If confluent trypsinize and split cells into T175 flasks at a dilution of 1:4 or 1:5, and culture for ~2 days. This is used for mitomycin C treatment in Subheading 3.2.2.

3.2.2 Mitotic Inactivation of EF Cells with Mitomycin C

EF cells are inactivated by mitomycin C treatment before using as a feeder layer for ES cells (*see* **Note 20**). Treat a small number of flasks at the same time (usually three flasks). If six flasks are treated, it should be done in a staggered way with an interval of 30 min to 1 h.

1. Prepare T175 flasks of EF cells (80–100 % confluent).
2. Remove medium and add 30 mL of EF medium and 1.5 mL of 20× mitomycin C.
3. Incubate in a tissue culture incubator for exactly 2 h.
4. Wash flasks twice with 20 mL of PBS and trypsinize with 5 mL of trypsin.
5. Add 10 mL of EF medium and triturate with a pipette to break cell clumps.
6. Pool the cell suspensions, if multiple flasks are treated at the same time, and pellet by centrifuging at 200 × *g* for 5 min.
7. Resuspend pellet in 40 mL of EF medium and pellet by centrifuging at 200 × *g* for 5 min.
8. Resuspend pellet at 100 cm^2 surface area of EF cells per 1 mL in freezing medium (*see* **Note 21**) and dispense into cryovials in 1 mL-aliquots.

9. Place cryovials in a styrofoam box at −80 °C and leave them overnight (*see* **Note 22**). Transfer the cryovials to liquid nitrogen.

3.3 General ES Cell Culture

ES cells are cultured on a monolayer of EF cells.

3.3.1 Thawing Frozen EF Cells for ES Cell Culture

1. Thaw vial(s) of frozen EF cells in a 37 °C water bath.
2. Transfer EF cells to 30 mL of EF medium (*see* **Note 23**) in a 50 mL conical tube and pellet cells by centrifuging at 200 × *g* for 5 min.
3. Resuspend pellet in the appropriate amount of EF medium and plate the cell suspension on a flask.
4. After 12 h, cells should form a monolayer covering almost the whole area of the flask bottom. If the coverage is too sparse, thaw more feeders as above and plate them on top of the existing one.

3.3.2 Thawing Frozen ES Cells

Prepare T12.5 flask of EF cells. ES cell colonies have an oval appearance with a smooth outline (Fig. 2a).

1. Thaw a vial of frozen ES cells (8 cm^2) in a 37 °C water bath.
2. Transfer ES cells to 10 mL of ES + LIF medium in a 15 mL conical tube. Pellet cells by centrifuging at 200 × *g* for 5 min.
3. Resuspend pellet in 5 mL of ES + LIF medium and plate on a confluent T12.5 flask of EF cells (*see* **Note 24**).

3.3.3 Passaging ES Cells

1. Check cells under the microscope for 70–80 % confluence.
2. Wash cells with PBS and add the appropriate volume of trypsin (*see* Table 1). Incubate in a tissue culture incubator for 5–10 min.
3. Tap the plate to detach the cells.

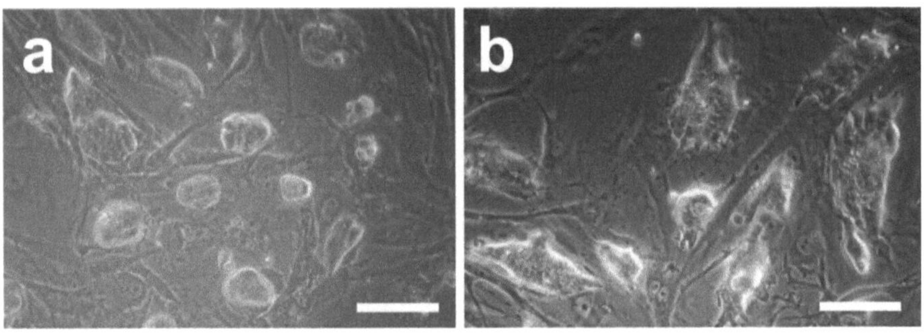

Fig. 2 Phase-contrast images of E14 ES cells. ES cell colonies on EF cells have an oval appearance with a smooth outline (**a**), while ES colonies appear more flattened when cultured on a layer of EF cells that have been plated at lower density (**b**). Scale bars, 100 μm

Table 1
Working volumes for cell culture vessels

Dish	Medium	Trypsin	Growth area
48 well plate	1.1 mL	100 μL	0.75 cm^2
24 well plate	1.5 mL	200 μL	2.00 cm^2
12 well plate	3 mL	300 μL	3.8 cm^2
6 well plate	6 mL	700 μL	9.6 cm^2
T12.5 flask	6 mL	700 μL	12.5 cm^2
T25 flask	10 mL	1 mL	25 cm^2

4. Add the appropriate volume of ES+LIF medium (*see* Table 1) (*see* **Note 25**) and triturate with a pipette to break cell clumps.
5. Plate on feeders at a dilution of 1:3 to 1:4.

3.3.4 Freezing ES Cells

The protocol below is for a T25 flask of ES cells into three cryovials (~8 cm^2 aliquots). Label cryovials with cell line, passage number, date, and plating area. Place freezing medium on ice next to the hood.

1. Trypsinize a T25 flask of ES cells.
2. Stop with 10 mL of ES+LIF medium and triturate with a 10 mL pipette to break cell clumps.
3. Pellet cells by centrifuging at 200×*g* for 5 min.
4. Resuspend pellet in 1.2 mL of freezing medium. Dispense into cryovials in 400 μL-aliquots, and quickly place the cryovials on ice.
5. Place the cryovials in a styrofoam box at −80 °C, and leave them overnight. Transfer the cryovials to liquid nitrogen.

3.4 Conventional Gene Targeting in ES Cells

The E14 ES cell line is used in combination with the *ACNf* cassette in the targeting vector and G418-resistant EF cells in this protocol.

3.4.1 Preparation of DNA for Electroporation

1. Isolate targeting construct DNA using a large scale plasmid preparation kit.
2. One day before electroporation, digest 35 μg of the DNA with the appropriate restriction enzyme (150 μL scale) to linearize, and assess the completion of the digestion by agarose gel electrophoresis using 0.5 μL of the digested DNA. Store at −20 °C.
3. On the day of electroporation, thaw the linearized DNA on ice.
4. Add 1.2 mL of 100 % ethanol and mix by inverting the tube (*see* **Note 26**). White DNA clump can be observed.

5. Remove the supernatant using a P200 Pipetman (*see* **Note 27**).
6. Wash DNA with 1.4 mL of 70 % ethanol and air-dry pellet for 10–15 min (*see* **Note 28**).
7. Dissolve DNA in 800 μL of PBS with Ca^{2+} and Mg^{2+} by incubating at 65 °C for 10 min. Mix well using a P1000 Pipetman.
8. Place on ice.

3.4.2 Electroporation of ES Cells

DNA can be transfected into ES cells by high-voltage electroporation.

1. Prepare three 60 mm dishes of EF cells and one T25 flask of ES cells (70–80 % confluent) after passaging recently thawed ES cells.
2. Feed ES cells with fresh ES + LIF medium 4 h before harvesting.
3. Wash ES cells twice with PBS. Add 1 mL of trypsin and incubate in a tissue culture incubator for 10 min.
4. Tap the flask to detach the cells. Add 10 mL of ES + LIF medium and triturate well by pipetting up and down five to ten times to break cell clumps into a single-cell suspension.
5. Pellet cells by centrifuging at 200 × *g* for 5 min.
6. Remove the supernatant and tap the pellet several times. Resuspend pellet in 800 μL of DNA/PBS solution from Subheading 3.4.1.
7. Transfer the cell suspension into precooled electroporation cuvette on ice (*see* **Note 29**).
8. Transfer the cuvette into the Bio-Rad electroporation chamber and apply the electric pulse (800 V and 3 μF).
9. Transfer the cell suspension to 10 mL of ES + LIF medium in a 15 mL conical tube.
10. Divide over three 60 mm dishes of EF cells. Drop the cell suspension evenly onto the dish. Culture for 15–20 h without changing medium.
11. Change to ES + LIF + G418 medium to start selection.
12. Change medium when the medium color begins to turn orange (*see* **Note 30**).
13. Pick colonies on 6–10th day of G418 selection when colonies are visible (*see* **Note 31**).

3.4.3 Picking Colonies

Use an inverted microscope to see ES colonies and pick them in a laminar flow hood. Prepare the Pipetman for picking colonies, P20- and P200-μL Multi-channel Pipetmans and their tips, and basins in the hood. For one targeting construct we usually pick 96 ES colonies per day for 3 consecutive days. Prepare EF cells in 96-well flat-bottom plates.

1. Prepare three 96-well round bottom plates with 15 μL of trypsin per well. Store at 4 °C until use (*see* **Note 32**).
2. Change ES + LIF + G418 medium in three 60 mm culture dishes of electroporated ES cells. Replace medium in a 96-well flat-bottom plate of EF cells with 100 μL of ES + LIF + G418 medium per well.
3. Transfer one 96-well trypsin plate to the hood.
4. Put one 60 mm dish of electroporated ES cells on the microscope in the hood.
5. Under the microscope dislodge the colony slightly with the pipette tip, and suck up the colony using the Pipetman set to 1 μL. Transfer the colony to a well of the 96-well plate containing trypsin.
6. Repeat picking colonies for no more than 30 min (*see* **Note 33**).
7. Incubate the plate in a tissue culture incubator for 10 min.
8. Tap the 96-well plate and add 100 μL of ES + LIF + G418 medium to each well by a Multi-channel Pipetman. Pipette up and down ten times.
9. Transfer the cell suspension to the 96-well plate of EF cells, and return the plate to the tissue culture incubator. Change medium daily.
10. Repeat **steps 5–9** using a new 96-well trypsin plate and another 60 mm dish of the ES cells until picking ~96 colonies per day.
11. Repeat **steps 1–10** on the second and third days (*see* **Note 34**).
12. When the 96-well plate is confluent overall (~3 days after picking colonies), proceed to the cell freezing step in Subheading 3.4.4 (*see* **Note 35**).

3.4.4 Freezing ES Cells in 96-Well Plates

Prepare a 96-well flat-bottom plate with 65 μL 2× freezing medium per well and store at 4 °C until use.

1. Feed ES cells with fresh ES + LIF + G418 medium 4 h before harvesting.
2. Wash twice with PBS.
3. Add 35 μL of trypsin and incubate in a tissue culture incubator for 10 min.
4. Tap the plate to detach the cells. Add 65 μL of ES + LIF + G418 medium and triturate well by pipetting up and down ten times (*see* **Note 36**).
5. Transfer 65 μL of the cell suspension to a well of the 96-well flat-bottom plate containing 2× freezing medium, and pipet up and down three times.

6. Label the plate with clone name and date. Seal the plate with parafilm around all the edges.
7. Place the plate in a styrofoam box and store at −80 °C.
8. Add 200 μL of ES + LIF + G418 medium to the remaining cell suspension (35 μL) at **step 5** and incubate the plate.
9. Change medium daily.
10. If the plate is mostly confluent, proceed to genomic DNA extraction in Subheading 3.4.5.

3.4.5 Genomic DNA Extraction from 96-Well Plate and Restriction Digestion for Southern Blotting

Prepare 70 % ethanol at −20 °C.

1. Wash cells in the 96-well plate from Subheading 3.4.4. with PBS.
2. Add 50 μL of ES lysis buffer per well.
3. Seal the 96-well plate completely with a sealing film. Place the plate in a moisture chamber and incubate at 60 °C for 24 h.
4. Add 100 μL of precipitation buffer per well and wrap the plate in plastic wrap (*see* **Note 37**), and leave the plate overnight at room temperature.
5. Discard the supernatant by turning upside down on paper towel.
6. Add 150 μL of prechilled 70 % ethanol per well.
7. Discard the supernatant by turning upside down on paper towel.
8. Place the plate at 55 °C for 20 min to air-dry the DNA (*see* **Note 38**).
9. Add the appropriate restriction enzyme mixture (30 μL/well) for Southern blotting and seal the plate with a sealing film. Incubate the plate for restriction reaction.
10. Perform standard Southern blotting using a ^{32}P-labeled or DIG-labeled probe for screening homologous recombinant clones (*see* **Note 39**).

3.4.6 Thawing and Expanding Mutant ES Cells from Frozen 96-Well Plates

1. Prepare a 48-well plate of EF cells with 400 μL of ES + LIF medium (*see* **Note 40**).
2. Remove the 96-well plate of ES cells from −80 °C and wipe it to remove ice.
3. Locate the clone to be thawed and warm with a finger under the well.
4. Add 100 μL of ES + LIF medium to the partially thawed ES cells. Triturate by pipetting up and down a few times and transfer to the 48-well plate of EF cells.
5. Repeat **step 4** until you transfer all cell suspensions. A total volume of 500 μL per well can be added.

6. Add ES+LIF medium if a total volume of 500 μL per well is not added, and return the 48-well plate to the tissue culture incubator.
7. Proceed to the next clone and repeat **steps 3–6** (*see* **Note 41**).
8. Rewrap the 96-well plate in parafilm and return to the –80 °C freezer.
9. Change medium after 12 h, and then change every 24 h.
10. If the plate is confluent, passage stepwise to 24-well plate, 12-well plate, 6-well plate, T25 flask, and three T25 flasks (*see* **Note 42**).
11. Make four cryovials of 8 cm^2 stock from two T25 flasks. Continue culture using the remaining ES cells in the T25 flask for genomic DNA extraction.
12. If the plate is confluent, trypsinize the cells, pellet by centrifugation, and resuspend in 2 mL of ES lysis buffer.
13. Incubate at 55 °C overnight.
14. Add 2 mL of isopropanol and mix well.
15. Fish genomic DNA, wash in 70 % ethanol and 100 % ethanol, and briefly air-dry.
16. Dissolve the DNA in 150 μL of TE buffer. This DNA is subject to Southern blotting for confirmation of the homologous recombination in the clones (*see* **Note 43**).

3.5 Deletion of a Receptor Gene Cluster by Chromosome Engineering

The AB2.2 ES cell line, which is deficient in hypoxanthine phosphoribosyl transferase (*Hprt*), has been used traditionally for chromosome engineering. Two complementary but nonfunctional fragments of a *Hprt* minigene are linked to the *loxP* site with either *neo* or *puro* in the targeting constructs. Cre-*loxP* mediated recombination in AB2.2 ES cells bearing these two *loxP* cassettes in a chromosome result in the reconstitution of functional *Hprt* gene structure allowing the selection in HAT medium (Fig. 1b).

3.5.1 Deletion of a Receptor Gene Cluster in ES Cells

1. Design the targeting vectors at both end points of a *VR* gene cluster as in Subheading 3.1. One targeting vector should contain *Hprt* (exons 1, 2)-*loxP-neo* from the *Hprt*Δ3′ plasmid for G418 selection and the other should contain *puro-loxP-Hprt* (exons 3–9) from the *Hprt*Δ5′ plasmid for puromycin selection. Make sure the *Hprt*-minigene fragments are placed in the correct order and orientation on the targeted chromosome(s) to reconstitute the functional *Hprt* after Cre-*loxP* mediated recombination. The correct order of DNA fragments is as follows: *Hprt* (exons 1, 2)/*loxP*/*neo*/*VR* cluster to be removed/*puro*/*loxP*/*Hprt* (exons 3–9) (Fig. 1b) (*see* **Note 44**).
2. Prepare G418-resistant EF cells and puromycin-resistant EF cells using embryos of TCRδ KO mouse and DR4 mouse, respectively, as in Subheading 3.2.

3. Perform gene targeting as in Subheading 3.4 using AB2.2 ES cells and a linearized targeting vector containing *Hprt* (exons 1, 2)-*loxP*-*neo* for electroporation. Use the G418-resistant EF cells as a feeder layer and ES + LIF + G418 medium for selection (*see* **Note 45**).
4. Perform gene targeting as in Subheading 3.4 using ES clones from **step 3** (*see* **Note 46**) and a linearized targeting vector containing *puro*-*loxP*-*Hprt* (exons 3–9) for electroporation. Use the puromycin-resistant EF cells as a feeder layer and ES + LIF + Puromycin medium for selection (*see* **Notes 47** and **48**).
5. Perform electroporation of ES cells and pick colonies as in Subheadings 3.4.2–3.4.3 using several independent double targeted ES clones from **step 4** and a Cre-expression vector, pOG231, for electroporation for transient expression of Cre. Preparation of pOG231 for electroporation is described in Subheading 3.5.2. The G418-resistant EF cells can be used as a feeder layer (*see* **Note 49**). At **step 10** in Subheading 3.4.2, plate ES cells at three different densities such as 3×10^6, 3×10^5, and 3×10^4 cells per 60 mm culture dish. The normal number of ES cells with 80 % confluency is ~1×10^7 per T25 flask (*see* **Note 50**). Use ES + LIF + HAT medium for selection and pick 6–10 HAT-resistant colonies. Passage HAT-resistant ES cells stepwise from 96-well plate to three T25 flasks without freezing the cells (*see* **Note 51**). Perform sib selection as in Subheading 3.5.3 during passaging the ES cells. Genomic DNA from the ES cells is subject to Southern blotting for confirmation of the correct Cre-*loxP* mediated recombination.

3.5.2 Preparation of pOG231 for Electroporation

For transient expression of Cre, use circular pOG231 plasmid for electroporation. Do not linearize the DNA to avoid increased probability of genomic integration.

1. Isolate pOG231 using a large scale plasmid preparation kit.
2. Put 25 μg of the plasmid in 1× restriction buffer as if it is digested (100 μL volume) (*see* **Note 52**).
3. Add 150 μL of 100 % ethanol and mix by inverting the tube (*see* **Note 53**).
4. Pellet DNA by centrifuging at 10,000 × *g* for 5 min.
5. Remove the supernatant.
6. Wash DNA with 1.4 mL of 70 % ethanol and air-dry pellet briefly.
7. Dissolve DNA in 800 μL of PBS with Ca^{2+} and Mg^{2+} by incubating at 65 °C for 10 min. Mix well using a P1000 Pipetman.

3.5.3 Sib Selection

Although most HAT-resistant clones should have the allele with the desired chromosomal deletion, there could be a few different genotypes on the other allele (Fig. 3). The genotype of ES clones can be distinguished by sib selection [6]. We perform sib selection during passaging HAT-resistant ES clones from 6-well plates to T25 flasks in Subheading 3.5.1.

1. Trypsinize HAT-resistant ES cells in a 6-well plate and stop with 10 mL of ES + LIF medium. Triturate with a 10 mL pipette.
2. Plate 0.3 mL of cell suspension on each of three wells of a 24-well plate containing 1 mL of ES + LIF (*see* **Note 54**). Incubate for 6–12 h.
3. Change medium of each of three wells to ES + LIF + HAT, ES + LIF + G418, or ES + LIF + Puromycin medium.
4. Change medium every 24 h and observe the growth of ES cells in each medium on the 4th day. Identify the genotype according to Fig. 3.

4 Notes

1. This is used for preparation of G418-resistant feeder cells expressing *neo*.
2. SLM-220-M can also be used as formulations of SLM-220-M and SLM-220-B are same.
3. LIF is a cytokine that can suppress differentiation of ES cells.
4. G418 is an analog of neomycin.
5. We use QIAfilter Plasmid Midi Kit (QIAGEN, Valencia, CA, USA; cat. no. 12243).
6. We use the Pipetman (Gray control button for 20 μL pipette tips, 0.5–10 μL, Eppendorf, Germany; cat. no. 022470051) and tips (epT.I.P.S. Biopur, 0.1–20 μL volume, Eppendorf, cat. no. 022491067).
7. Although EF cells prepared from the DR4 transgenic mouse strain is resistant to concentrations of the drug G418, 6-thioguanine, puromycin, and hygromycin, we use the DR4 EF cells only for puromycin selection in this protocol.
8. DR4 EF cells are resistant to 1 μg/mL, but sensitive to 2 μg/mL puromycin.
9. AB2.2 ES cells can be requested through the Welcome Trust Sanger Institute Web site: http://www.sanger.ac.uk/technology/clonerequests/resources/.
10. We do not usually gelatinize tissue culture plates. Alternatively, gelatinization of tissue culture plates can be performed with

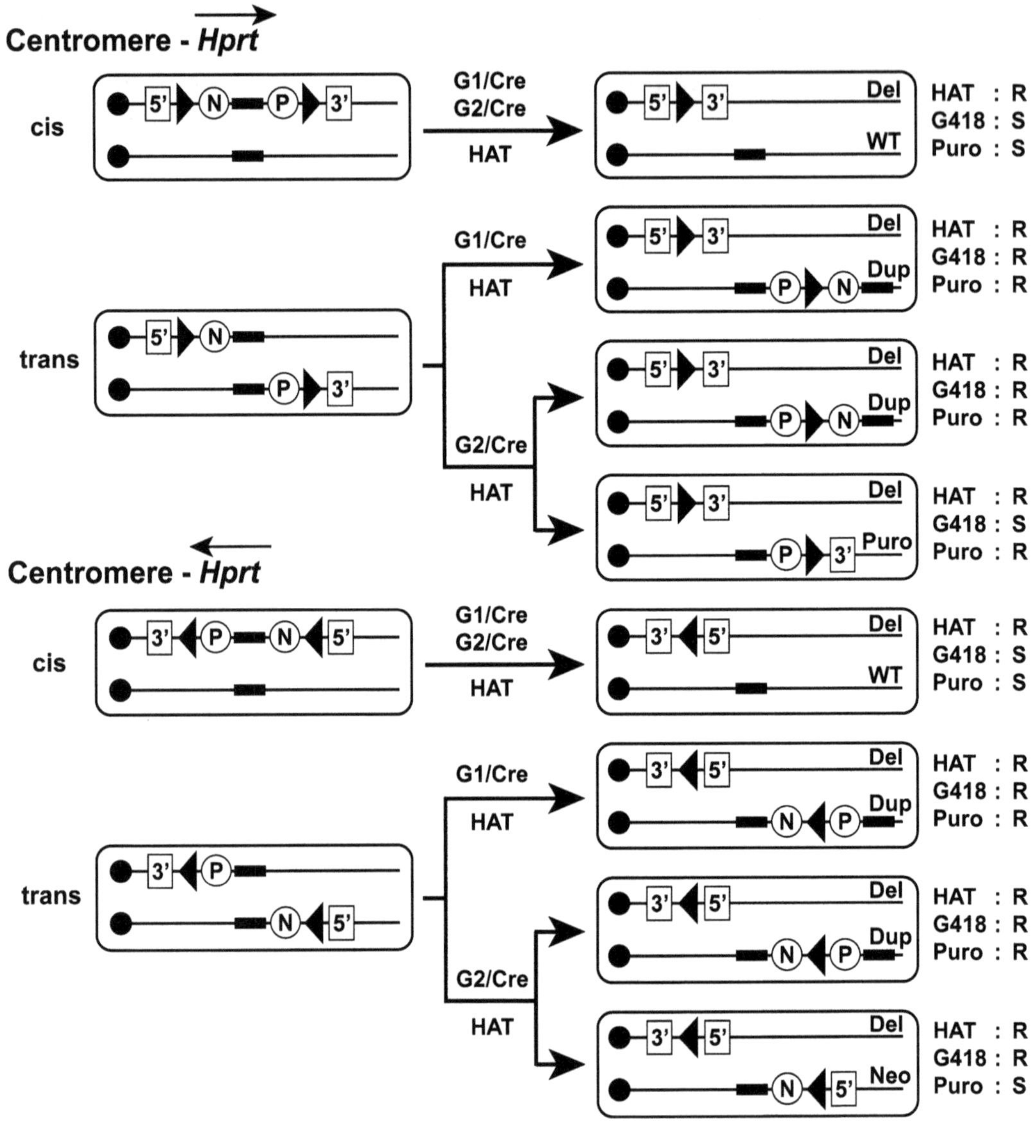

Fig. 3 HAT-resistant ES clones. HAT-resistant clones should harbor a *VR* cluster deletion if the targeting vectors are integrated in the correct order and orientation. Among the HAT-resistant clones there could be G418/Puromycin-sensitive, G418/Puromycin-resistant, G418-sensitive/Puromycin-resistant, and G418-resistant/Puromycin-sensitive clones. It depends on integration of two targeting vectors into the same chromosome (*in cis*) or different chromosomes (*in trans*), G1 or G2 phase of the cell cycle during which Cre-*loxP* mediated recombination occurs, and orientation to the centromere. *Closed circle* centromere, *triangle loxP* site, *closed box VR* cluster, *5′ Hprt* (exons 1, 2), *3′ Hprt* (exons 3–9), *N neo*, *P puro*, *Del* cluster deletion, *Dup* cluster duplication, *R* resistant, *S* sensitive, *G1/Cre, G2/Cre* phases of the cell cycle in which Cre-recombination occurs. The various DNA sequences involved in chromosome engineering are not drawn to scale

0.1 % gelatin solution (Millipore, cat. no. ES-006-B) for 10 min at room temperature followed by aspirating off the solution and allowing it to dry for 5 min.

11. Possible other Web sites include Mouse Genome Resources at NCBI: http://www.ncbi.nlm.nih.gov/projects/genome/guide/mouse/; UCSC genome browser: http://genome.ucsc.edu/; Ensembl genome browser: http://www.ensembl.org/Mus_musculus/Info/Index.
12. A possible other Web site is Ensembl genome browser: http://www.ensembl.org/Mus_musculus/Info/Index.
13. Alternatively, BAC clones (RP21 and RP22) from 129S6/SvEvTac strain can be obtained by screening high-density microarray filters to acquire the BAC ID and purchase the clone from Children's Hospital Oakland Research Institute (CHORI).
14. The length of left and right homology arms is 3–5 kb and the total length of homology is 8–10 kb in most of our targeting vectors.
15. *ACNf* is a *neo* expression cassette for G418 selection with self-excising property during germ line transmission of the targeted gene.
16. G418-resistant EF cells are also commercially available from Millipore (Millipore, cat. no. PMEF-N) and gene targeting in E14 ES cells using this EF cells is successful.
17. Avoid making air bubbles.
18. Although EF cells can be expanded in EF medium with G418 to select for resistant cells, we usually do not use the drug.
19. Alternatively make EF cell stocks (75 cm^2 area of EF cells in 1 mL of freezing medium per vial) for future mitomycin C treatment.
20. Work carefully with gloves and protective clothing because mitomycin C is toxic and teratogenic.
21. Remove all of the supernatant by first sucking off all bubbles, and then slowly inverting the tube while aspirating medium. Tap the pellet several times and add freezing medium.
22. The cells should be frozen slowly at ~1 °C per min to −80 °C.
23. EF cells can also be plated in ES + LIF medium.
24. The recovery may vary from 25 to 75 %. Cells should be plated and left alone for 12–24 h as they take a while to recover.
25. Trypsin action should be stopped by adding more than 10× volume of ES + LIF medium.
26. We do not purify the digested DNA by phenol/chloroform extraction.

27. Avoid touching the Pipetman tip to DNA clump because DNA is sticky.
28. Do not over-dry DNA because it might become harder to dissolve.
29. Avoid making air bubbles. Do not touch the metal plate of the cuvette. Using plastic wrap, avoid the surface of the cuvette touching ice directly.
30. Watch the color of medium carefully. We change medium every 8 h on the first and second day, and every 12 h on the third-fifth day of G418 selection.
31. Try to pick round ES colonies with a smooth outline.
32. We usually add trypsin to 36 wells in each 96-well plate because all wells are not used.
33. 24–36 colonies can be picked for 30 min.
34. 288 ES colonies in total are sufficient to obtain homologous recombinant clones in most cases. If you expect higher or lower efficiency of homologous recombination, adjust the number of colonies to be picked.
35. Cell growth will vary significantly among wells.
36. Avoid making air bubbles during pipetting.
37. Do not mix by pipetting or shaking.
38. If you perform Southern blotting later, you can seal the dried plate with a sealing film and store at –20 °C.
39. It is convenient to use high-throughput gel system allowing multi-channel pipetting.
40. It is not necessary to use G418 for selection any more.
41. We usually thaw six to eight clones in total from three frozen 96-well plates.
42. We do not plate EF cells on these plates. ES cells cultured without EF cells form more flattened colonies than cells cultured on EF cells (Fig. 2b). When the ES cells are plated on EF cells again, ES cell colonies revert to their original oval appearance with a smooth outline.
43. Analyze the homologous recombination on the side of the homology arm of the targeting vector different from the one analyzed during ES clone screening.
44. A library of ready-made targeting vectors, incorporating a drug resistance gene for gene targeting, the 5′ or 3′ portion of an *Hprt* minigene, and *loxP* site for chromosome engineering, is available and clones can be requested through the MICER Web site [23]. http://www.sanger.ac.uk/resources/mouse/micer/. It is worth searching for clones.

45. It is advisable to perform karyotyping to confirm that selected ES clones have maintained the correct number of chromosomes prior to **step 4**.
46. It is advisable to use two or more independent clones.
47. The duration of selection varies depending on the drug. Puromycin selection is faster than G418 selection.
48. It is advisable to perform karyotyping to confirm that selected ES clones have maintained the correct number of chromosomes prior to **step 5**.
49. G418-resistance in EF cells is not necessary for HAT selection.
50. The frequency of Cre-*loxP* mediated recombination varied from 3×10^{-6} to 5×10^{-3} in our hand, and could probably be lower depending on the genes or distance to be deleted [5]. The frequency is lower in case of two *loxP* sites on chromosome in *trans* than those in *cis*. Therefore plate more ES cells in case the frequency should be very low.
51. Most HAT-resistant clones should have desired chromosome deletion if the design of targeting vector is appropriate.
52. Do not add enzymes.
53. Many small white clumps are observed.
54. The 24-well plate does not contain EF cells. The remaining cell suspension is used to expand the clone.

Acknowledgments

I am grateful to P. Mombaerts for critical reading the manuscript, M. Omura for photos of ES cells and valuable comments, S. Nishimura for comments on chromosome engineering, and members of the Mombaerts laboratory for improvements in the methods. This work was supported in part by JSPS KAKENHI Grant Number 24500456.

References

1. Martin GR (1981) Isolation of a pluripotent cell line from early mouse embryos cultured in medium conditioned by teratocarcinoma stem cells. Proc Natl Acad Sci USA 78: 7634–7638
2. Evans MJ, Kaufman MH (1981) Establishment in culture of pluripotential cells from mouse embryos. Nature 292:154–156
3. Vasquez KM, Marburger K, Intody Z et al (2001) Manipulating the mammalian genome by homologous recombination. Proc Natl Acad Sci USA 98:8403–8410
4. Zhang H, Hasty P, Bradley A (1994) Targeting frequency for deletion vectors in embryonic stem cells. Mol Cell Biol 14:2404–2410
5. Ramirez-Solis R, Liu P, Bradley A (1995) Chromosome engineering in mice. Nature 378:720–724
6. Yu Y, Bradley A (2001) Engineering chromosomal rearrangements in mice. Nat Rev Genet 2:780–790
7. Belluscio L, Koentges G, Axel R et al (1999) A map of pheromone receptor activation in the mammalian brain. Cell 97:209–220

8. Rodriguez I, Feinstein P, Mombaerts P (1999) Variable patterns of axonal projections of sensory neurons in the mouse vomeronasal system. Cell 97:199–208
9. Del Punta K, Leinders-Zufall T, Rodriguez I et al (2002) Deficient pheromone responses in mice lacking a cluster of vomeronasal receptor genes. Nature 419:70–74
10. Del Punta K, Puche A, Adams NC et al (2002) A divergent pattern of sensory axonal projections is rendered convergent by second-order neurons in the accessory olfactory bulb. Neuron 35:1057–1066
11. Ishii T, Mombaerts P (2008) Expression of nonclassical class I major histocompatibility genes defines a tripartite organization of the mouse vomeronasal system. J Neurosci 28: 2332–2341
12. Haga S, Hattori T, Sato T et al (2010) The male mouse pheromone ESP1 enhances female sexual receptive behaviour through a specific vomeronasal receptor. Nature 466:118–122
13. Leinders-Zufall T, Ishii T, Mombaerts P et al (2009) Structural requirements for the activation of vomeronasal sensory neurons by MHC peptides. Nat Neurosci 12:1551–1558
14. Ishii T, Mombaerts P (2011) Coordinated coexpression of two vomeronasal receptor V2R genes per neuron in the mouse. Mol Cell Neurosci 46:397–408
15. Ringwald M, Iyer V, Mason JC et al (2011) The IKMC web portal: a central point of entry to data and resources from the International Knockout Mouse Consortium. Nucleic Acids Res 39:D849–D855
16. Skarnes WC, Rosen B, West AP et al (2011) A conditional knockout resource for the genome-wide study of mouse gene function. Nature 474:337–342
17. Guan C, Ye C, Yang X et al (2010) A review of current large-scale mouse knockout efforts. Genesis 48:73–85
18. Mouse Genome Sequencing Consortium, Waterston RH, Lindblad-Toh K, Birney E et al (2002) Initial sequencing and comparative analysis of the mouse genome. Nature 420: 520–562
19. Adams DJ, Quail MA, Cox T et al (2005) A genome-wide, end-sequenced 129Sv BAC library resource targeting vector construction. Genomics 86:753–758
20. Mombaerts P, Wang F, Dulac C et al (1996) Visualizing an olfactory sensory map. Cell 87:675–686
21. Bunting M, Bernstein KE, Greer JM et al (1999) Targeting genes for self-excision in the germ line. Genes Dev 13:1524–1528
22. Bozza T, Feinstein P, Zheng C et al (2002) Odorant receptor expression defines functional units in the mouse olfactory system. J Neurosci 22:3033–3043
23. Adams DJ, Biggs PJ, Cox T et al (2004) Mutagenic insertion and chromosome engineering resource (MICER). Nat Genet 36: 867–871
24. Hooper M, Hardy K, Handyside A et al (1987) HPRT-deficient (Lesch-Nyhan) mouse embryos derived from germline colonization by cultured cells. Nature 326:292–295
25. Itohara S, Mombaerts P, Lafaille J et al (1993) T cell receptor delta gene mutant mice: independent generation of alpha beta T cells and programmed rearrangements of gamma delta TCR genes. Cell 72:337–348
26. O'Gorman S, Dagenais NA, Qian M et al (1997) Protamine-Cre recombinase transgenes efficiently recombine target sequences in the male germ line of mice, but not in embryonic stem cells. Proc Natl Acad Sci USA 94:14602–14607
27. Tucker KL, Wang Y, Dausman J et al (1997) A transgenic mouse strain expressing four drug-selectable marker genes. Nucleic Acids Res 25:3745–3746

Part III

Methods in Measuring Neural Responses to Pheromones

Chapter 11

Electroantennogram and Single Sensillum Recording in Insect Antennae

Shannon B. Olsson and Bill S. Hansson

Abstract

Electrophysiology is an invaluable technique to quickly and quantitatively assess the response of the olfactory system to odor stimuli. For measuring the response of the insect antenna, two basic techniques exist, electroantennography and single sensillum recording. Here, we describe the general practice of both methods in terms of equipment used, insect preparation, recording technique, and basic analysis.

Key words Invertebrate, Olfaction, Electrophysiology, Odor stimulation, GC-EAD, EAG, SSR

1 Introduction

For most insects, the chemical senses are indispensable for survival. Insects rely on chemical cues to locate food, hosts, predators, kin, and, of course, potential mates. It is perhaps, then, no surprise that the first pheromone identified was the sex attractant of the female silkworm moth, *Bombyx mori*, in 1959 [1].

Given that electrical potentials are the *modus operandi* for transmitting information within the nervous system, electrophysiology provides an incredibly efficient and effective means for assessing chemical signaling. In insects, the primary sensory structure for detecting volatile chemical cues is the antenna. Indeed, the first physiological recordings from the antenna occurred nearly simultaneously with the identification of the first pheromone [2, 3]. From that time until now, electrophysiological studies have been invaluable for detecting, identifying, and assessing pheromone communication.

Insect antennae vary widely in structure, from the long pectinate (feathered) antenna of the silkworm moth to the lamellate club antenna of scarab beetles (*see* descriptions in [4]). Adult pterygote (winged) insect antennae have three sections: the basal scapus, attached to the head capsule, the pedicellus, housing the Johnston's

Kazushige Touhara (ed.), *Pheromone Signaling: Methods and Protocols*, Methods in Molecular Biology, vol. 1068, DOI 10.1007/978-1-62703-619-1_11, © Springer Science+Business Media, LLC 2013

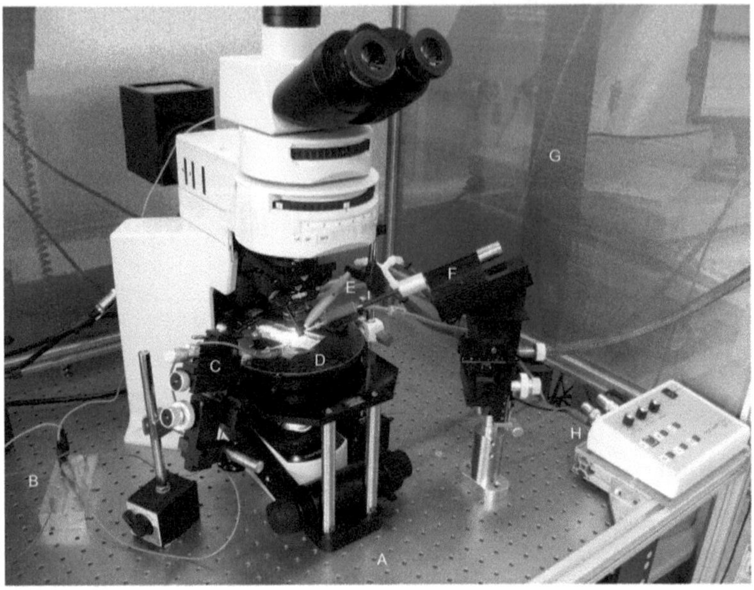

Fig. 1 Electrophysiology setup showing basic equipment, *clockwise from bottom center*: (**a**) anti-vibration table, (**b**) grounding block with ground wires, (**c**) manual micromanipulator with ground electrode, (**d**) microscope with long working distance objectives and stage, (**e**) odor stimulus outlet with tubing, (**f**) motorized micromanipulator with recording electrode and amplifier, (**g**) Faraday cage, (**h**) motorized micromanipulator control board

organ, and the flagellum housing the majority of the olfactory sensory neurons (OSNs) [4]. The OSNs are housed in cuticular structures called sensilla, which also range from long hair-like sensillum trichodea to plate-like sensillum chaetica. These sensilla house one or several neurons depending on species and sensillum type (up to 50 in some basiconic sensilla of locusts [5]). It is thus the flagellum that is the focus of most olfactory studies.

Despite the incredible morphological diversity of insect antennae and the large body of electrophysiological studies over the past 50 years, the basic equipment required for electrophysiological recording has remained relatively unchanged from Schneider's initial experiments ([6], Fig. 1). The basic setup consists of: conductive metal electrodes, an amplifier, an anti-vibration table, a microscope with long working distance objectives, and a Faraday cage to reduce external electrical noise. Likewise, there are mainly two types of antennal recording, both of which were developed over 50 years ago: electroantennography [3], which records electrical activity across the antennal flagellum, and single sensillum recording [7, 8], which measures the electrical activity within single sensilla. Most current methods are variations of these two techniques.

Electroantennography (EAG) is an excellent technique for quickly assessing the receptive range of an insect's antenna. EAGs are performed by measuring the change in electrical potential between distal and proximal sections of the antennal flagellum caused by olfactory stimulation. EAGs are generally believed to measure the sum of electrical potentials created by activated OSNs on the insect antenna (see review in [9]). Kaissling [10] suggested that the EAG is likely the summed potential of several OSNs lying in series, and indeed the amplitude of the response potential to pheromones in *Ostrinia nubilialis* was directly proportional to antennal length [11]. Nagai further surmised that the EAG signal is a result of the initial fast negative drop in potential caused from OSNs situated between the recording and ground electrodes combined with the slower electrotonic potential spreading from neighboring regions of the recording electrode [11]. The electrotonic spread also occurs more effectively in proximal, rather than distal directions of the antenna [12]. Nevertheless, the nature of the EAG potential has been questioned in antennal flagella whose OSNs do not lie in series (e.g., flies [6, 13]). In addition, EAG amplitude is subject to change depending on connection strength, insect vitality, and even position of the electrode (e.g., [13, 14]). Although EAGs show a concentration–response relationship with stimulus concentration [9], the EAG response should in general be treated as a qualitative, rather than quantitative indicator of olfactory reception.

For more quantitative measurement of insect olfactory response, single sensillum recording (SSR) is suggested. SSR is a form of extracellular electrophysiology, where action potentials generated by OSNs within single sensilla on the insect antenna can be measured by an electrode in contact with the extracellular receptor lymph. Individual OSNs can often be separated by action potential amplitude, which is believed to result from the diameter of the nerve cell generating the signal [15]. In this manner, detailed assessment of the sensitivity and selectivity of individual OSNs can be obtained. As OSNs expressing the same receptor type converge onto the same region of the insect brain [16, 17], single sensillum electrophysiology can be a very effective technique for mapping the receptive range of the olfactory receptor proteins expressed within the OSNs [18, 19].

From these two basic techniques, a variety of additional analyses can be performed. For example, both electroantennography (reviewed in [20]) and single-sensillum recording (reviewed in [21]) can be combined with gas chromatography to measure antennal responses to natural headspace extracts. The two techniques can also be combined for simultaneous recording [14]. Sensillum incision allows manipulation of the internal sensillum environment during SSR [22–24]. In the absence of tissue incision, diffusion [25, 26] or the recently established microinjection technique [27] can allow the study of pharmacological effects on OSN response kinetics.

In this chapter, we outline the basic methodology for establishing EAG or SSR. As electrode placement and insect preparation vary widely with chosen species (*see* included figures), we attempt a general overview of important considerations rather than a description of specialized techniques. As with all electrophysiology, effective recording requires practice and experience with the chosen organism. We consider this chapter a primer to introduce the reader to the technique. However, researchers wishing to perform these analyses for the first time are highly encouraged to practice under supervision of an experienced electrophysiologist who can give additional tips specific to the species being assessed.

2 Materials

2.1 Insect Preparation

1. Animals: Use healthy, young insects with undamaged antenna. Generally, insects older than 1 day and less than 1 week generate the best electrophysiological signals.
2. Mounting equipment: *For excised antennae*: electrode gel or glass capillary electrodes filled with hemolymph ringer (*see* Subheading 2.2). *For whole animal preparations*: microscope slide, soft dental wax or modeling clay, parafilm, double-sided tape, glass capillaries, magnetic stand, test tube clamp.
3. Other materials: For long-term preparations (particularly for single-sensillum recording), moistened cotton or paper towels can be used on the body of the preparation to maintain humidity.

2.2 Electrodes

1. Electrodes: conductive metal (usually tungsten wire, silver wire and/or two-pronged fork electrode; choice depends on procedure, *see* Subheading 3), glass capillaries with filament, laboratory tissue.
2. Electrode sharpening/etching device: *For glass*: micropipette puller (e.g., Sutter Instrument Co). *For tungsten*: approximately 10 V voltage source, electrolytic solution (e.g., saturated potassium nitrite, KNO_2), cathode (carbon electrode), manual micromanipulator.
3. Hemolymph Ringer: 6.4 mM KCl, 20 mM KH_2PO_4, 12 mM $MgCl_2$, 1 mM $CaCl_2$, 9.6 mM KOH, 354 mM glucose, 12 mM NaCl, Osmotic pressure 450, pH 6.5 after [28, 29].
4. Sensillum lymph ringer: 171.9 mM KCl, 9.2 mM KH_2PO_4, 10.8 mM K_2HPO_4, 3 mM $MgCl_2$, 1 mM $CaCl_2$, 1.5 mM HCl, 22.5 mM glucose, 25.0 mM NaCl, Osmotic pressure 475, pH 6.5 after [28].
5. Electrode holders (e.g., from Ockenfels Syntech GmbH; *see* **Note 1**).

2.3 Electrophysiological Recording Equipment

1. Air/anti-vibration table (recommended but not necessary for well-isolated environments or EAG in most cases).
2. Faraday cage (recommended but not necessary for well-isolated environments or EAG in most cases).
3. Upright compound binocular microscope with up to 1,000× magnification and as long working distance as possible (e.g., Olympus BX-51 with 4×, 10×, and 50× objectives, Fig. 1).
4. Adjustable light source (preferably both regular and epi-luminescence sources).
5. Manual (e.g., Narshige) and/or motorized micromanipulators (e.g., Märzhäuser-Wetzlar PM-10 piezo micromanipulator; *see* **Note 2**).
6. Magnetic or other appropriate mounting stands for manipulators.
7. Insulated wire with alligator clips and/or clamps for grounding.
8. Extracellular 10–100× amplifier with high impedance (10^{12} Ω [9]; e.g., Syntech universal AC/DC 10× probe will perform both EAG and single sensillum recording; *see* **Note 3**).
9. Signal acquisition and analog to digital converter with computer inputs (e.g., Syntech IDAC systems).
10. Audio output (for single sensillum recording only).
11. Desktop or laptop computer and peripherals.
12. Signal acquisition software (e.g., Autospike by Syntech).

2.4 Odor Stimulation

1. Filtered air source (e.g., compressed air with pressure regulator, *see* **Note 4**).
2. Humidifier (e.g., gas washing bottle filled with distilled water).
3. 3-way directional-control valve (e.g., Lee solenoid control valves) with controller.
4. Silicon tubing, between 6 and 10 mm I.D. (*see* **Note 5**).
5. Y or T-shaped silicon connectors.
6. 0–5 L/min adjustable gas flow meter.
7. Stimulus outlet (stainless steel or glass tube roughly 10 cm long, 6–10 mm I.D., with small hole roughly 1 cm from end large enough for Pasteur pipette tip to pass through).
8. Clamps, stand, or other holders to support outlet within 1 cm or preferably closer to antenna.
9. Disposable glass Pasteur pipettes.
10. Filter paper cut into 1 cm diameter disks.
11. Appropriate volatiles dissolved preferably in mineral oil (*see* **Note 6**).

3 Methods

3.1 Insect Preparation

3.1.1 For Excised Antennae

1. For many insects (e.g., with large and/or filiform antenna), an excised antenna can be used for recording (*see* **Notes 7** and **8**).
2. Remove the antenna at the base, making sure not to pinch or close the antennal shaft.
3. Immediately attach antenna to each side of a two-pronged electrode "fork" recording probe (base of antenna at the ground position) using a small amount of electrode gel (*see* **Note 9**, Fig. 2c). Alternatively, two glass micropipettes with Ag/AgCl electrodes can be used (*see* Subheading 3.5).
4. Make sure that the antenna is straight, but not stretched, between the two prongs/electrodes. If necessary, the antennal tip can be cut or opened if insufficient electrical contact is obtained.

3.1.2 For Whole Animal Preparations

1. Immobilize the insect's abdomen and thorax with parafilm. Smaller insects can be placed headfirst into a disposable pipette tip cut so that the antennae and eyes can protrude (*see* **Notes 10** and **11**).
2. *Option 1 (for single sensillum or small working distances)*: Fix the prostrate insect onto a microscope slide with a small amount of dental wax or modeling clay. For many insects, placement with the ventral side exposed provides the best access to antennal sensilla.
3. *Option 2 (for EAG or large working distances)*: Fix the insect upright directly under the microscope objective via magnetic stand and test tube clamp. Make sure both antennae and eyes are accessible.
4. Immobilize antenna either with a small amount of wax or clay or with a glass capillary placed gently over the antennal shaft. Do not cover antenna completely or restrict airflow to antennal surface.

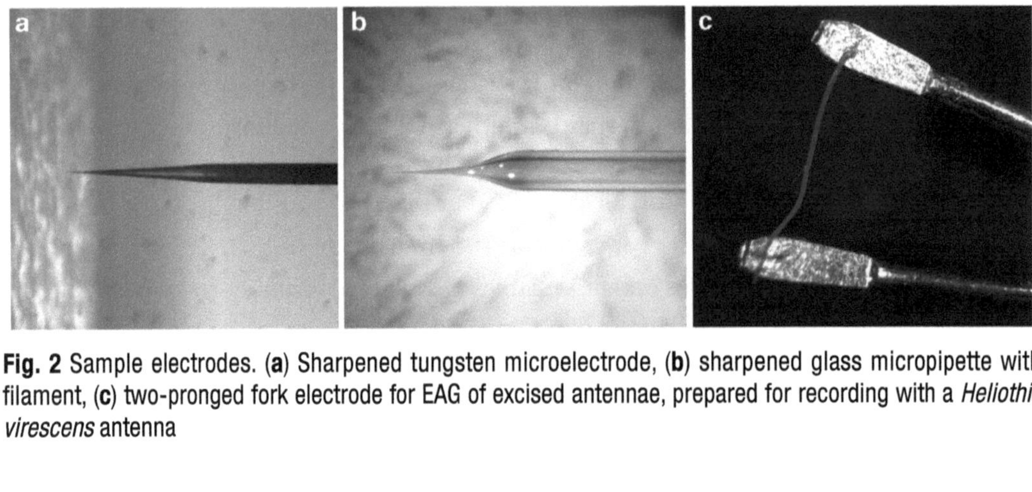

Fig. 2 Sample electrodes. (**a**) Sharpened tungsten microelectrode, (**b**) sharpened glass micropipette with filament, (**c**) two-pronged fork electrode for EAG of excised antennae, prepared for recording with a *Heliothis virescens* antenna

3.2 Electrode Preparation

1. *For Tungsten Electrodes*: Cut tungsten wire of appropriate length for electrode holder. Assemble an electrolytic sharpener by obtaining a 10 V electrical source and connecting the positive via insulated wire to a carbon cathode. Insert the cathode into a saturated potassium nitrite (KNO_2) solution and place the solution under a dissecting microscope (*see* **Note 12**). Connect the negative to the tungsten electrode placed in an appropriate holder and turn on the electrical source at lowest voltage setting. Slowly dip the electrode into the solution. Increase voltage as needed to produce bubbles within the solution when contact with tungsten is made, indicating that the nitrite is reacting with the tungsten metal. Repeated movement of the electrode into and out of the solution will produce a fine tip (approximately 1 μm tip; Fig. 2; *see* **Note 13**). Unless using a high-power objective, electrode tip size will not be apparent. Rinse electrodes with water before use (the nitrite is toxic). Make sure electrodes are completely dry before use to avoid electrical shorts. For immediate use, rinse electrode with ethanol and acetone for quick drying. Be careful to avoid touching the tip.
2. *For Silver Electrodes*: Silver wire of appropriate size for insertion into the glass micropipette shaft should be cleaned and cut to appropriate length. In most cases, the tip of the silver electrode should also be chloridized prior to use. Chloridization (i.e., electrochemical reaction with HCl) produces a silver–silver chloride (Ag/AgCl) electrode that is non-polarizable and exhibits reduced low-frequency noise. This process can be performed electrolytically in a manner similar to tungsten sharpening (*see* **step 1**). In this case, an AA battery provides enough electricity for chloridization. Connect the positive end of the battery to a carbon cathode in 0.1 N HCl, and the negative end to the silver wire itself. Leave the silver wire in the solution with the cathode for 0.5–2 h until wire turns black indicating proper chloridization (*see* **Note 14**). This process should be repeated periodically to maintain proper chloridization. Ag/AgCl electrodes require a salt bridge such as a saline solution (*see* Subheading 2.2), and should therefore be placed in glass micropipettes filled with saline.
3. *For Glass Micropipettes*: Using an appropriate Flaming-Brown or laser pipette puller, follow manufacturer's instructions to pull short tapered electrodes of approximately 2–3 MΩ resistance (*see* **Note 15**; Fig. 2b). For EAG, gently break electrode tips by lightly tapping electrode tip against a laboratory tissue.

3.3 Electrophysiological System Setup

1. Fix microscope in center of anti-vibration table (if used) or on a surface in a relatively electrically isolated area away from large electrical equipment (e.g., freezers, refrigerators). If possible, use an isolated circuit for all setup equipment.
2. Setup Faraday cage (if needed) around table and connect to central ground with grounding wire (*see* **Step 5**).

3. Place two micromanipulators (one for ground and one for recording) on either side of microscope. Place recording manipulator on same side that signal acquisition equipment will be placed.
4. Attach recording amplifier/probe/holder to recording manipulator and ground holder to ground manipulator.
5. Connect an insulated grounding wire (with clips or hooks as needed) from ground holder to a central ground plate that is earthed (*see* **Note 16**).
6. Connect the recording electrode probe/amplifier to the signal acquisition system and/or analog-digital converter.
7. Connect signal acquisition system to computer.
8. Install appropriate signal acquisition software for your chosen amplifier/system (e.g., Autospike or EAG software for Syntech systems).
9. Connect audio output to audio port on signal acquisition system or computer as needed.
10. Confirm that all signal acquisition hardware and software is functional and installed correctly. An oscilloscope can alternatively be used for system check.
11. To test the system, connect appropriate electrodes (e.g., Ag/AgCl wire in sharpened glass micropipettes filled with saline) to both ground and recording holders. Prepare a small Petri dish of saline and carefully insert both probes into the liquid. Check that a signal is obtained when both probes are in the liquid.
12. Observe the noise level. Sources of noise can be systematically isolated and reduced by connecting an insulated ground wire to various locations (e.g., microscope, Faraday cage, light source) until the noise is sufficiently reduced. Make sure to connect all ground wires to a central ground terminal and use as few grounding wires as possible. Too many wires can create ground loops and actually increase noise (*see* **Notes 17** and **18**).

3.4 Odor Stimulation Setup

1. Securely connect pressure-regulated, filtered air source to Y or T shaped connector.
2. Connect silicon tubing of appropriate size between one outlet of the connector and valve inlet.
3. Connect silicon tubing (at least 6–10 mm ID) from other outlet of connector to humidifier filled with distilled water, and then to flow meter regulated to 0.5–1.0 L/min.
4. Mount stainless steel or glass stimulus outlet so that end is as close to the recording site as possible (*see* Figs. 1, 3 and 4). Make sure that small hole used for odor presentation is easily accessible.

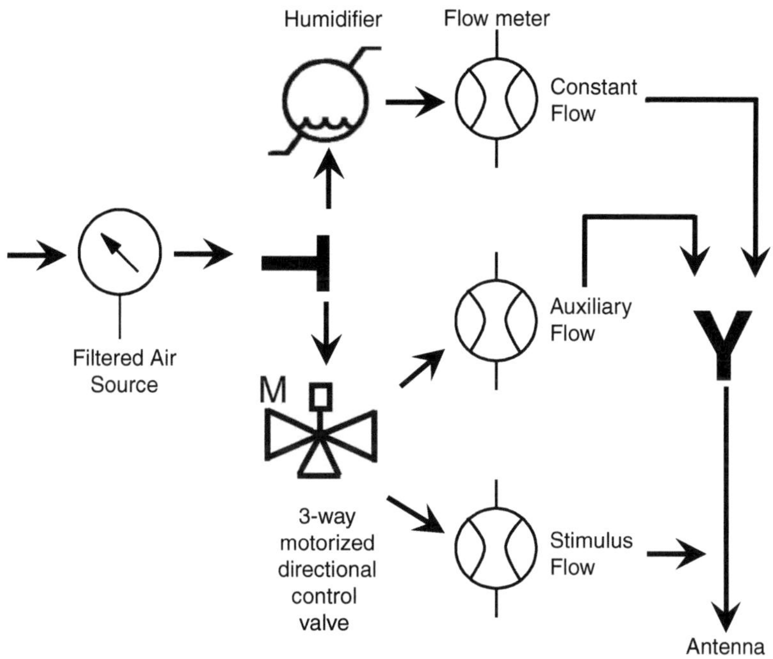

Fig. 3 Overview of odor stimulation equipment and setup. *Arrows* indicate direction of airflow

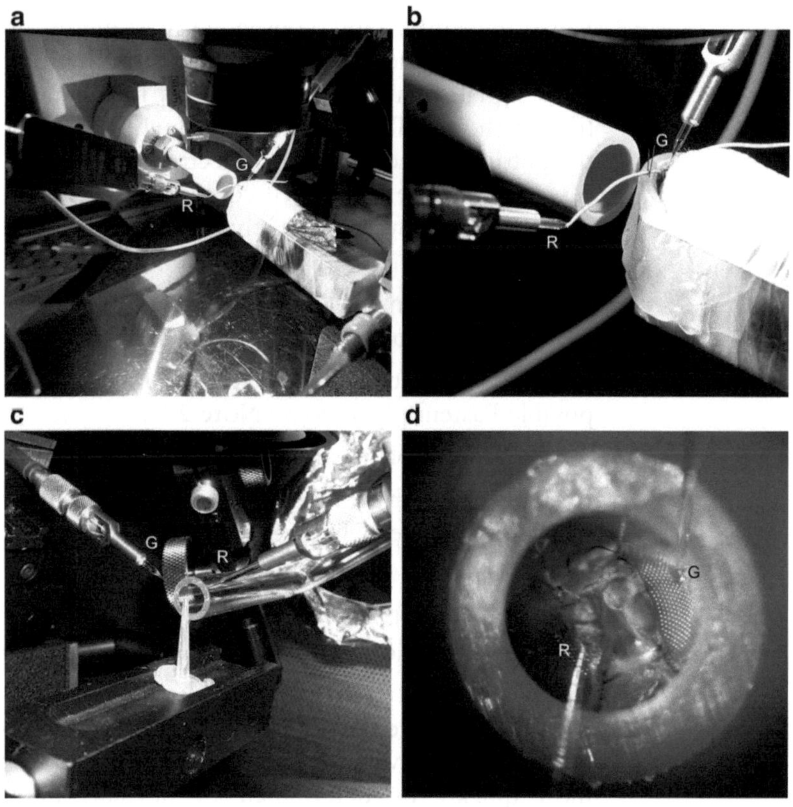

Fig. 4 Electroantennography at different scales. Electroantennography with *Manduca* (**a**, **b**, Hawk moth) and *Drosophila* (**c**, **d**, fruit fly). *G and R* indicate ground and recording electrode placement, respectively. Both preparations are undergoing gas chromatography coupled with electroantennographic detection (GC-EAD). Images kindly provided by Dr. Andreas Reinecke and Jeanine Linz

5. Connect a Y-shaped connector to opposite end of stimulus outlet, with base of Y attached to outlet (*see* **Notes 19** and **20**, Fig. 3).
6. Connect silicon tubing from flow meter to one arm of Y connector at stimulus outlet.
7. Connect silicon tubing of equal length (at least 6–10 mm ID) to both valve outlets and two flow meters. Adjust both flows equally to $\leq$0.5 L/min.
8. Connect tubing of equal length from both flow meters. Make sure tubing is long enough to reach stimulus outlet placed in **step 4**.
9. Connect tube with auxiliary flow (i.e., has airflow when valve is off) to either the other branch of the Y-tube connector at outlet or stationary, empty Pasteur pipette (*see* **Note 20**).
10. Connect tube with stimulus flow (i.e., has airflow when valve is on) to stimulus pipette (*see* **Note 21**).
11. Adjust valve opening time according to manufacturer's instructions (i.e., manually or via software). 0.5–1.0 s is a standard stimulus presentation time.
12. Valve activation (and consequently stimulus presentation) should be recorded along with the electrophysiological signal. Most signal acquisition systems allow for digital input from a trigger such a TTL pulse. Subsequent valve operation can then trigger recording of electrophysiological responses. Alternatively, the valve itself could be triggered by the signal acquisition software. The choice depends on the type of hardware used.
13. Prepare stimuli in solvent (preferably mineral oil) at behaviorally and/or ecologically appropriate dilutions.
14. Pipette 10 μl onto filter paper circles inserted into clean, disposable Pasteur pipettes (*see* **Note 22**).
15. Always prepare blank stimuli with solvent only.
16. Keep pipettes refrigerated or sealed with Parafilm or plastic tips in a ventilated area when not in use to avoid contamination or loss of stimuli.
17. Refill stimuli frequently after use. *See* [30] for details.
18. Be aware that odor concentration and timing is highly dependent on stimulus presentation [31].
19. For most accurate physiological analysis, stimulus delay should be measured between the valve and the preparation. This can most quickly be determined by portable photoionization detection of a volatile standard (e.g., [31, 32]). Stimulus delay should then be subtracted from all responses if measuring response latency.

3.5 Electro-antennography

3.5.1 For Excised Antennae

1. *For two-pronged electrode*: Make sure base of antenna is contacting ground arm of electrode (Fig. 2c). Connect electrode to appropriate holder and amplifier. In this method, a separate ground manipulator is not necessary and should be disconnected and removed to avoid additional noise. Instead, the ground wiring should connect to the ground supplied on the "fork" electrode holder.
2. *For glass electrodes*: Prepare two sharpened glass micropipettes (Fig. 2b) filled with hemolymph ringer, each containing an Ag/AgCl wire extending as far into the glass micropipettes as possible. Break tips of micropipettes with laboratory tissue until they are just large enough to envelop the antenna. Place base of antenna into ground electrode. Connect electrodes to the holders (one ground and one recording), making sure that wire has sufficient contact with holder base. Move ground electrode with antenna so that it is in field of view under the microscope. Move recording electrode until it is one antennal length from the ground electrode. Using fine forceps, gently insert tip of antenna into recording electrode, making sure to keep base inside recording electrode and leaving antenna taught, but not stretched, between both points.

3.5.2 For Whole Animal Preparations

1. Mount insect preparation, either prostrate (*Option 1*) or upright (*Option 2*) under microscope field of view.
2. Prepare and mount microelectrodes as in **step 2** for excised antenna.
3. Insert ground electrode near recording site (usually eye, Fig. 4).
4. Connect recording electrode with antennal tip (*see* **Note 23**, Fig. 4).

3.5.3 Signal Acquisition

1. Confirm contact is made between ground, antenna, and recording electrodes. A deflection should appear in the signal trace when contact is made or broken.
2. Make sure the selected amplification of the signal in the acquisition software matches the amplifier chosen (i.e., usually 10–100×).
3. Set the high pass filter to <0.5 Hz.
4. Observe the signal baseline. An unstable baseline indicates poor electrical contact with the preparation and/or excessive external noise. After confirming proper contact is made with the antenna and ground (e.g., through replacement of the electrodes), systematically test for noise as described in Subheading 3.3, **step 12**.
5. A low pass filter is not generally needed for low-frequency EAG signals.

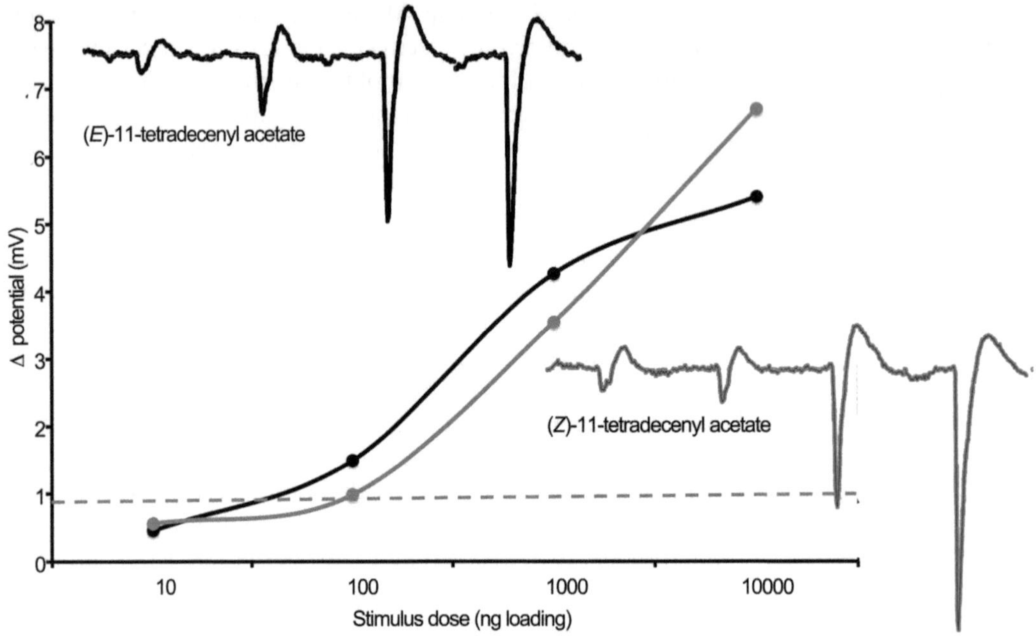

Fig. 5 Electroantennographic analysis. EAG response of an E-strain European corn borer (*Ostrinia nubilalis*) excised antenna to its two major pheromone components, *E*-11-tetradecenyl acetate (*black*) and *Z-11-tetracenyl* acetate (*gray*). Five-second EAG response traces are shown in series for four concentrations of the two components. *Line graph* depicts maximum change in potential in traces displayed, dashed line indicates blank response level

6. Present a known physiologically active stimulus to the antenna. If no specific compounds are known, an extract/headspace collection made of the insect's food or host make good first candidates.
7. An odor response is indicated by a sharp drop in baseline followed by a slow decay (Fig. 5). A rise in baseline may indicate that ground and recording electrodes are reversed. However, some compounds are known to elicit such responses.
8. A weaker (or ideally) absent drop in baseline should be observed for the blank stimulus.
9. Preferably, responses should produce >1 mV change in baseline potential. However, signal strength varies with preparation and species. For this reason, results should always be reported in comparison to a standard response so that inter-preparation differences due to age, health, and mounting can be avoided (*see* **Note 24**).
10. Record stimulus responses for at least 1–2 s before and after response (at least 5 s total) to allow for accurate measurement of response.
11. Responses can be measured as change in potential between baseline and maximum deflection. Time to maximum response

and/or return to baseline are also good parameters to measure.

12. Responses should be reported as % difference from blank or standard to avoid inter-individual differences. However, raw mV change from baseline can also be included with standard deviations.
13. Cut antennal preparations can generally be assessed for <2 h. Whole animal preparations can survive for several hours if humidity is maintained and antenna is relatively undamaged.

3.6 Single-Sensillum Electrophysiology

3.6.1 For Sensillum Penetration Recording

1. Make sure that antenna is well stabilized and immobilize with wax, tape, or microcapillary (*see* Subheading 2.1).
2. Mount insect preparation on microscope stage so that antenna is roughly parallel to and facing the recording electrode. This positioning generally provides the best resistance for electrode penetration of the sensillum cuticle.
3. Insert ground electrode as close to recording site as possible (often the eye; *see* **Note 25**, Fig. 6).
4. Prepare an electrolytically sharpened tungsten electrode or glass microelectrode (with Ag/AgCl wire) with approximately 1 μm tip. Before recording, clarify under high magnification that a sufficiently sharp tip has been created to penetrate the sensillum cuticle (*see* Subheading 3.2, **Note 26**, and Fig. 2a).
5. Mount recording electrode on holder.
6. Check that signal acquisition equipment is on and make sure that audio output is functional (*see* **Note 27**).
7. Slowly lower electrode to antenna. To avoid damaging the electrode, it is recommended to first move the electrode at least 6–10 mm directly above the antenna and slowly lower the electrode to the preparation, rather than moving it horizontally at the level of the preparation.
8. Move electrode a few μm in front of targeted sensillum. Slowly and carefully place electrode in base and/or shaft of sensillum (*see* **Note 28**, Fig. 6). Careful and slow movement is required to avoid damaging the electrode tip. Some thick or large sensilla may require extra force (such as via piezoelectric step action) to allow cuticular penetration.
9. Generally, initial contact with cuticle will cause a significant increase in signal noise, which also indicates that preparation is both alive and properly grounded.

3.6.2 For Cut Sensillum Recording

1. Make sure that the antenna is stabilized and immobilized with wax, tape, or microcapillary (*see* Subheading 2.1).
2. Prepare a glass microelectrode with Ag/AgCl wire in sensillum lymph ringer. Tip should be <10–15 μm diameter.

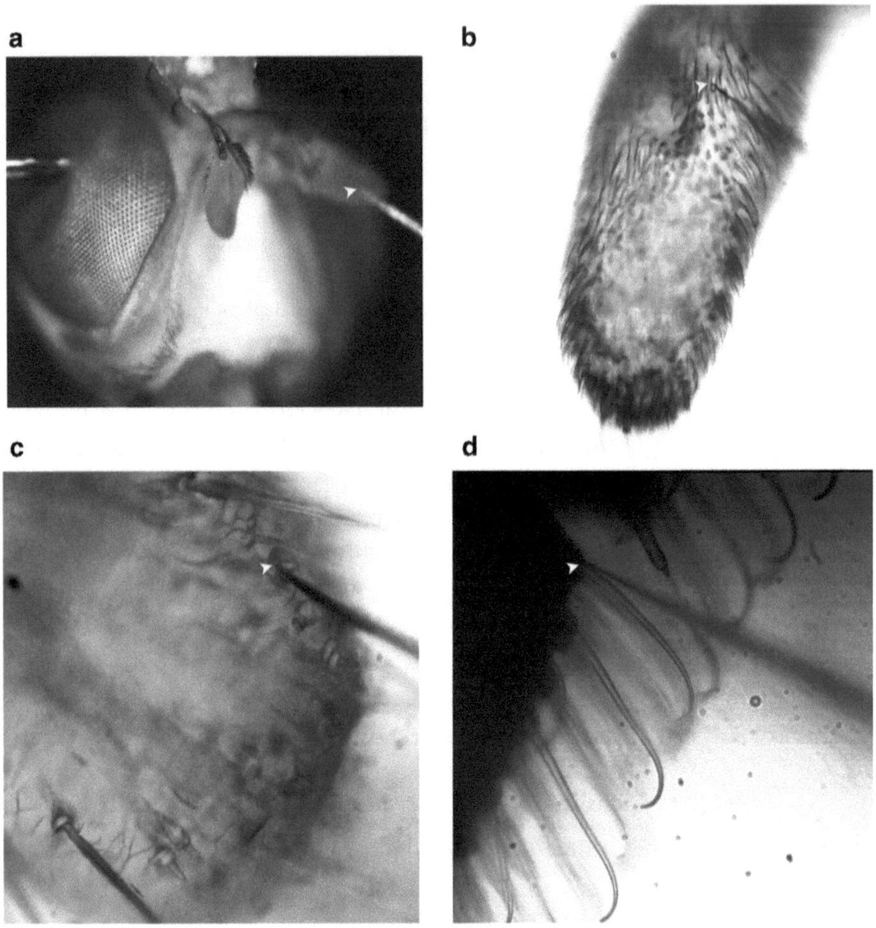

Fig. 6 Single sensillum electrophysiology across insect taxa. Recording sites indicated with *white arrows*. (**a**) Ground and recording electrode placement in *Rhagoletis pomonella* (Tephritid fruit fly), (**b**) SSR in an ab2 *Drosophila melanogaster* basiconic sensillum, (**c**) SSR in a *Tetropium fuscum* basiconic sensillum (brown spruce longhorn beetle, courtesy of Colin MacKay), (**d**) SSR in a male *Ostrinia nubilalis* trichoid pheromone sensillum

3. Mount recording electrode on holder (*see* **Note 29**).
4. Carefully cut a few targeted sensilla using fine forceps [33], glass knives [34], or vibrating glass stylets [35, 36].
5. To prevent sensillum from sealing closed, Vaseline can be used on forceps, knives, or stylet. Polyvinylpyrrolidone can also be added to the electrode ringer to prevent leakage (*see* ref. 37 for details).
6. Immediately move recording electrode so that it sheathes a single cut sensillum tip. Do not cover sensillum completely.
7. If no electrical contact is apparent, make sure sensillum and electrode are not sealed and in contact. In addition, be careful that electrode tip is not too large.

3.6.3 Signal Acquisition

1. Proper contact with the extracellular OSN space will reveal spontaneous action potential activity. In some preparations (particularly pheromone sensilla), very low spontaneous activity will occur.
2. If significant noise is present, first confirm that appropriate contact is made with internal sensillum lymph. If necessary, reposition electrode. Additional noise may indicate improper grounding (*see* Subheading 3.3).
3. Set the high pass filter between 30–100 Hz and the low pass at 1,000–3,000 Hz. Filters can be adjusted to reduce noise, and an additional 50/60 Hz filter can be used to reduce noise from electronic equipment (*see* **Note 30**).
4. Action potentials should be easily distinguished from background noise (Fig. 7). Electrode replacement and/or contacting other sensilla can improve signal strength.
5. Many insects house more than one sensory neuron in a single sensillum. These neurons can often be distinguished by spontaneous top-to-top action potential amplitude, which should remain relatively consistent in a stable recording (Fig. 7). Nevertheless, some sensilla housing many neurons (e.g., ants, locusts) or sensory neurons of similar size may be indistinguishable. In such cases, only "sensillum" response, rather than single cell response, should be reported.
6. Present a known physiologically active stimulus to the antenna. If no specific compounds are known, an extract/headspace collection made of the insect's food or host make good first candidates.
7. A response is indicated by a sharp increase (or decrease) in action potential activity.
8. A weaker (or ideally) absent change in spontaneous activity should be observed for the blank stimulus.
9. Record stimulus responses for at least 1–2 s before and after response (at least 5 s total) to allow for accurate measurement of change in spike activity.
10. Use appropriate spike sorting software to separate sensory neurons by amplitude (Fig. 7).
11. Responses can be measured in several ways. Responses can be recorded as raw spike count (in Hz) during stimulation, change in spike count with spontaneous activity, or through a peristimulus time histogram (instantaneous spike activity over time).
12. Whole animal preparations can survive for several hours (or even days) if humidity is maintained and antenna is relatively undamaged. Single sensillum contacts can also be maintained for several hours.

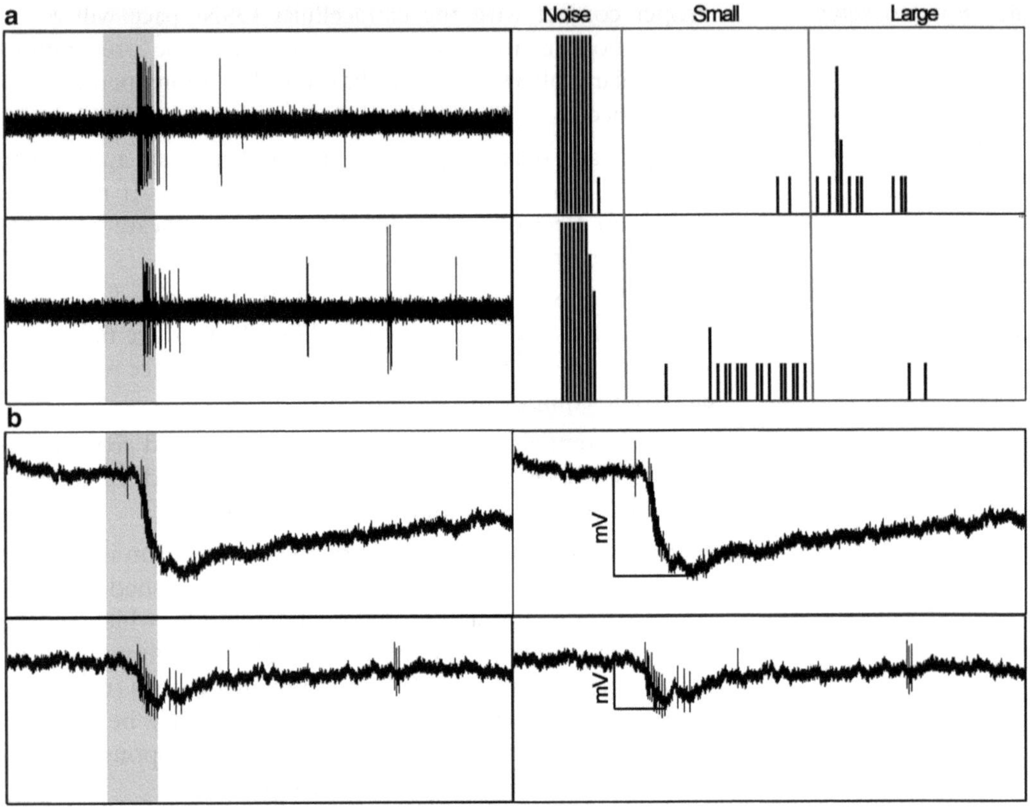

Fig. 7 Single sensillum electrophysiology analysis. Action potential (**a**) and receptor potential (**b**) activity in a single trichoid pheromone sensillum of a male silkworm moth, *Bombyx mori*. Traces were measured in AC mode (**a**, 30/3,000 Hz filter settings, 2 mV display) and DC mode (**b**, DC/3,000 Hz filter settings, 5 mV display) using an Ag/AgCl glass electrode penetrating the base of the sensillum near the cell bodies. (*Left*) Five-second traces with stimulus window as *gray bar*. Top traces in each panel show the response to the main pheromone component, bombykol, and bottom traces to the minor component, bombykal. (*Right, top*) Separation of the large, bombykol-responsive OSN and smaller, bombykal-responsive OSN by amplitude. (*Right, bottom*) measurement of receptor potential from beginning of stimulus onset to maximum change in potential (mV)

4 Notes

1. Holders should be chosen to interface properly with probe and amplifier used. Syntech (http://www.syntech.nl) provides a selection of compatible holders and probes/amplifiers.
2. For EAG and ground electrode placement, manual micromanipulators are sufficient. For single sensillum recording, a motorized micromanipulator is strongly recommended for positioning the recording electrode.
3. For best signal to noise ratio, amplifier should be placed as close to recording site as possible.

4. An "all-in-one" stimulus system can be purchased that generates, regulates, and controls airflow (e.g., Syntech stimulus controller). Such a system can be advantageous for setups that do not have access to compressed air.
5. Make sure to use tubing made of a relatively inert material, such as silicon. Nevertheless, all tubing will require occasional solvent cleaning (e.g., ethanol, water) and heat drying at 80–100 °C for 4 h (as allowed by tubing material) to prevent contamination.
6. Odor cartridge preparation and maintenance is essential to avoid nonspecific responses due to contamination, and/or inaccurate concentration–response measurement due to depleted odor sources.
7. Excised antennae provide rapid and efficient screening for electroantennography. However, the preparation is short-lived (<2 h), and is not generally practical for long-term single-sensillum recording (but see, e.g., [22]).
8. Many insects such as moths are amenable to excised antennal recording. If quick qualitative results are needed on a species new to electrophysiology, excised antenna can be tested before attempting whole animal mounting.
9. Be careful not to add too much electrode gel to an excised antennal prep, which can reduce contact with the electrode. Add only enough gel to ensure proper adhesion to the electrode surface.
10. Ensure that all appendages are secure. Leg, wing, and obviously antennal movement can cause significant noise during recording.
11. If using disposable pipettes to immobilize the animal, make sure to cut the pipette tip only enough to allow the head to protrude and not the prothorax. In many cases, only part of the head needs be exposed to allow placement of both recording and ground electrodes. Partial head exposure also helps prevent movement.
12. Generally, it is easier to place the nitrite solution horizontally under the microscope in, for example, a plastic syringe. 10 % sodium nitrite ($NaNO_2$) or 1 M potassium hydroxide (KOH) can also be substituted for potassium nitrite as the electrolytic solution.
13. Electrode sharpening takes practice to acquire the most efficient technique. The most favorable shape for an electrode is pencil-like—a broad electrode coming to a sharp and fine point at the end (Fig. 2). This creates a strong and sturdy electrode for penetration into the insect cuticle. This shape can be made by slow and deep movements at the beginning of sharpening

to begin to narrow the tip, followed by small and quick movements just at the tip to create a short and fine point.

14. Silver wires chloridized for too long will become brittle.
15. In general, short electrode tips are more sturdy for both single sensillum and EAG analyses. However, be careful not to make electrodes too short as their tips will quickly become too large following inevitable breakage from cuticular contact.
16. Anti-vibration tables are generally grounded internally, so central connection to the tabletop is sufficient. If using a bench top or table, connect the grounding wire to a stainless steel grounding plate that is then earthed.
17. Essentially, only one grounding wire is needed—from the ground electrode to the earth. All other wires are simply to reduce electrical noise created by electronic and metal equipment. Be very frugal with the use of grounding wires and in general, a "working rig" should not need re-grounding unless new equipment is added.
18. Despite best efforts, some preparations do not produce sufficient signal-to-noise ratio due to unforeseen physiological or environmental conditions. If no signal is obtained, first confirm that a complete circuit from the recording electrode through the animal to the ground is present. Then, clean and dry all electrode holders to remove potential salt, water or debris buildup. After retesting setup with saline (*see* Subheading 3.3, **step 11**), confirm that animal vitality is maintained. If there is still no signal, establish another animal preparation.
19. To avoid backflow, it is best to orient the Y connector so that the two airflows proceed in the same direction (as in both short arms of the Y-connector). Also, the main airflow through the humidifier should be much higher than the auxiliary flow to avoid backflow during valve operation.
20. As an alternative to the Y connector, an additional hole can be drilled in the stimulus outlet itself next to the stimulus inlet hole for the auxiliary airflow to pass through an empty Pasteur pipette fixed permanently at the outlet. This, in fact, produces a better auxiliary flow as it replicates the stimulus flow more precisely, and is closer to the stimulus inlet. However, as in **Note 19**, make sure to angle the pipette so that its airflow proceeds in the same direction as the main flow, and is at a lower flow rate (generally ≤ 0.5 L/min).
21. Be aware that the line proceeding to the stimulus cartridges can quickly become contaminated. To avoid this, disposable connectors can be made from 1,000 μL disposable plastic pipette tips cut slightly at tip with base connected to the stimulus

pipette base. In addition, be careful not to allow any chemicals to contact the stimulus pipette base. *See* also **Note 5**.

22. Pipette preparation and maintenance is very important for accurate physiological assessment.
23. For many flies and moths, the antennal tip can be placed into the recording electrode with no additional tissue incision. However, for insects with particularly thick cuticle (such as beetles), a small incision can be made at the antennal tip to allow sufficient contact with internal lymph for recording electrical potentials.
24. There is no rule for what is a "sufficient EAG signal". The major determining factor is signal-to-noise ratio. If your stimulus response is clearly distinguishable from both noise and the blank, then you have a sufficient signal.
25. Generally, it is best to use the same material for both ground and recording electrodes. Thus, if recording with tungsten, use a tungsten electrode as the ground.
26. Electrode shape and sharpness depend heavily on the preparation. Animals with large and thick sensilla may require much larger and thicker electrode tips. Generally, proper electrode shape is determined through experience with the preparation.
27. Many researchers agree that it is very beneficial to hear the sensory neuron activity during single sensillum recording. Audio output from signal acquisition allows the researcher to focus on the preparation during electrode placement rather than a computer screen. Proper contact will produce the familiar "popping" sound caused by spontaneous neural activity.
28. Obtaining electrical contact with the extracellular sensory neuron space initially requires tremendous patience and practice. Prepare for many bent and damaged electrodes until you obtain the fine motor skills required for accurate and precise electrode placement. In time, you will be able to obtain many contacts without requiring electrode sharpening.
29. Cut tip and sensillum penetration require amplifiers with different impedances. In the authors' experience, Syntech universal probes are not suitable for cut sensillum recording.
30. As an alternative to measuring action potential activity, the "receptor potential" (i.e., the negative deflection of the transepithelial potential of the sensillum; [38]) can be measured using an Ag/AgCl electrode in a sharpened glass micropipette to contact the internal sensillum lymph (Fig. 7b). To measure receptor potentials, the high pass filter should be set to DC mode.

Acknowledgments

This work was supported by the Max Planck Society. Special thanks to Dr. Andreas Reinecke, Jeanine Linz, and Colin MacKay for images.

References

1. Butenandt A, Beckmann R, Stamm D, Hecker E (1959) Über den sexual-lockstoff des seidenspinners *Bombyx mori.* Z Naturforshg 14b:283–284
2. Schneider D (1957) Electrophysiological investigation on the antennal receptors of the silk moth during chemical and mechanical stimulation. Experientia 13:89–91
3. Schneider D (1957) Elektrophysiologische untersuchungen von chemorezeptoren und mechanorezeptoren der antenne des seidenspinners *Bombyx mori*-L. Z Vergl Physiol 40:8–41
4. Keil TA (1999) Morphology and development of the peripheral olfactory organs. In: Hansson BS (ed) Insect olfaction. Springer-Verlag, Berlin, p 457
5. Ochieng SA, Hallberg E, Hansson BS (1998) Fine structure and distribution of antennal sensilla of the desert locust, *Schistocerca gregaria* (Orthoptera: Acrididae). Cell Tissue Res 291:525–536
6. Ignell R, Hansson B (2004) Insect olfactory neuroethology: an electrophysiological perspective. In: Christensen TA (ed) Methods in insect sensory neuroscience (frontiers in neuroscience). CRC Press, Boca Raton, p 435
7. Schneider D, Boeckh J (1962) Rezeptorpotential und nervenimpulse einzelner olfaktorischer sensillen der insektenantenne. Z Vergl Physiol 45:405–412
8. Boeckh J (1962) Elektrophysiologische untersuchungen an einzelnen geruchsrezeptoren auf den antennen des Totengrabers (Necrophorus, Coleoptera). Z Vergl Physiol 46:212–248
9. Roelofs WL (1984) Electroantennogram assays: rapid and convenient screening procedures for pheromones. In: Hummel HE, Miller TA (eds) Techniques in pheromone research. Springer-Verlag, Berlin, p 464
10. Kaissling K-E (1971) Insect olfaction. In: Beidler LM (ed) Handbook of sensory physiology, vol IV, Chemical senses 1, olfaction. Springer-Verlag, Berlin, p 518
11. Nagai T (1981) Electroantennogram response gradient on the antenna of the European corn borer, *Ostrinia nubilalis.* J Insect Physiol 27:889–894
12. Nagai T (1983) Spread of local electroantennogram response of the European corn borer, *Ostrinia nubilalis.* Pestic Biochem Physiol 19:291–298
13. Crnjar R, Scalera G, Liscia A, Angioy A-M, Bigiani A, Pietra P, Barbarossa IT (1989) Morphology and EAG mapping of the antennal olfactory receptors in *Dacus oleae.* Entomol Exp Appl 51:77–85
14. Nagai T (1983) On the relationship between the electroantennogram and simultaneously recorded single sensillum response of the European corn borer *Ostrinia nubilalis.* Arch Insect Biochem Physiol 1:85–91
15. Hansson B, Hallberg E, Löfstedt C, Steinbrecht R (1994) Correlation between dendrite diameter and action potential amplitude in sex pheromone specific receptor neurons in male *Ostrinia nubilalis* (Lepidoptera: Pyralidae). Tissue Cell 26:503–512
16. Vosshall LB, Wong AM, Axel R (2000) An olfactory sensory map in the fly brain. Cell 102:147–159
17. Gao Q, Yuan B, Chess A (2000) Convergent projections of *Drosophila* olfactory neurons to specific glomeruli in the antennal lobe. Nat Neurosci 3:780–785
18. Hallem E, Ho M, Carlson J (2004) The molecular basis of odor coding in the *Drosophila* antenna. Cell 117:965–979
19. Hallem E, Carlson J (2006) Coding of odors by a receptor repertoire. Cell 125:143–160
20. Struble DL, Arn H (1984) Combined gas chromatography and electroantennogram recording of insect olfactory responses. In: Hummel HE, Miller TA (eds) Techniques in pheromone research. Springer-Verlag, Berlin, p 464
21. Wadhams LJ (1984) The coupled gas chromatography-single cell recording technique. In: Hummel HE, Miller TA (eds) Techniques in pheromone research. Springer-Verlag, Berlin, p 464
22. Kaissling K, Keil T, Williams J (1991) Pheromone stimulation in perfused sensory hairs of the moth *Antheraea polyphemus.* J Insect Physiol 37:71–75
23. van den Berg M, Ziegelberger G (1991) On the function of the pheromone binding protein

in the olfactory hairs of *Antheraea polyphemus*. J Insect Physiol 37:79–85
24. Pophof B (2004) Pheromone-binding proteins contribute to the activation of olfactory receptor neurons in the silkmoths *Antheraea polyphemus* and *Bombyx mori*. Chem Senses 29:117–125
25. Laughlin J, Ha T, Jones D, Smith D (2008) Activation of pheromone-sensitive neurons is mediated by conformational activation of pheromone-binding protein. Cell 133: 1255–1265
26. Flecke C, Dolzer J, Krannich S, Stengl M (2006) Perfusion with cGMP analogue adapts the action potential response of pheromone-sensitive sensilla trichoidea of the Hawkmoth *Manduca sexta* in a daytime-dependent manner. J Exp Biol 209:3898–3912
27. Olsson SB, Getahun MN, Wicher D, Hansson BS (2011) Piezo controlled microinjection: an in vivo complement for in vitro sensory studies in insects. J Neurosci Methods 201:385–389
28. Kaissling K-E, Thorson J (1980) Insect olfactory sensilla: structural, chemical and electrical aspects of the functional organization. In: Sattelle DB, Hall LM, Hildebrand JG (eds) Receptors for neurotransmitters, hormones, and pheromones in insects: Proceedings of the workshop in neurotransmitter and hormone receptors in insects held in Cambridge, 10–12 September, 1979. Elsevier/North Holland Biomedical Press, Amsterdam, p 310
29. Kaissling K-E (1995) Single unit and electroantennogram recordings in insect olfactory organs. In: Spielman AI, Brand JG (eds) Experimental cell biology of taste and olfaction: Current techniques and protocols. CRC Press, Boca Raton, p 464
30. Andersson MN, Schlyter F, Hill SR, Dekker T (2012) What reaches the antenna? How to calibrate odor flux and ligand-receptor affinities. Chem Senses 37:403–420
31. Vetter RS, Sage AE, Justus KA, Cardé RT, Galizia CG (2006) Temporal integrity of an airborne odor stimulus is greatly affected by physical aspects of the odor delivery system. Chem Senses 31:359–369
32. Olsson SB, Kuebler LS, Veit D, Steck K, Schmidt A, Knaden M, Hansson BS (2011) A novel multicomponent stimulus device for use in olfactory experiments. J Neurosci Methods 195:1–9
33. Kaissling K-E (1974) Sensory transduction in insect olfactory receptors. In: Jaenicke L (ed) Biochemistry of sensory functions. Springer, Berlin, p 641
34. van der Pers JNC, den Otter CJ (1978) Single cell responses from olfactory receptors of small ermine moths to sex-attractants. J Insect Physiol 24:337–343
35. Gödde J (1989) Vibrating glass stylets: Tools for precise microsurgery on cuticular structures. J Neurosci Methods 29:77–83
36. Hillier NK, Kleineidam C, Vickers NJ (2006) Physiology and glomerular projections of olfactory receptor neurons on the antenna of female *Heliothis virescens* (Lepidoptera: Noctuidae) responsive to behaviorally relevant odors. J Comp Physiol A 192:199–219
37. Kaissling K, Meng L, Bestmann H (1989) Responses of bombykol receptor-cells to (Z, E)-4,6-hexadecadiene and linalool. J Comp Physiol A 165:147–154
38. Kaissling K (1986) Chemo-electrical transduction in insect olfactory receptors. Annu Rev Neurosci 9:121–145

Chapter 12

Calcium Imaging of Pheromone Responses in the Insect Antennal Lobe

Susy M. Kim and Jing W. Wang

Abstract

Calcium imaging is a powerful technique that permits the visual monitoring of neural responses to pheromones and other odors in large ensembles of neurons. Here, we describe a method that permits the monitoring of *Drosophila* antennal lobe responses to odors using the genetically encoded calcium monitor GCaMP.

Key words *Drosophila melanogaster*, Antennal lobe, *cis*-Vaccenyl acetate (cVA), GCaMP, Olfactory system, Calcium imaging, Pheromone

1 Introduction

Elucidating how neural circuits process pheromone information is crucial towards understanding the role of chemical communication in the social interactions between conspecific insects. While electrophysiological methods permit monitoring of neural activity with unparalleled resolution, they are labor intensive and not amenable to monitoring large groups of specific neuronal types. Continuous technical improvements of genetically expressed calcium sensors and other activity reporters in recent years [1–3], however, bypass some of these problems. A genetic approach allows one to easily target calcium sensors to specific populations of neurons [4, 5], thus making calcium imaging a useful approach for monitoring pheromone responses at different levels of the fly brain.

Two-photon calcium imaging of odor responses in the antennal lobe is a useful technique in the study of how odor information is processed at this second relay in the insect nervous system [4]. In calcium imaging studies of antennal lobe odor processing in *Drosophila melanogaster*, a genetically encoded calcium indicator such as GCaMP [2] is typically expressed in either the odorant receptor neurons or the projection neurons of this circuit [6].

Kazushige Touhara (ed.), *Pheromone Signaling: Methods and Protocols*, Methods in Molecular Biology, vol. 1068,
DOI 10.1007/978-1-62703-619-1_12,

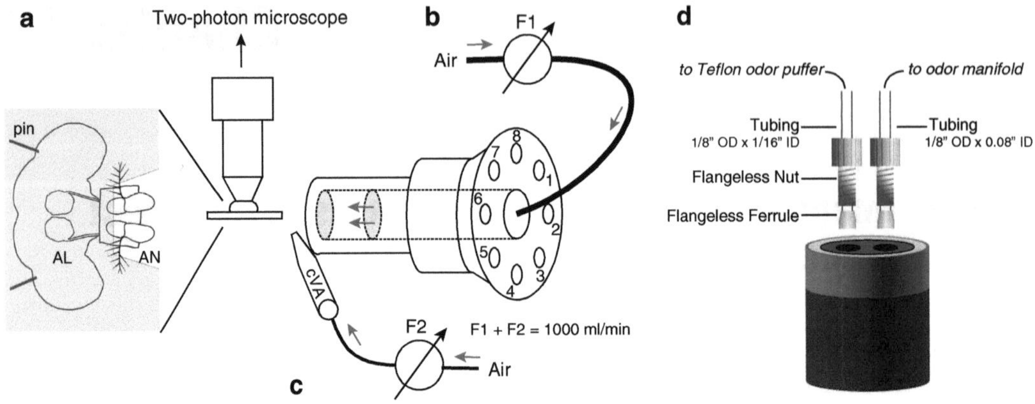

Fig. 1 Schematic diagram depicting *Drosophila* antennal lobe preparation and odor delivery system. (**a**) Close-up view of antennal lobe prep. Note that the antennae are exposed to the open air while the rest of the brain is submerged in agarose. A rectangular coverslip is placed on top of this prep and lies between the objective and the brain. (**b**) Teflon odor puffer carrying humidified airstream. The total airflow is kept constant at 1,000 mL/min. (**c**) Truncated glass Pasteur pipette containing a small slip of filter paper with 1 μL of cVA. The solenoid valve controller diverts a small percentage of the total airflow through the glass pipette containing cVA upon receipt of a signal from the stimulator box. (**d**) Vacuum fittings for odor delivery system consist of inert tubing, a flangeless ferrule, and a flangeless nut. As the nut is screwed into threaded fittings (¼-28) in the bottle cap, the ferrule seals the vacuum line at the juncture

Other methods for monitoring neural activity such as synaptopHluorin have also been successfully used in this preparation [3, 7]. In this technique, the fly brain is dissected out from the rest of the head capsule with the antennae and the antennal nerve connections to the antennal lobe still intact. Highly volatile and relatively inexpensive chemical odorants are typically delivered to the intact antennae via odor streams redirected through bottles containing odorant [6–9] (Fig. 1). However, many important insect pheromones such as 11-*cis* vaccenyl acetate (cVA) exhibit lower volatility than many chemical odorants due to relatively high molecular weights (>300 g/mol). In addition, these lipids must be purified and therefore can be expensive reagents. Standard odor delivery techniques, therefore, must be modified to accommodate these limitations. We describe here a simple system for delivering odor streams of pheromone using minute quantities of the pheromone cVA.

2 Materials

2.1 Antennal Lobe Preparation Materials

1. Adult hemolymph-like (AHL) saline: 108 mM NaCl, 5 mM KCl, 2mM $CaCl_2$, 8.2 mM $MgCl_2$, 4 mM $NaHCO_3$, 1 mM NaH_2PO_4, 5 mM trehalose, 10 mM sucrose, 5 mM HEPES, pH 7.5, 265 mOsm. Weigh and transfer to a 1-L graduated flask:

6.312 g NaCl.

0.373 g KCl.

0.337 g $NaHCO_3$.

0.120 g NaH_2PO_4.

1.892 g Trehalose·$2H_2O$.

3.423 g Sucrose.

1.192 g HEPES.

Add 8.2 mL of 1 M $MgCl_2$. Add deionized and ultrafiltered (D.I.U.F.) water (Fisher W2-20) to a volume of 990 mL. Mix and adjust pH with 1 M NaOH. Transfer to a 1 L volumetric flask and make up to 1 L with water. Filter sterilize 500 mL to get calcium-free AHL saline. Add 1 mL of 1 M $CaCl_2$ to the remaining solution and filter sterilize to get AHL saline with calcium. Store at 4 °C (*see* **Note 1**).

2. 2.5 % Agarose solution: Weigh out 0.5 g low melting point agarose (Type VII-A; Sigma A0701) and dilute with 20 mL AHL saline containing 2 mM Ca^{2+}. Heat agarose solution in microwave just until agarose is completely dissolved. Keep the solution warm and in a liquid state on a hot plate set to 37 °C.
3. Two #55 fine forceps (Dumont).
4. Scalpel.
5. Tungsten wire pins: California Fine Wire Company, Grover Beach, CA 93433.

   Tungsten 99.95 %, diameter = 0.001 in.

   Tungsten 99.95 %, diameter = 0.0005 in.
6. Dissection dish with Sylgard layer: Prepare Sylgard according to instructions. Fill small petri dish (Falcon, 35 × 10 mm) with approximately 3–5 mm layer of Sylgard and let it cure for 24–48 h until hardened.
7. Glass slide (25 × 75 × 1 mm) with small square (~5 mm × 5 mm × 2 mm) of Sylgard placed near one of the edges.
8. Coverslip 11 × 22 mm, 0.13–0.17 mm thick (Thomas Scientific).

2.2 Odor Delivery System

The odor delivery system described here permits quantitative control of odor stimulation by diluting saturated vapor concentration up to a 1:1,000 ratio (Fig. 1). Two independent humidified airstreams are generated. One airstream enters the solenoid valve system, which can be switched from an empty vial (air) to the pheromone odor cartridge (Fig. 2). The constant airflow before, during, and after the 2-s odor stimulation ensures minimum mechanical change for odor stimulation. The other airstream is used to dilute the saturated vapor. The sum of flow rates for the

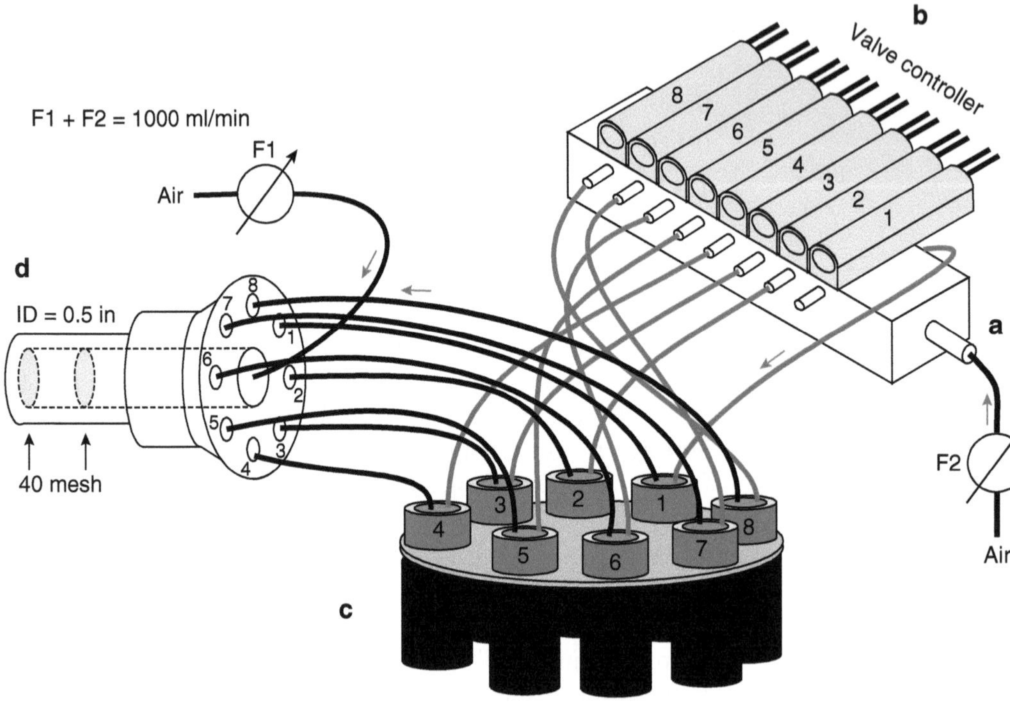

Fig. 2 Schematic diagram depicts the flow of air from the (**b**) manifold/solenoid valve controller, (**c**) glass vials containing odor samples, and (**d**) the odor puffer. A small fraction of the total airstream flows into the manifold through a side port (**a**). The bulk of the total airstream flows through the central port in the odor puffer (**d**). Upon receipt of a signal from the stimulator box, the solenoid valve controller (**b**) switches off the flow of air through valve 1/bottle 1 which is empty and switches airflow into selected valves (2–8) or through the glass pipette containing cVA. Air flows from selected open valves in the manifold (**b**) into bottles containing odor (**c**). Air containing odor is then sent into the odor puffer and delivered to the antennal lobe prep

two airstreams is kept constant at 1,000 mL/min. When a new odor is used in the system all tubing and odor vials downstream of the solenoid valves must be replaced to eliminate any potential contamination from previously used odors.

1. 2 port bottle cap (SV-435, Western Analytical, CA).
2. Flangeless Fitting for 1/8" OD tubing, P-300× ETFE ferrule (Western Analytical).
3. Delrin Nut, P-305 flangeless nut (Western Analytical).
4. 1/8″ OD × 1/16″ ID Teflon Tubing from bottlecap to odor puffer (PTFE2062, Western Analytical).
5. 1/8″ OD × 0.08″ ID Teflon Tubing from bottlecap to manifold (FEP2080, Western Analytical).
6. Custom manifold (*see* **Note 2**).
7. Odor puffer: 7-Port PTFE Manifold with 6 ¼–28 ports and 1 ¼″ NPT port (MA-014, Western Analytical; *see* **Note 3**).

8. Pipe Hex Nipple, ½″ MNPT × ½″ MNPT (PN-008, Western Analytical) with steel mesh placed at both openings (Fig. 2).
9. Stimulator box (Grass Instruments).
10. Custom eight-channel solenoid valve controller system.
11. *cis*-vaccenyl acetate (neat, Cayman).
12. Whatman filter paper #8.
13. Pasteur glass pipette (5 ¾″) truncated at both ends and fire polished such that ~1 cm of the tip and 1 ½″ of the barrel remain.

3 Methods

Carry out all procedures at room temperature unless otherwise specified.

3.1 Antennal Lobe Dissection

1. Anesthetize fly on ice or with CO_2 (*see* **Note 4**).
2. Decapitate fly with scalpel.
3. Transfer head with paintbrush or forceps and place with anterior side facing upwards on the dissection dish with antennae positioned in front.
4. Pin down proboscis with large tungsten pin (dia. 0.001 in.).
5. Cover head with cold calcium-free saline (*see* **Note 5**).
6. Remove air bubbles by gently coaxing them away from the surface of the head with the sides of forceps. This prevents the brain from floating upwards towards the surface of the saline during the dissection.
7. With a pair of forceps in either hand, begin by inserting the tips of both forceps under one corner of the cuticle plate between the eyes. Hold the cuticle firmly with one set of forceps and use the other pair to tear away pieces of the cuticle and air sacs away from the brain (*see* **Note 6**). Leaving the proboscis and antennae intact, continue peeling off the outer cuticle and compound eyes. The brain itself can be clearly distinguished from the outside cuticle by its opaque and white appearance.
8. Flip the head and reattach the large pin through the proboscis such that its posterior aspect now faces upwards. Remove the back panel of cuticle attached to posterior side of the brain.
9. Position the dissection dish so that the proboscis points away from you. Gently pull the brain back towards yourself. Observe the white connective tissue and nerves lying in between the brain and proboscis. Begin gently clearing away this material by pulling it away with your forceps. Utmost caution, however, must be taken to avoid severing the nerves connecting the antennae to the brain.

10. Stop once the nerves connecting the antennae to the brain become visible. These should be located within a short distance on either side of the brain's midline and must be intact in order for information about odors detected by the antennae to reach the antennal lobe in your experiments.
11. A slender pair of motor nerves may also be visible even closer to the brain's midline than the antennal nerves. These may be severed at this time with your forceps to reduce spontaneous proboscis muscle contractions.
12. Remove the pin holding down the proboscis on the Sylgard plate. Gently suction up and transfer the brain with a glass pipette to a slide with square piece of Sylgard. The brain should rest on the Sylgard in a drop of saline (*see* **Note 7**).
13. Arrange the brain such that its anterior aspect is facing upwards and the antennae are pointing out towards the edge of glass slide (*see* **Note 8**).
14. Replace the calcium-free saline with regular saline containing 2 mM calcium.
15. Draw up agarose solution in glass pipette. Make a thin ring of agarose around brain prep to create a well. Then cover brain with agarose. Mix the agarose and saline by gently pipetting the solution in and out of the glass pipette until it is uniform (*see* **Note 9**).
16. Place the slide in a humidified chamber [a petri plate with a wet kimwipe] to prevent dehydration. Allow the agarose approximately 20 min or more to solidify (*see* **Note 10**).
17. When ready, use a scalpel to cut away chunks of agarose from areas immediately adjacent to the antennae and proboscis while leaving the brain intact and submerged in agarose. Next using a scalpel, remove the layer of agarose lying on top of the proboscis (Fig. 1). Next, insert forceps into agarose above antennae and gently peel agarose layer away from the antennae. Do not disturb either the antennal nerves or the brain itself (*see* **Note 11**). Use a kimwipe to dry off the antennae and rub off any bits of agarose that may still be clinging to the cuticle.
18. Drop coverslip on top of the prep.

3.2 Antennal Lobe Stimulation and Imaging

1. Pipette 1 μL of cVA on a small slip of filter paper and place in glass pipette. Push filter paper towards the tip and then add small wisps of cotton to prevent it from falling back out. Secure the pipette to the pheromone tubing system.
2. Position slide in the imaging microscope setup. Position the tip of the glass pipette delivering the pheromone such that it is resting on the slide or the Sylgard slab approximately 5 mm

away from the antennae. Position the odor puffer as close to the antennae as possible without disturbing the glass pipette.

3. After locating the prep with a low magnification objective (e.g., 4×), add a drop of water on top of the coverslip. Add a second drop to the tip of a higher magnification water immersion objective (e.g., 40×). Position the high magnification objective until the water droplets touch before adjusting the focus to locate the brain.
4. Starting at a zero setting, increase the laser power slowly just until brain structures expressing GCaMP are visible. Focus and zoom in on the antennal lobe (*see* **Note 12**). To maximize the dynamic range of the signal and to minimize phototoxicity, adjust the laser power to set basal fluorescence to about 10–15 % of the full dynamic range of the light detector.
5. Begin stimulation and image acquisition. Stimulate for 2 s with at least 2-min interval between odor presentations to allow any desensitized odor receptors to recover and to permit odors to clear from the fly's headspace. For antennal lobe imaging, we typically acquire images at ~4 frames per second (256 ms per frame) with a resolution of 128 × 128 pixels (0.75 × 0.75 μm for each pixel; Fig. 3). After all odor stimulations are finished, acquire a high-resolution image stack of the antennal lobe at 512 × 512 pixels (1–2 μm step in *z*-axis) for glomerular identification.

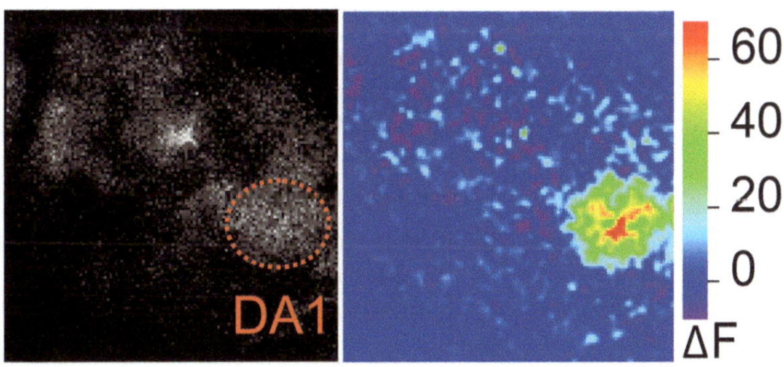

Fig. 3 Anterior view of antennal lobe from a female fly bearing the transgenes GH146-Gal4 and UAS-GCaMP3. Grayscale (*left*) and pseudocolored (*right*) images depict odor-evoked activity in DA1 projection neurons in response to delivery of cVA at 2 % SV (saturated vapor pressure)

4 Notes

1. For best results, saline should be made fresh every week. A freezing-point osmometer can be used to measure the osmolarity of the AHL saline, which should be about 265 mOsm for adult *Drosophila melanogaster* [10].
2. The custom manifold with eight odor ports is under the control of the custom solenoid valve controller. Switches on the solenoid valve controller determine which of the eight odor ports is open at the time of odor stimulation.
3. The custom-made odor puffer in our lab contains nine ports. A 7-port version is available commercially from Western Analytical.
4. Antennal lobe preps from younger flies are more robust and easier to dissect out than older flies. Flies, however, should be old enough to permit sufficient levels of GCaMP to accumulate in the glomeruli. Therefore, flies approximately 2–3 days old are ideal.
5. Dissection with saline containing calcium may lead to reduced odor responses. Cold calcium-free saline serves to preserve the integrity of the fly brain and minimize the loss of synaptic strength.
6. Using two sets of forceps in this manner—one to hold the cuticle, and the other to tear away pieces of it—minimizes damage to the brain by reducing the exertion of tensile forces on the tissue.
7. Brain tissue is highly charged and tends to stick to the side of a glass pipette. Prevent the loss of your tissue by first suctioning up a small volume of saline into your glass pipette. Expel a tiny volume of the saline before suctioning up the brain. The brain should hang just at the tip of the pipette in a drop of saline.
8. Orient the fly brain such that the antennae are fairly close to the edge of the slide. This later ensures that the antennae may be positioned as close to the odor puffer as possible.
9. Be sure that the saline around the brain and the agarose are thoroughly mixed. Otherwise, the agarose may fail to solidify around the antennae, making its complete removal messy and difficult in a later step.
10. To determine whether agarose has sufficiently dried, one can make a small incision into the agarose with the scalpel. If it is properly set, saline should not appear to seep out from the gel.
11. Use a Kimwipe to dry off the antennae and rub off any bits of agarose that may still be clinging to the cuticle.

12. If the image of the antennal lobe is not clear, there may be too much agarose in between the brain and the objective. Strive to make the layer of agarose a little thinner to improve image quality.

Acknowledgement

This work was supported by NIH (R01DK092640) and NSF (0920668) grants to J.W.W.

References

1. Tian L et al (2009) Imaging neural activity in worms, flies and mice with improved GCaMP calcium indicators. Nat Methods 6:875–881
2. Nakai J, Ohkura M, Imoto K (2001) A high signal-to-noise Ca(2+) probe composed of a single green fluorescent protein. Nat Biotechnol 19:137–141
3. Ng M et al (2002) Transmission of olfactory information between three populations of neurons in the antennal lobe of the fly. Neuron 36:463–474
4. Wang JW et al (2003) Two-photon calcium imaging reveals an odor-evoked map of activity in the fly brain. Cell 112:271–282
5. Ruta V et al (2010) A dimorphic pheromone circuit in Drosophila from sensory input to descending output. Nature 468:686–690
6. Root CM et al (2007) Propagation of olfactory information in Drosophila. Proc Natl Acad Sci U S A 104:11826–11831
7. Root CM et al (2008) A presynaptic gain control mechanism fine-tunes olfactory behavior. Neuron 59:311–321
8. Root CM et al (2011) Presynaptic facilitation by neuropeptide signaling mediates odor-driven food search. Cell 145:133–144
9. Semmelhack JL, Wang JW (2009) Select Drosophila glomeruli mediate innate olfactory attraction and aversion. Nature 459:218–223
10. Singleton K, Woodruff RI (1994) The osmolarity of adult Drosophila hemolymph and its effect on oocyte-nurse cell electrical polarity. Dev Biol 161:154–167

Chapter 13

Live Cell Calcium Imaging of Dissociated Vomeronasal Neurons

Angeldeep Kaur, Sandeepa Dey, and Lisa Stowers

Abstract

Sensory neurons in the vomeronasal organ (VNO) are thought to mediate a specialized olfactory response. Currently, very little is known about the identity of stimulating ligands or their cognate receptors that initiate neural activation. Each sensory neuron is thought to express 1 of approximately 250 variants of Vmn1Rs, Vmn2Rs (A, B, or D), or FPRs which enables it to be tuned to a subset of ligands (Touhara and Vosshall, Annu Rev Physiol 71:307–332, 2009). The logic of how different sources of native odors or purified ligands are detected by this complex sensory repertoire remains mostly unknown. Here, we describe a method to compare and analyze the response of VNO sensory neurons to multiple stimuli using conventional calcium imaging. This method differs from other olfactory imaging approaches in that we dissociate the tightly packed sensory epithelium into individual single cells. The advantages of this approach include (1) the use of a relatively simple approach and inexpensive microscopy, (2) comparative analysis of several hundreds of neurons to multiple stimuli with single-cell resolution, and (3) the possibility of isolating single cells of interest to further analyze by molecular biology techniques including in situ RNA hybridization, immunofluorescence, or creating single-cell cDNA libraries (Malnic et al., Cell 96:713–723, 1999).

Key words Vomeronasal, Calcium imaging, Neurons, Dissociated

1 Introduction

In mice, the vomeronasal organ (VNO) has been implicated in mediating reception of a variety of ligands informing both interspecies and intraspecies interactions [1]. Sensory neurons within the VNO can each express 1 of approximately 250 receptors. With such a complex receptor family of interest, ligand identification and receptor deorphanization have become important in order to study the mechanisms of sensory detection and function. Calcium imaging of dissociated sensory neurons is a relatively simple technique to identify and analyze ligands detected by the VNO [2]. Moreover, because the closely packed neurons are dissociated into individual sensory neurons, single cells can be easily isolated following calcium imaging for analysis by conventional molecular biology [3].

Kazushige Touhara (ed.), *Pheromone Signaling: Methods and Protocols*, Methods in Molecular Biology, vol. 1068, DOI 10.1007/978-1-62703-619-1_13, © Springer Science+Business Media, LLC 2013

Activation of the vomeronasal sensory neurons (VSNs) occurs upon binding of ligands by the cognate receptors, followed by a signaling cascade that leads to opening of a Trpc2-associated calcium channel [4]. This in turn results in a high influx of extracellular calcium ions [5], a property that has been exploited to identify activated cells using chemical dyes or genetically encoded marker proteins that detect changes in intracellular calcium (reviewed by Grienberger, 2012 [6]). This principle can be combined with simple microscopy to study the changes in calcium concentration within individual VSNs. The densely packed VSNs must first be separated by physical dissociation followed by mild protease digestion. The separated neurons are plated on coverslips at a density that enables analysis of individual neurons. This process is referred to as a "dissociated preparation" of VSNs. The attached VSNs are incubated in a solution containing a calcium-sensing chemical dye to "load" the neurons with the fluorescent indicator. A dissociated preparation allows each neuron in the field of view to be independently selected as "regions of interest" or areas on the coverslip for commercially available software to monitor changes in calcium concentration. The recorded change in fluorescence of the dye reflects the changes in intracellular calcium within the selected VSNs. Rapid changes in calcium can be recorded and the timing of each rise in calcium can be correlated to specific ligands being introduced in the cell environment. The collected data can be studied for an in-depth comparative analysis of ligands that may broadly or specifically activate populations of VSNs.

Dissociation of the VSNs allows for a relatively rapid and even uptake of chemical dyes within each cell. Consequently, this preparation results in a highly reliable resolution of ligand responses by sensory neurons normally present in both superficial and deep layers. The preparation is highly efficient; from three mice, up to four coverslips can be prepared, each containing between 150 and 500 cells that can be imaged. Since the timing of each stimulant delivery is known, the window of time during which a neural response may occur can be precisely defined. This allows us to parse out spontaneous changes in intracellular calcium from real responses to presented ligands. Additionally, the preparation is easy to learn and does not require highly specialized equipment for imaging or analysis of gathered data. Following imaging of a dissociated preparation, it is possible to physically isolate single cells of interest from the coverslip for molecular analysis, for instance RNA in situ hybridizations, immunofluorescence, and creation of single-cell cDNA libraries. The ability to do subsequent analysis on individual activated cells is a distinct advantage of imaging dissociated neurons over whole epithelium or slice preparations.

The most prominent advantage of using chemical dyes over genetic markers is their high sensitivity in that they undergo a large change in fluorescence in response to single-action potentials [7].

Without this added sensitivity, a subset of relevant and functional neuronal activities would go undocumented, further impairing the ability to fully understand vomeronasal function. We use UV excitable ratiometric dye Fura-2-AM. Once bound to Ca^{2+}, the absorption of Fura-2-AM shifts, which can be observed by scanning the excitation spectrum from 300 to 400 nm, whereas the emission spectrum stays at ~510 nm (product information—Invitrogen). In the calcium-free state, fluorescence at 380 nm is higher than at 340 nm whereas in the bound state, the fluorescence at 380 nm decreases and that at 340 nm increases. The simultaneous change in absorption spectra results in a more reliable measurement of calcium concentration. By measuring fluorescence at both 340 and 380 nm a relative quantification of calcium concentrations can be made based on the 340/380 fluorescence values (product information—Invitrogen). The use of the ratio also aids in reducing artifactual information by accounting for differences in cell thickness, uneven dye loading, loss of dye, and photobleaching [8]. Several other commercially available fluorescent dyes may also be used; however an advantage of Fura-2-AM over other similar dyes is that it can be used in combination with and does not interfere with the fluorescence spectrum of GFP. Addition of chemical modifications, such as the acetoxymethyl (AM) ester [9, 10], to calcium indicators enhances their permeability and facilitates more uniform uptake by cells. Greater uptake of the dye reduces the number of dye micelles observed, which can produce a background signal as flecks of fluorescence during imaging. Noise can be further reduced by using pluronic acid, which aids in dispersion of Fura-2-AM, increasing dye uptake efficiency and breaking up fluorescent aggregates.

Despite the simplicity and sensitivity of using a dissociated preparation and calcium-sensing dye-based fluorescence imaging, the technique is not without disadvantages. Protease digestion, albeit mild, followed by dissociation of the closely packed neurons reduces overall cell health compared to vibratome slices or whole-epithelium preparations where cell connections are not entirely disrupted. A limited number of sequential ligand pulses can be perfused over the neurons before their reduced cell health affects reliable cell responses. Therefore, each step of the protocol must be carried out within a minimal amount of time so as to preserve cell health. UV phototoxicity may also reduce the longevity of the neurons during imaging; thus designing shorter experiments yields more reliable results. For each experiment, the time of stimulant or buffer delivery must be precisely noted so that data analysis can be carried out with reliable knowledge of which stimuli are responsible for cell depolarization events. Most calcium dye indicators when used in large concentrations have a calcium buffering effect whereby the peak amplitude of calcium transients is significantly reduced [11], so indicator loading must be carried out carefully to maintain a good signal-to-noise ratio. Lastly, since the topological structure of the VNO is lost during dissociation, all information on the spatial arrangement of the neurons is irretrievable, though molecular characterization can inform on whether imaged cells are from the apical or the basal layer of the VNO. By designing experiments while keeping these caveats in mind, a vast amount of information can be gathered about the neurons of interest.

The combined use of chemical dye Fura-2-AM and a dissociated cell preparation enables characterization of the response profile of a single neuron when presented with particular ligands. With these tools, stimuli relevant to the VNO can be identified and the specificity and tuning of vomeronasal neurons can be more fully understood.

2 Materials

2.1 Dissection of VNO

1. 1× Phosphate buffer saline (PBS): pH 7.4 (*see* **Note 1**).
2. Forceps and surgical scissors.

2.2 Dissociation of Neurons

1. DL-Cysteine-HCl (1 M): Dissolve 15.7 mg DL-cysteine-HCl in 100 μL distilled water (*see* **Note 2**).
2. Ethylene diamine tetra acetate (EDTA, 100 mM).
3. Papain: Dissolve papain in water to 0.22 U/μL concentration, sterile filter, and divide into 11 μL aliquots for single use (*see* **Note 3**).
4. Dissociation buffer (in 1× PBS): 1.1 mM EDTA; 5.5 mM DL-cysteine-HCL; papain 2.2 U/mL. Mix 5.5 μL DL-cysteine-HCl solution (1 M), 11 μL EDTA solution (100 mM), and 10 μL papain (0.22 U/μL) in 1 mL 1× PBS (*see* **Note 4**).
5. DNAse solution: 10× DNAse buffer; DNase 1; 1× PBS. Mix 200 μL of 10× DNAse buffer and 800 μL 1× PBS, add 50 U DNase 1, and mix well (*see* **Note 5**).

2.3 Plating of Neurons

1. Plating media: 10 % supplemented Dulbecco's modified Eagle medium (D-MEM). D-MEM High Glucose (1×)+ GlutaMAX+Sod.Pyr, heat-inactivated fetal bovine serum (FBS). Mix 90 mL D-MEM with 10 mL heat-inactivated FBS and divide into 12 mL single-use aliquots. Pre-warm aliquots to 37 °C before use (*see* **Note 6**).
2. Concanavalin A (0.5 μg/μL)-coated coverslips: Solid Concanavalin A (type V), distilled water. Weigh out 10 mg Concanavalin A and dissolve in 20 mL distilled water to obtain a 0.5 μg/μL solution. Divide into 1–2 mL single-use aliquots. Incubate coverslips overnight in Concanavalin A solution at

4 °C, with continuous shaking. Place them in a coverslip rack to dry in a laminar flow chamber for 2–7 h and store at 4 °C (*see* **Note 7**).

3. Stacks (optional): Prepare stacks by cutting off a center portion of P-200 tips (Fig. 2). Autoclave the stacks before use.

2.4 Calcium Imaging

1. Hank's Balanced Salt Solution (HBSS) 10×, anhydrous calcium chloride 1,400 mg/L, magnesium chloride-6H_2O 1,000 mg/L, magnesium sulfate-7H_2O 1,000 mg/L, potassium chloride 4,000 mg/L, potassium phosphate monobasic 600 mg/L, sodium chloride 80,000 mg/L, sodium phosphate dibasic-7H_2O 900 mg/L, dextrose 10,000 mg/L (*see* **Note 8**).
2. 4-(2-Hydroxyethyl)-1-piperazineethanesulfonic acid (HEPES) buffer, 1 M (*see* **Note 9**).
3. Imaging buffer: Mix 50 mL HBSS (10×), 5 mL HEPES (1 M), and 445 mL of distilled water for a final volume of 500 mL (*see* **Note 10**).
4. Pluronic acid 20 %: Add 20 mg pluronic acid to 100 μL DMSO in an Eppendorf tube, warm in 37 °C water bath for 10–15 min with intermittent vortexing to dissolve (*see* **Note 11**).
5. Fura-2-AM dye (1 μg/μL): Dissolve 50 μg Fura-2-AM in 50 μL DMSO by pipetting up and down several times (*see* **Note 12**).
6. Loading solution: Mix 5 μL Fura-2-AM solution with 2.5 μL pluronic acid solution by pipetting up and down several times. Add 1,000 μL imaging buffer (*see* **Note 13**).

2.5 Equipment

1. 37 °C cell culture incubator with 5 % carbon dioxide.
2. Certified class II biological safety cabinet with laminar flow.
3. Centrifuge machine with swinging bucket rotor for 15 mL conical tubes.
4. Dissection microscope.
5. Phase contrast microscope with 10× and 20× objective lenses.
6. Inverted deconvolution microscope with a 20× fluar 0.75 objective or an upright microscope with water immersion objective.

3 Methods

All reagents and chemicals should be of cell culture grade. For best results, all solutions should be prepared using sterile technique and stored at the appropriate temperature for future use. Each step must be carried out within minimal time to ensure optimal health of the neurons.

3.1 VNO Dissection

1. Transfer 1 mL 1× PBS to a well in a 4-well plate and set on ice. Transfer 20 mL 1× PBS each into two petri dishes and chill on ice.
2. Prepare papain-containing dissociation buffer immediately prior to starting dissections. Incubate on ice.
3. Isolate both VNO lobes from a total of three mice transferring each upon removal to a petri dish containing chilled PBS on ice from **step 1**. For best results, this step should be completed within 10–15 min.
4. Move VNOs to the second petri dish containing chilled PBS one at a time. Remove cartilage from each VNO lobe and transfer tissue to chilled PBS in the 4-well plate prepared in **step 1**.

3.2 Dissociation of Vomeronasal Neurons

1. Once all VNOs have been removed from their cartilage, carefully aspirate PBS in the well and replace with 1 mL ice-cold dissociation solution prepared in Subheading 3.1, **step 2**.
2. Dissociate each lobe of the VNO by tearing with fine forceps into minute pieces (Fig. 1). For best results, complete within 10 min.
3. Transfer solution containing dissociated VNO tissue to a 15 mL Falcon tube and incubate at 37 °C for 15–20 min with continuous shaking (at approximately 225 rpm).
4. Prepare the DNAse solution.
5. Add DNAse solution to dissociated VNO tissue and gently triturate until all aggregates are dispersed (*see* **Note 14**).

3.3 Plating Vomeronasal Neurons

1. Add 10 mL of pre-warmed D-MEM–FBS media. Tap several times to mix and spin at 1,000 rpm (200g) for 5 min, at room temperature.

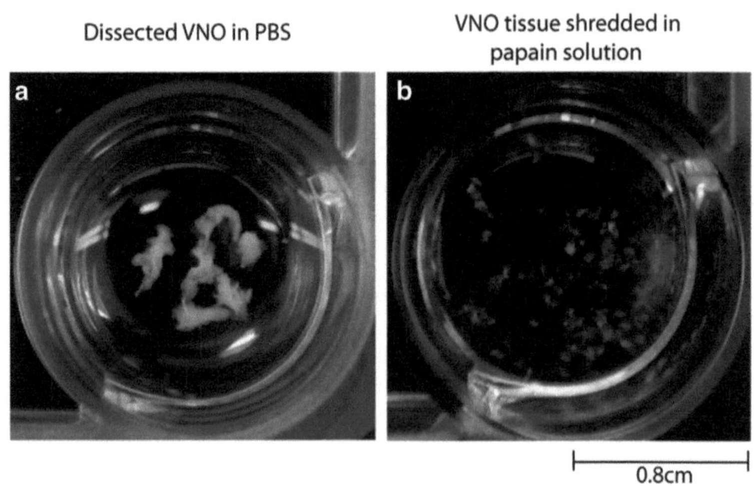

Fig. 1 Dissociating VNO tissue in protease solution (**a**) before and (**b**) after. Three whole VNOs are depicted before and after

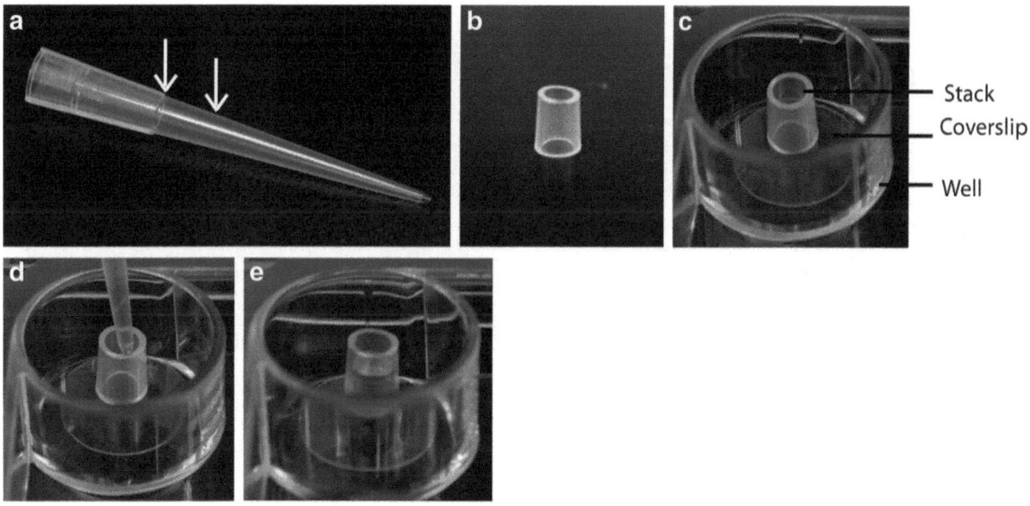

Fig. 2 (**a** and **b**) Preparing a stack. (**a**) Cut a 200 μL pipette tip at the positions indicated by *white arrows*. (**b**) A stack obtained from the pipette tip. (**c**) Stack placed on a Concanavalin A-coated glass coverslip in a well. (**d**) Pipetting cell suspension in media inside the stack. (**e**) Media containing dissociated neurons pipetted in stacks, ready to be incubated at 37 °C

2. Place one Concanavalin A-coated coverslip per well in a new 4-well plate. Place a stack in the center of each coverslip to concentrate application of cells (Fig. 2).
3. Aspirate supernatant without disturbing the pellet. Gently resuspend in 100 μL of D-MEM–FBS media.
4. Transfer 25 μL of resuspended VSNs onto each coverslip inside the stack.
5. Incubate at 37 °C for 45–60 min. Check to see if cells are attached to the coverslip using a phase contrast microscope (density similar to Fig. 3).

3.4 Loading Vomeronasal Sensory Neurons with Calcium-Sensing Dye

1. Thaw Fura-2-AM at room temperature.
2. Prepare dye loading solution, cover in aluminum foil, and store in the dark through experiment.
3. Remove stack from the coverslip, transfer the coverslip to a well, and gently layer with 250 μL dye loading solution. Incubate the coverslip for 15–30 min at room temperature, covered in aluminum foil.
4. Remove coverslip from dye solution and assemble it on the stage. Gently cover the cells with 250 μL of imaging buffer to prevent drying (*see* **Note 15**). Coverslip is now ready for imaging.

3.5 Calcium Imaging

1. Prior to imaging, wash stimulant/buffer delivery system and accessory tubes thoroughly with distilled water. Other wash solutions may also be used based on the solubility of stimulants used.

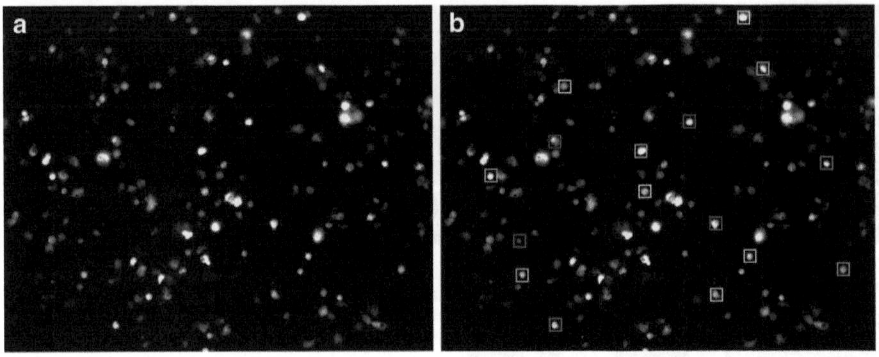

Fig. 3 (**a**) A field of view of dissociated cells under 380 nm. (**b**) Individual cells selected, shown in *colored squares* as "regions of interest"

2. Dilute stimulants in imaging buffer to desired strength.
3. Set up dye-loaded coverslip on microscope and set perfusion system up to wash cells with imaging solution.
4. Using a data collection software such as MetaFluor, select each dissociated cell as an individual "region of interest" (Fig. 3). Select an empty region of the coverslip for background subtraction during data collection (*see* **Note 16**).
5. Apply stimulant(s) for desired duration followed by imaging buffer for an intervening duration to wash cells before perfusing the next stimulant (*see* **Note 17**). Following the last test stimulant, perfuse a pulse of a known activator of the cells of interest (for a positive control). If such a positive control ligand is not defined for cells of interest, pool individual stimulants to pulse at the end (*see* **Notes 18–20**).

3.6 Data Analysis

1. Identify the cells that show an increase of the 340/380 ratio measuring more than 1.5 times the baseline signal during the time window(s) of positive control or pooled stimulant application. Graphing software such as Microsoft Excel may be used to do this (*see* **Note 21**). Plot the 340/380 fluorescence ratio for each region of interest that shows the increase in calcium coinciding with the delivery of the positive control or pooled stimulant.
2. Precisely define the time during the experiment when stimulus was perfused over the cells. In case the stimulants are delivered through a tube, determine the length of time required for the stimulant front to reach the coverslip so that the time of delivery of stimuli is defined precisely (*see* **Note 22**). If analyzing by a trace plot, annotate the plot from the time the stimulus reaches the coverslip to the time the subsequent wash reaches the coverslip.

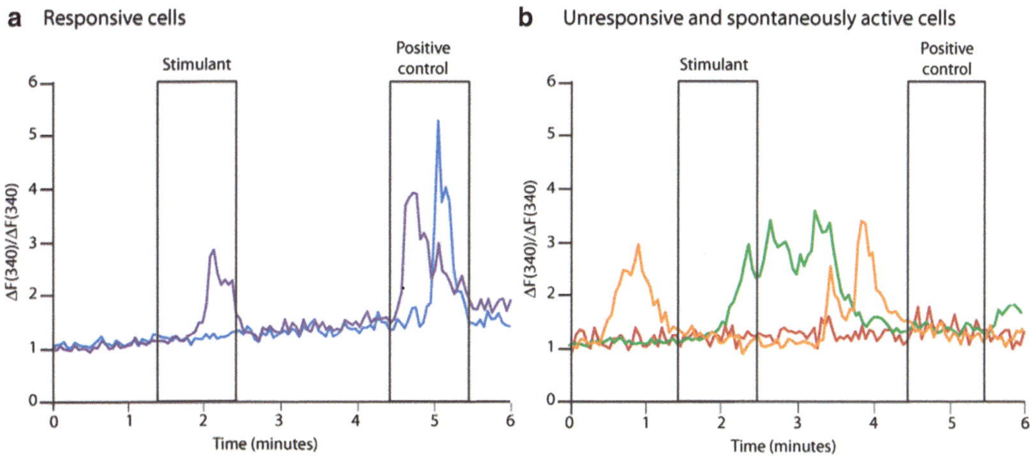

Fig. 4 Example of output data. Each *colored line* represents the calcium trace of a single neuron, the *black rectangles* represent time windows for application of stimulus. (**a**) Responsive cells. *Purple*: Cell showing rise in intracellular calcium on application of test stimulus and positive control. *Blue*: Cell showing rise in intracellular calcium on application of positive control only. (**b**) Unresponsive and noisy cells. *Red*: Unresponsive cell. *Orange*: Cell showing intracellular calcium increase before application of test stimulus and positive control. *Green*: Cell showing intracellular calcium increase randomly on application of test stimulus but not to positive control. (*X*-axis: Ratio of fluorescence change measured at 340 nm and that measured at 380 nm, *Y*-axis: Time (in min))

3. To define a cell of interest as "responsive," we use the following criteria (Fig. 4):
   - Cells must respond to the positive control stimulus pulse at the end.
   - An increase in fluorescence ratio measuring at least 1.5 times the baseline during the time of interest is counted as a response.
   - Cells must only respond during the time that stimuli are present on the coverslip (*see* **Note 23**).
   - Cells must not show an increase in calcium greater than 1.5 times the baseline signal outside of the stimulus windows.

4 Notes

1. PBS used for dissection may be kept chilled on ice as VNOs are dissected. Chilled PBS is helpful in keeping the neurons alive if dissection takes longer than 15 min.
2. Prepare fresh each day.
3. Papain aliquots dissolved in distilled water may be stored at −20 °C up to several months. To avoid repeated freeze–thaw

cycles, they are best divided into single-use aliquots. Thaw before preparing dissociation buffer.

4. Prepare fresh each day. Once prepared, keep dissociation buffer on ice until use to avoid losing enzyme activity.
5. Prepare DNase solution right before use while neurons are incubating in 37 °C shaker. If DNAse buffer is not available, this may be substituted by other buffers in which DNAse is fully active, for instance 10× PCR buffer or 10× endonuclease buffer.
6. Prepare 10 % (by volume) FBS-supplemented DMEM in the laminar flow chamber to keep it contamination free and store aliqouts at 4 °C. Pre-warm each aliquot at 37 °C at the start of the experiment to avoid adding cold media to cells.
7. Of the available reagents, we have found that the use of Concanavalin A to coat coverslips allows for the maximum number of cells to adhere after plating. Aliquots of 0.5 μg/μL Concanavalin A may be stored at −20 °C and thawed before use.
8. Store 10× HBSS at 4 °C after opening.
9. Store 1 M HEPES at 4 °C.
10. Make fresh each day. Keep at room temperature after preparation.
11. Make fresh every 7–10 days. Store at room temperature.
12. Fura-2-AM is light sensitive. Once dissolved, keep Fura-2-AM solution wrapped in aluminum foil at −20 °C. Thaw before use each day.
13. Prepare loading solution fresh each day. Wrap loading dye in aluminum foil and store at room temperature while experiment is in progress.
14. At the end of protease digestion, the pieces of tissue appear to aggregate together. After addition of DNAse, the pieces of tissue appear to be more disintegrated: fewer and smaller pieces of tissue should be seen in suspension. The solution should appear more turbid than that in **step 2** (Subheading 3.2).
15. Do not pipet the solution directly on the coverslip. Instead pipet the solution on the side of the stage and tip the stage to cover the cells. This will prevent cells from being forcibly dislodged from the coverslip.
16. For urinary proteins and total urine, we have empirically determined that setting the 340–380 gain controls at a ratio of 3:1 obtains the optimal signals to differentiate neuronal responses from noise in our setup. This may have to be adjusted for different ligands and instrumental setups.
17. We apply stimuli such as mouse urinary proteins or total urine for 1 min alternating with buffer for 2 min. These durations of

stimulus and buffer pulses may have to be varied according to the nature and concentrations of different ligands.

18. A positive control pulsed at the end of the experiment helps to identify cells that survived through the entire experiment. Only these cells can be investigated for their response to all the ligands presented during the course of the experiment. Another possible positive control for neurons is potassium chloride [12].
19. In our setup, a perfusion rate of approximately 5 mL per minute is optimal to obtain laminar flow of buffer over the neurons. This may have to be optimized differently for each setup.
20. To obtain the most accurate results, it is necessary to image each coverslip quickly. Depending on the quality of the preparation, and the nature and concentration of ligands, up to five total pulses may be applied, spanning approximately 15 min of imaging time. To maximize speed, subsequent coverslips should be incubated in dye solution while the previous is being imaged such that one can complete imaging four coverslips in 90 min or less. If the cells are imaged for longer than 15 min, the constant exposure to pulses of ligands and UV light may adversely affect cell viability, leading to less accurate output. As previously noted, one preparation yields four coverslips. If each coverslip is imaged within 15–20 min, there is no substantial loss of cell health between neurons from the first and last coverslip. Cell health may not be optimal if neurons are not imaged within 3 h of plating.
21. We have empirically determined, for mouse urine and urinary proteins, that a 1.5-fold increase in the 340–380 nm ratio is a good indicator of neuronal response. This may have to be optimized for various ligands and instrument setup. We use this cutoff to sort our data, separating cells that show an increase in calcium during presentation of the positive control from the rest of the imaged cells. This way the analysis can be focused on responsive cells instead of every single imaged cell, many of which show no changes in calcium during the experiment.
22. Delivery time may be determined by introducing a bubble or a colored dye in the delivery tube and then monitoring the time taken by the bubble or the dye front to reach the end of the tubing with a stopwatch or a timer.
23. There are spontaneous changes in calcium in cells, a natural phenomenon noticed in many types of preparations [12]. Being able to differentiate spontaneous activity and ligand-based activity is important in order to use this technique to follow ligand activity. A precise notation of the window of stimulus perfusion and repetitive pulses of the same stimulus enables correlation between stimulus perfusion and calcium changes within the cell.

References

1. Touhara K, Vosshall LB (2009) Sensing odorants and pheromones with chemosensory receptors. Annu Rev Physiol 71:307–332
2. Malnic B, Hirono J, Sato T, Buck LB (1999) Combinatorial receptor codes for odors. Cell 96(5):713–723
3. Munger SD, Leinders-Zufall T, Zufall F (2009) Subsystem organization of the mammalian sense of smell. Annu Rev Physiol 71:115–140
4. Chamero P, Marton TF, Logan DW, Flanagan K, Cruz JR, Saghatelian A et al (2007) Identification of protein pheromones that promote aggressive behaviour. Nature 450(7171):899–902
5. Kaupp UB (2010) Olfactory signalling in vertebrates and insects: differences and commonalities. Nat Rev Neurosci 11(3):188–200
6. Grienberger C, Konnerth A (2012) Imaging calcium in neurons. Neuron 73(5):862–885
7. Smetters D, Majewska A, Yuste R (1999) Detecting action potentials in neuronal populations with calcium imaging. Methods 18(2):215–221
8. Becker PL, Fay FS (1987) Photobleaching of fura-2 and its effect on determination of calcium concentrations. Am J Physiol 253(4 Pt 1): C613–C618
9. Tsien RY (1988) Fluorescence measurement and photochemical manipulation of cytosolic free calcium. Trends Neurosci 11(10):419–424
10. Tsien RY (1981) A non-disruptive technique for loading calcium buffers and indicators into cells. Nature 290(5806):527–528
11. Saftenku EE, Teslenko VI (ed) (1995) Effect of fura-2 on calcium transients and its dependenece on kinetics and location of endogenous buffers (a model study). Kluwer Academic-Plenum, New York
12. Holy TE, Dulac C, Meister M (2000) Responses of vomeronasal neurons to natural stimuli. Science 289(5484):1569–1572

Chapter 14

Whole-Mount Imaging of Responses in Mouse Vomeronasal Neurons

Pei Sabrina Xu and Timothy E. Holy

Abstract

Imaging permits the visualization of neural activity from the whole-mount vomeronasal sensory epithelium with single-cell resolution. The preparation preserves an intact tissue environment, enabling the robust detection of cellular responses upon chemical stimulation and study of the precise 3D mapping of vomeronasal sensory neuron (VSN) functional types within the epithelium. Using objective-coupled planar illumination (OCPI) microscopy to perform fast volumetric imaging, we routinely record the responses of thousands of VSNs for hours from a single intact vomeronasal organ preparation. Here we document the preparation of the whole-mounted vomeronasal epithelium, multichannel stimulus delivery, and three-dimensional calcium imaging by OCPI microscopy.

Key words In situ, Calcium imaging, Three-dimensional, Whole-mount VNO, Neural population, OCPI microscopy

1 Introduction

Mouse vomeronasal sensory neurons (VSNs) are situated in the sensory epithelium of the vomeronasal organ (VNO). Within the sensory epithelium, each bipolar VSN extends a single apical dendrite through the supporting cells to the surface, where the dendrite terminates as a knob covered by a tuft of microvilli [1]. It is the vomeronasal receptor proteins, localized on the dendritic tips in the microvilli layer, that bind ligands (often considered pheromones) and initiate activation of particular VSNs. To maintain the microvilli layer, an intact whole-mount preparation of the VNO was used, at first for electrical recording of VSN responses [2]. A similar preparation, mounting the neuroepithelium on a nitrocellulose membrane, was later used for imaging [3].

Calcium imaging is widely used for imaging the neural activity, including dissociated VSNs [4–6] and VNO acute slices [7]. However, for the whole-mount VNO preparation, some form of

Kazushige Touhara (ed.), *Pheromone Signaling: Methods and Protocols*, Methods in Molecular Biology, vol. 1068,
DOI 10.1007/978-1-62703-619-1_14,

optical sectioning (collecting only in-focus light) is necessary. However, for sampling many hundreds or thousands of neurons, conventional techniques such as two-photon microscopy are limited in speed. We therefore developed a form of light sheet-based microscopy called objective-coupled planar illumination (OCPI) microscopy, which enables the fast three-dimensional fluorescence imaging of activity in thousands of neurons [3]. With OCPI microscopy, we have been able to image from at least one-fifth of the whole VNO epithelium. Because OCPI microscopy is also less phototoxic than most alternatives, a single VNO preparation can be repetitively scanned for at least 1.5 h without significant loss of responsiveness. This enabled us recently to test up to 13 different stimuli on a single VNO preparation with four or more repeats of each stimulus [8]. Consequently, responsivity of VSNs can be measured quantitatively and analyzed by systematic statistical criteria. Given that VSN physiology is often perceived to be a "needle in a haystack" problem, and VSNs show abundant spontaneous activity with characteristics disturbingly close to those of stimulus-driven responses [9], such systematic approaches are among the few tools available for reducing the risk of false positives.

2 Materials

2.1 Reagents

1. Ringer's solution: 115 mM NaCl, 5 mM KCl, 2 mM $CaCl_2$, 2 mM $MgCl_2$, 25 mM $NaHCO_3$, 10 mM HEPES, and 10 mM D-(+)-glucose. Solution is freshly prepared on the experiment day from 20× Ringer's A, 20× Ringer's B, and glucose. Prepare 20× Ringer's A: Weigh 67.17 g NaCl, 3.73 g KCl, 2.94 g $CaCl_2$ (dehydrate), and 4.07 g $MgCl_2$ (hexahydrate); add water; mix; and make up to a volume of 500 mL. Prepare 20× Ringer's B: Weigh 21.00 g $NaHCO_3$ and 23.86 g HEPES, add water to dissolve, and make up to 500 mL. Water is prepared by purifying deionized water to attain a sensitivity of 18 MΩ cm at 25 °C. 20× Ringer's A and B can be stored at 4 °C for 3 months.
2. Stimulus stocks: Will be diluted to desired concentration by freshly prepared carboxygenated Ringer's solution on the experiment day (*see* **step 2** of Subheading 3.1). For water-insoluble stimuli, they need to be dissolved in appropriate solvent (e.g., methanol, DMSO) as high-concentration stock (*see* **Note 1**).

2.2 Animals and VNO Dissection Tools

1. Mice with calcium indicator-labeled VSNs, e.g., OMP-GCaMP2 mice [7, 8], mice with Oregon-Green BAPTA-labeled VNO by surgical injection [3].
2. VNO dissection tools: Decapitation scissors, Ring forceps, scalpel, #3 forceps, #5 forceps, Vannas spring scissors (3 mm cutting edge),

25 mm Petri dish, Sylgard-coated 25 mm Petri dish, ice bucket with cover, and dissection stereomicroscope.

3. VNO mounting tools: Nitrocellulose membrane (Millipore), custom glass dropping pipet (cut out the rear ~5 cm part of a disposable borosilicate glass Pasteur pipets, and cap the cutting end with a rubber head) (*see* **Note 2**), custom imaging chamber (can hold the nitrocellulose membrane at the bottom; is big enough to immerse the objective and light sheet optics inside), kimwipes tissue.

2.3 Equipment and Software

1. Microscopy: Custom OCPI microscopy [3] (US patent application 20090174937 pending) controlled by custom imaging acquisition software *Imagine* (http://holylab.wustl.edu/software.htm).
2. Superfusion equipment: 16-channel superfusion system (AutoMate Scientific, Inc. CA, USA) pressurized with gas cylinder (40 % O_2, 3 % CO_2, balance helium). We use 35 mL plastic syringe reservoirs. The switchable pinch valves are computer controlled by *Imagine* to synchronize with the imaging system.
3. Liquid heating equipment: PH01 heatable perfusion probe and TC02 temperature controller (Multi Channel Systems MCS GmbH, Reutlingen, Germany).

3 Methods

3.1 Prepare Solution

1. Mix 950 mL water, 50 mL 20× Ringer's buffer stock, 50 mL 20× Ringer's salt stock, and 1.8 g D-(+)-glucose in a 1 L bottle to prepare fresh Ringers' solution. Carboxygenate the Ringer's solution by bubbling with 95 % O_2/5 % CO_2 in 37 °C water bath for at least 30 min.
2. Dilute the stimulus stock solutions into carboxygenated Ringer's solution in 35 mL falcon tubes to desired concentration. Dilute the same amount of solvent as the Ringer's control. Prepare at least two tubes of Ringer's controls (*see* **Note 3**). Mix the stimuli solution gently by inverting the tubes back and forth several times (*see* **Note 4**).
3. Load the stimulus solution into the perfusion system. First, switch on the gas to a pressure level allowing flow rate of 2.2 mL/min. Then, for each stimulus, pour the stimulus solution into the syringe reservoir, and briefly blow the unfilled space of the syringe with pressurized gas before sealing with the rubber cap. Keep the stimulus solution pressurized throughout the experiment.

4. Test each stimulus line by briefly switching on the valves one by one. Chase the air bubbles and make sure that flow rate is consistent among all lines.
5. Save ~200 mL carboxygenated Ringer's solution on ice for VNO dissection. Transfer the remaining Ringer's solution into a 2 L bottle reservoir in the superfusion system and check its flow as in the previous step. It will provide Ringer's flush to the VNO tissue following each stimulus solution.
6. Before moving to the next step, make sure that the following list is done:
   - Ensure that the optical table is floating.
   - Launch the imaging acquisition software *Imagine.*
   - Power on the laser and laser shutter controller (but keep the shutter closed).
   - Empty the container for liquid waste collected by suction.
   - If the camera is equipped with water-cooling system, switch on the water cooling and shut down the internal fan (to minimize vibration). If necessary, set the optimized working temperature for the camera (we set −55 °C for the Andor ixon 885 EMCCD camera).

3.2 Prepare VNO Sample

1. Prior to performing the VNO dissection, get all the following items ready around the dissection microscope, including all dissection and mounting tools (*see* **items 2** and **3** of Subheading 2.2). In addition, prepare the following three Petri dishes: a 25 mm Petri dish filled with ice-cold carboxygenated Ringer's solution (dish #1), a Sylgard-coated Petri dish in the same way (dish #2), an empty and dry Sylgard-coated Petri dish (dish #3), a piece of kimwipes tissue that is rolled up into a compact stick, and a piece of nitrocellulose membrane (*see* **Note 5**). If fixing the nitrocellulose membrane into the imaging chamber involves assembly, get assembly tools ready.
2. Euthanize the mouse by CO_2 and decapitate. Remove the VNO quickly and place into dish #1. Under the stereomicroscope, using #3 forceps carefully remove the bony capsule from one VNO (*see* **Note 6**), and transfer the VNO into dish #2. With #5 forceps or spring scissors, separate the VNO neural epithelium from the blood vessel by cutting along the edge of the VNO. More detail about VNO dissection is described in [10].
3. Gently suck up the dissected VNO together with small amount of liquid into the tip of glass pipet, and then place down VNO with a drop of liquid at the center of the precooled dish #3. Under the microscope, adjust the VNO position that (1) it is in the center of the liquid drop and (2) the basal lamina side faces up.

4. Use the tip of the kimwipes tissue stick to carefully dip around the VNO in order to get rid of the liquid drop. To best preserve the VNO activity, it is recommended to complete the following three actions within 5 s: (1) obtain a well-flat VNO tissue (*see* **Note 7**) with basal lamina side up in minimal liquid volume (*see* **Note 8**); (2) hold the nitrocellulose membrane with a #3 forceps, and carefully put down the membrane in a way that the center of the membrane touches the VNO; and (3) add a drop of fresh ice-old oxygenated Ringer's solution to the back of the nitrocellulose membrane.
5. Add more drops of fresh Ringer's solution until the nitrocellulose membrane floats. The VNO should be firmly adhered. Transfer the membrane into the imaging chamber. Fill the imaging chamber with fresh Ringer's solution as quickly as possible to preserve the tissue health.

3.3 Recording

1. Place the imaging chamber under the OCPI microscope. Position the heater probe above the VNO tissue. Superfuse the tissue with heated Ringer's solution at 35 °C (*see* **Note 9**). Position the suction probe on the liquid surface at a distant corner of the imaging chamber.
2. Lower the OCPI microscope objective, until the objective immerses into the chamber. Turn on the laser shutter and start preview in *Imagine*. Adjust the chamber position by translating stage control knob while further lowering down the objective until the VNO images appear in the preview on the computer screen (*see* **Note 10**).
3. Adjust the translating stage to position VNO in the center of the filed of view. Sweep the piezo to scan the whole VNO and check if the tissue is intact. Once done, close the laser shutter and preview.
4. Since the imaging chamber has been repositioned, adjust the location of heater probe and suction probe accordingly (*see* **Note 11**).
5. Turn off the room light. Leave the VNO tissue being superfused for at least 30 min to acclimate (*see* **Note 12**).
6. Set up the recording parameters in *Imagine*, which includes the following: (1) generate the randomly ordered stimulus file by custom software and load into *Imagine* (*see* **Note 13**); (2) set the following parameters (values from our typical experiment are listed as reference):
   - Piezo travel range: 200 μm.
   - Exposure time: 50 ms/frame.
   - Frame number in a stack: 40.
   - Number of stacks to acquire: Set according to the loaded stimulus file. For a typical experiment with 15 stimuli and 4 repeats,

duration of each stimulus is 5 stacks, duration of interval flush is 10 stacks, and pre-leading flush is 10 stacks. The total number of stacks for such a recording is 910. We acquire one stack every 5 s, so the typical experiment with 910 stacks lasts for ~75 min.

7. After 30 min, dump the first 20-s stimulus solution for each line one by one before recording (*see* **Note 14**). Flush the tissue for 3 min.
8. Adjust the laser power to obtain the appropriate fluorescence level by monitoring the intensity profile in *Imagine* (*see* **Note 15**).
9. Begin recording. Imaging data are saved to disk. Keep the room dark and uninterrupted during recording.

3.4 Data Analysis

1. Analyze the imaging data offline. The resulting data sets can be used to address many different questions. For example, from the raw trace intensity (*F*) of each cell (*see* Fig. 1), one can measure the $\Delta F/F$ of recorded VSNs in response to particular stimuli, and also get the 3D position of the VSNs. Because of large 3D volume as well as long hours of recording, a data set is large in size. Analyzing such data sets requires programming skills (C, Matlab, R, Python, or similar). In addition, during long hours of recording, VNO tissue undergoes detectable deformity, such as swelling, shrinking, and wrapping. Thus, 3D image registration is required in the data analysis.

4 Notes

1. Be aware that solvents like methanol and DMSO may cause tissue damage or artificial signal during recording. Adding equivalent solvent in control groups is recommended. In our experiments, solvent concentration never exceeds 0.01 % in the final stimulus solutions.
2. The resulting glass dropper has a wide opening with smooth glass walls, which generates minimal damage to the tissue during VNO transfer.
3. Because fluorescence intensity can change even for small differences in carboxygenation, one Ringer's control will be used to subtract any artifact response. The other Ringer's control will serve as the negative control for tested stimuli (and is treated exactly as if it is one of the stimulus solutions).
4. To preserve the health of the VNO tissue in the experiment, care should be taken to preserve oxygen inside the solution: avoid rigorous shaking, minimize the time of preparing stimulus solution before loading into the pressurized superfusion system, etc. Handle each stimulus tube consistently to ensure consistent carboxygenation level.

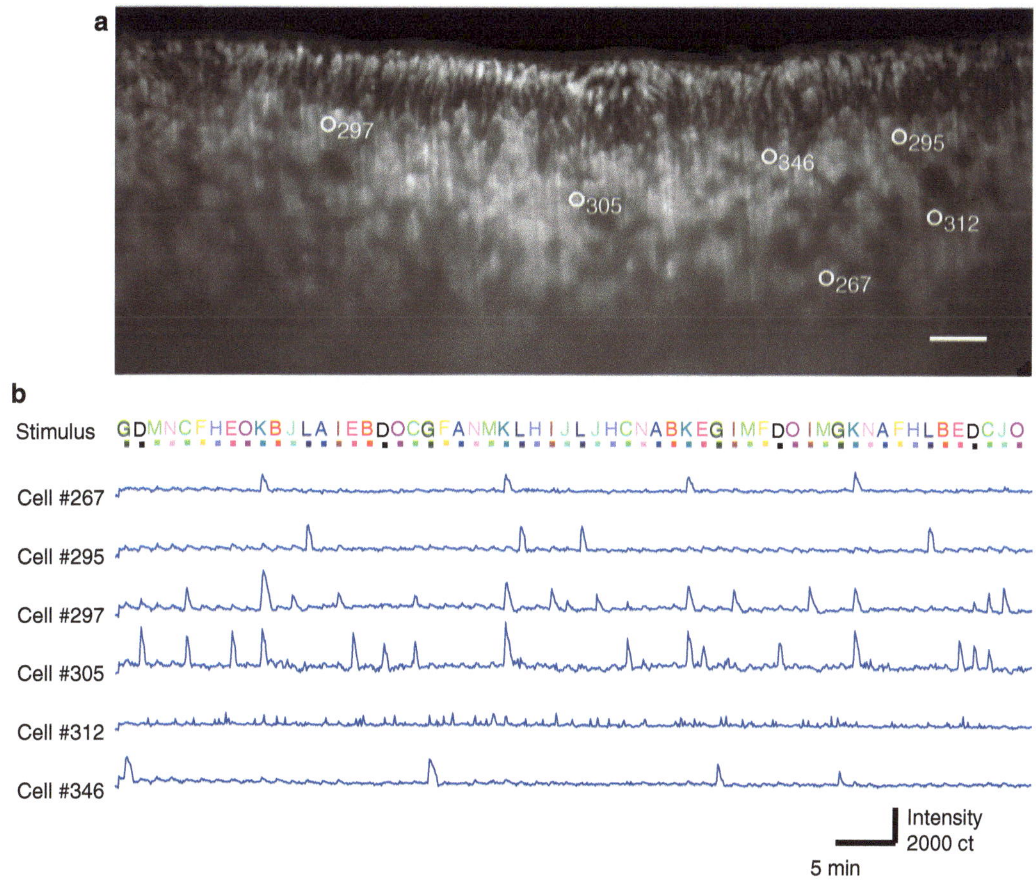

Fig. 1 Raw images and representative raw traces of individual VSNs labeled by GCaMP2. (**a**) Shows one of the frames in a three-dimensional stack image of a single VNO preparation. Six example VSNs are noted in *white circles* and labeled by *numbers*. Scale bar = 50 μm. (**b**) Shows the fluorescent intensity traces of example VSNs throughout a 75-min recording. The *top color bars* and *letters* represent 15 different stimuli (B–N: 13 different sulfated steroids [8], A and O: Ringer's negative controls). Each stimulus was tested four times, in randomized order. Each example VSN responds to particular sulfated steroids with high reproducibility in all four trials. Example neurons were chosen nearly at random from regions of interest within this optical section, and are typical of the majority of neurons in the recording

5. If the nitrocellulose membrane breaks easily when mounting in the imaging chamber, one can attach a piece of black electrical tape to the back of the nitrocellulose membrane in advance.
6. The other VNO tissue from the same mouse can be dissected and mounted in the same way. Keep it (after adhering it to the nitrocellulose membrane) in the carboxygenated Ringer's on ice. Cover the ice bucket to prevent the bleaching of fluorescent calcium indicator. Changing the top half of the medium with fresh carboxygenated Ringer's 1–2 times per hour is recommended. The VNO tissue can last for at least 4 h in the ice bucket without obvious compromised responsiveness.

7. It is important to make the VNO flat. Normally when the whole dissection procedure is performed quickly (within 12–18 min), the VNO is relatively flat; and it automatically flattens as the liquid drop is wicked away. However, when the dissection takes too long, the VNO starts to fold. In that case, try to flatten the VNO using #5 forceps under the stereomicroscope.
8. If too much liquid remains, VNO tissue will not attach to the nitrocellulose membrane firmly. Tissue may float up during later recording.
9. At room temperature, VNO tissue may die in minutes without carboxygenated Ringer's solution, so be certain to keep the Ringer's flush flowing throughout the experiment. If you temporarily stop the superfusion, remember to switch off the heater, and keep in mind that the solution sitting inside the heater probe will be substantially overheated—you will likely need to dump this solution without flowing it onto living tissue.
10. Never bump the objective onto the tissue or any part of the imaging chamber. The most straightforward approach is to find the tissue and/or bottom of the chamber using reflected light, which, because it can be performed with very-low-intensity illumination, has the further advantage of limiting any photobleaching while finding the tissue. Set the illumination to very low power (you may need to reduce the laser intensity by adding an extra neutral density filter in the light path) and temporarily take out the emission filter. Lower the objective until you see a "line of light" corresponding to reflection of the light sheet off the bottom of the chamber. Move the stage until the tissue (which will not show cellular detail due to laser speckle) is in the field of view. You can then restore the emission filter, adjust the laser intensity, and check the fluorescent image. Keep the light power low when the emission filter is out of the imaging pathway, to avoid any risk of damaging the camera.
11. With little space between the objective and light sheet optics, heater probes can easily hit the objective or the light sheet optics when one lowers the objective. To avoid this, initially position the heater probe higher up and off to the side of the VNO itself. Once the tissue has been positioned in the field of view, one can safely reposition the heater probe directly above the VNO. After that, one should avoid changing the position of the heater probe, especially after the tissue has situated under the constant superfusion pressure. Otherwise, the tissue may change shape due to the changed superfusion direction, pressure, etc.
12. Be patient when waiting. A good VNO tissue will well maintain responsiveness to stimuli in the superfused chamber for over 4 h.

On the other hand, even a perfectly mounted VNO tissue will shift by at least 100 μm in the first 30 min (probably because of the new superfusion environment). Such large shifts are difficult to correct at the analysis step. Normally, after 30 min, VNO tissue will keep relatively stable geometry if the heater probe remains in the same position. The small amount of remaining warping may be addressed by image registration during data analysis.

13. Stimuli are randomly ordered, and repeated 4–5 times each over the duration of the experiment.
14. Solutions sitting in the Teflon tubing and the silicon pinch tubing over prolonged periods may lose oxygen and CO_2, causing artificial fluorescence increases in the whole VNO tissue. Thus, one should dump solutions in all lines right before the recording.
15. One should carefully choose the laser power. There are two factors to consider: first, to achieve a decent signal-to-noise ratio, one needs to collect enough photons from each cell; secondly, one should minimize the laser power to avoid photobleaching and phototoxicity of the tissue. To balance the two factors, calculate how many photons (N) are needed to achieve your target signal-to-noise ratio. For example, suppose you need to be sensitive to $\Delta F/F$ of 1 % for each cell. When the dominant source of noise is shot noise, the noise is $\sqrt{N}$. So for $\sqrt{N}/N <= 1$ %, one requires $N >= 10{,}000$ photons/cell. If the camera gain is 15 (one photon gives a signal of 15 counts), then counts required for each cell are $10{,}000 \times 15 = 150{,}000$. If a cell occupies 100 pixels, each pixel needs 1,500 counts. Thus, one should adjust the laser power to get an average intensity no less than 1,500 counts above the camera bias.

Acknowledgements

This work was supported by NIH/NIDCD 005964, NIH/NINDS 068409, and NIH/NDPA 006437 (T.E.H.).

References

1. Døving KB, Trotier D (1998) Structure and function of the vomeronasal organ. J Exp Biol 201:2913–2925
2. Holy TE, Dulac C, Meister M (2000) Responses of vomeronasal neurons to natural stimuli. Science 289:1569–1572
3. Holekamp TF, Turaga D, Holy TE (2008) Fast three-dimensional fluorescence imaging of activity in neural populations by objective-coupled planar illumination microscopy. Neuron 57:661–672
4. Boschat C, Pélofi C, Randin O et al (2002) Pheromone detection mediated by a V1r vomeronasal receptor. Nat Neurosci 5:1261–1262
5. Leinders-Zufall T, Brennan P, Widmayer P et al (2004) MHC class I peptides as chemosensory signals in the vomeronasal organ. Science 306:1033–1037

6. Chamero P, Marton TF, Logan DW et al (2007) Identification of protein pheromones that promote aggressive behaviour. Nature 450:899–902
7. He J, Ma L, Kim S et al (2008) Encoding gender and individual information in the mouse vomeronasal organ. Science 320:535–538
8. Turaga D, Holy TE (2012) Organization of vomeronasal sensory coding revealed by fast volumetric calcium imaging. J Neurosci 32: 1612–1621
9. Arnson HA, Holy TE (2011) Chemosensory burst coding by mouse vomeronasal sensory neurons. J Neurophysiol 106:409–420
10. Arnson HA, Fu X, Holy TE (2010) Multielectrode array recordings of the vomeronasal epithelium. J Vis Exp. doi: 10.3791/1845

Chapter 15

Calcium Imaging of Vomeronasal Organ Response Using Slice Preparations from Transgenic Mice Expressing G-CaMP2

C. Ron Yu

Abstract

The vomeronasal organ (VNO) in vertebrate animals detects pheromones and interspecies chemical signals. We describe in this chapter a Ca^{2+} imaging approach using transgenic mice that express the genetically encoded Ca^{2+} sensor G-CaMP2 in VNO tissue. This approach allows us to analyze the complex patterns of the vomeronasal neuron response to large number of chemosensory stimuli.

Key words Vomeronasal organ, Pheromone, Urine, Slice preparation, G-CaMP2, Calcium imaging

1 Introduction

The vomeronasal organ (VNO) detects pheromones; chemical cues that carry information about the social, sexual, and reproductive status of the individuals within the same species [1, 2]; as well as signals emitted by animals and other species [3]. These signals activate the vomeronasal sensory neurons (VSNs) with high levels of specificity and sensitivity [4]. At least three distinct families of G-protein-coupled receptors, V1R, V2R, and FPR [5–14], are expressed in VNO neurons to mediate the detection of the chemosensory cues. To understand how pheromone information is encoded by the VNO, it is critical to analyze the response profiles of individual VSNs to various stimuli and identify the specific receptors that mediate these responses.

The neuroepithelia of VNO are enclosed in a semi-blind tubular structure, with one open end (the vomeronasal duct) connecting to the nasal cavity. VSNs extend their dendrites to the lumen part of the VNO, where the dendritic knobs and microvilli interact with pheromone cues to generate electrical currents. The currents transmit pheromone activation to the cell body to evoke action potentials. Various approaches have been developed to detect responses

Kazushige Touhara (ed.), *Pheromone Signaling: Methods and Protocols*, Methods in Molecular Biology, vol. 1068,
DOI 10.1007/978-1-62703-619-1_15,

of VSNs to sensory stimuli, including electrophysiological recordings and calcium imaging [4, 12, 15–19]. Sensory stimulation elicits strong Ca^{2+} influx in VSNs that is indicative of receptor activation [4, 20]. Imaging calcium signals, therefore, provides a means to monitor large number of neurons.

Here I describe a method using acute slice preparation from transgenic mice expressing the genetically encoded calcium sensor G-CaMP2 to conduct calcium imaging. Traditional calcium imaging utilizes synthetic calcium dyes. The dye loading processes are usually invasive and often cause damage to tissues. The detection of the calcium signals can also be affected by the uneven loading and the limited penetration of the dye in the tissue. The use of genetically encoded sensor provides a sensitive readout of the neuronal responses. The specific expression of the sensor in targeted tissue and cell type make it feasible to monitor response from specific population of neurons. These sensors also make it possible for chronic imaging in live animals. The G-CaMP proteins are calcium sensors that have been used in a variety of animal systems. They provide some of the highest sensitivity and temporal response to calcium transients [21–24].

We have developed transgenic mice that express G-CaMP2 in the olfactory sensory neurons, including the VSNs [15, 25]. The sensitivity and the genetic nature of the probe greatly facilitate Ca^{2+} imaging experiments by eliminating the dye loading process [4, 20]. We also employ a ligand delivery system that enables application of various stimuli to the VNO slices. The combination of the two techniques allows us to monitor multiple neurons simultaneously in response to large number of stimuli. These procedures are described in this chapter.

2 Materials

2.1 Solutions, Stimuli, and Embedding Materials

1. Prepare three stock 10× stock solutions (R1, R2, and R3) according to the table. They will be used to prepare the mouse artificial cerebrospinal fluid (mACSF) and Ringer's solutions. The solutions should be kept at 4 °C.

   10× R1 stock solution: In 0.8 L water, dissolve 73.05 g NaCl 1.86 g KCl, 2.94 g $CaCl_2 \cdot 2H_2O$, and 1.72 g $NaH_2PO_4 \cdot H_2O$. Add 10 mL of 1 M stock $MgCl_2$ solution. Add water to a final volume of 1 L.

   10× R2 stock solution: In 0.8 L water, dissolve 21 g $NaHCO_3$. Add water to a final volume of 1 L.

   10× R3 stock solution: In 0.8 L water, dissolve 73.05 g NaCl 1.86 g KCl, 2.94 g $CaCl_2 \cdot 2H_2O$, and 1.72 g $NaH_2PO_4 \cdot H_2O$. Add 20 mL of 1 M stock $MgCl_2$ solution. Add 50 mL of 1 M stock HEPES solution. Add water to a final volume of 1 L.

2. On the day of the experiment make fresh mACSF: In 0.7 L double-distilled water, add 100 mL of 10× R1 and 100 mL of 10× R2. Add 1.8 g of dextrose. Add water to a final volume of 1 L. The mACSF should have osmolarity of 310–315 mOsm/L. Adjust the osmolarity with dextrose or water if necessary. Aerate the solution with carboxygen gas, containing 95 % oxygen and 5 % carbon dioxide, for at least 30 min before adjusting the pH of the solution to 7.2–7.4. This method prevents calcium carbonate precipitation at high pH.
3. On the day of the experiment make 1 L of Ringer's solution: In 0.7 L water, add 100 mL of 10× R2 and 100 mL of 10× R3. Add 1.8 g of dextrose. Add water to final volume of 1 L. Adjust if necessary to the same osmolarity and pH as the mACSF. Aerate the solution with carboxygen gas.
4. Prepare 4 % low-melting agarose (LMA) in Ringer's solution: Weight 0.4 g LMA and add to 10 mL Ringer's solution. Melt the LMA and aliquot the agarose in Eppendorf tubes and store at 4 °C till use.
5. Prepare pheromone stimuli in Ringer's solution. For mouse urine, 1:100 dilution gives robust response. For chemical compounds, peptides, and proteins, dissolve them at the desired concentration.

2.2 Mice

Compound heterozygotic mice containing both the *OMP-IRES-tTA* allele and the *tetO-G-CaMP2* allele are used. The two lines have been deposited at the Jackson laboratories. Stock numbers are no. 017754 (*OMP-IRES-tTA*) and no. 017755 (*tetO-G-CaMP2*), respectively.

2.3 Vibratome and Microscope

1. Any vibratome for live brain tissues can be used. Our laboratory uses the VF300 tissue slicer (Precisionary Instruments, Greenville, NC) to make sections.
2. We use the Zeiss AxioSkope FS2 microscope with a 10× or a 20× water-dipping lens for time-lapse imaging. Standard GFP band-pass filter (450–490 nm) is used for G-CAMP2 signals. The epifluorescent images are acquired by a CCD camera (Zeiss HRM) with 1 × 1 or 2 × 2 binning depending on the expression levels of G-CaMP2.

2.4 Perfusion System

1. Set up a perfusion system on the microscope stage. The perfusion system includes three main parts: a delivery system providing constant flow of oxygenated mACSF, a stimulus delivery system, and a suction system to remove excessive liquid (Fig. 1a).
2. Place a perfusion chamber (Siskiyou, Grants Pass, OR) on the stage (Fig. 1b).

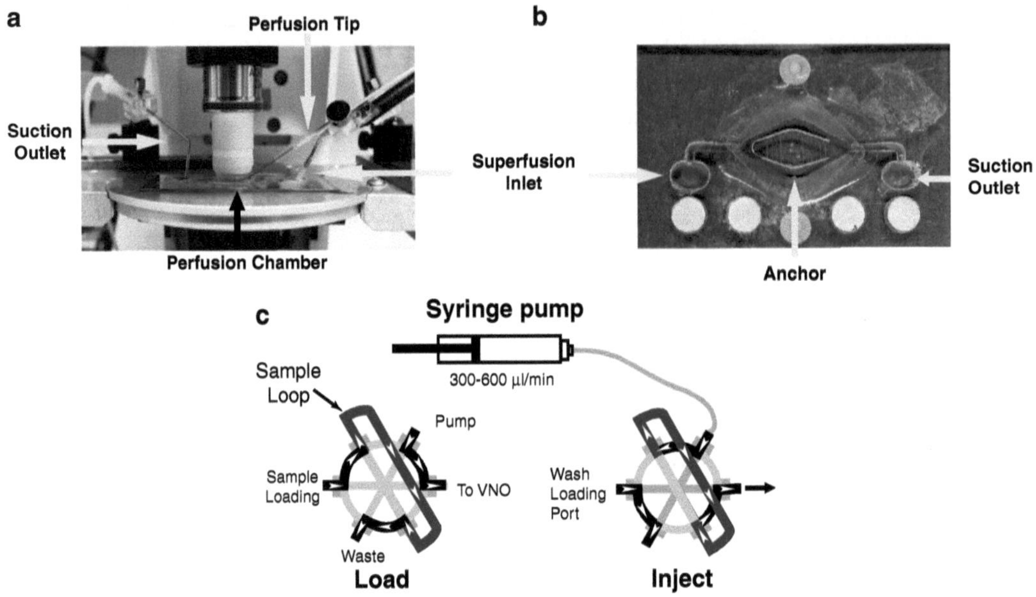

Fig. 1 Illustration of perfusion system setup. (**a**) A typical perfusion chamber with inlet and outlet. The perfusion chamber is placed on microscope stage under the dipping objective. The mACSF inlet, suction outlet, and perfusion tip are indicated. (**b**) Top view of the perfusion chamber. A VNO slice is positioned in the center of the chamber and pressed down with a tissue anchor. (**c**) Schematic illustration of the flow directions at the "load" and "inject" positions. The single-barrel syringe pump provides continuous flow of Ringer's through the perfusion tip. *Light gray contour* illustrates the ports in the HPLC injection port. The *dark gray line* illustrates the injection loop. *Black lines* illustrate the flow of liquid within the port at different positions. *Arrowheads* indicate the directions of liquid flow

3. Place two perfusion port holders (MPIOH-S ALA scientific, Inc.) on both sides of the chamber, one holding the mACSF superfusion inlet and the other holding the suction needle.
4. Position a micromanipulator alongside the superfusion inlet port to hold the perfusion tip such that it can be position to deliver the stimuli.
5. Place a reservoir of oxygenated mACSF above the imaging setup and connect the outflow tubing to the superfusion inlet. A gravity-driven continuous flow is fed into the imaging chamber at a speed of ~1 mL/min.
6. Set up stimulus delivery system. We use an HPLC injection loop (V-451, Chromtech, Apple Valley, MN). The injection loop (PEEK sample loop, 1803, Chromtech) has two flow routes controlled by a switch (Fig. 1c). At either position, a constant flow of Ringer's solution is injected by a syringe pump (NE-300, New Era Pump Systems, Farmingdale, NY). At the "load" position, the flow bypasses the sample loop and goes directly to the outlet connected to the perfusion tip. Stimulus solution can be injected into the sample loop using a Hamilton

precision syringe (80630, Chromtech). The liquid contained in the sample loop will be the volume delivered. Excess sample exits the sample loop through waste outlet. The volume being injected can be controlled by the size of the loop. We use a 20 μL loop in our experiments. When switched to the "injection" position, the pump solution flows through the sample loop and pushes the stimuli into the outlet. Other commercial or custom-made injection systems can also be implemented for stimulus delivery.

3 Methods

3.1 Slice Preparation

1. Before sacrificing the animal, melt two tubes of LMA on a heat block at >60 °C. As soon as the gel liquefies, transfer the tubes to a 37 °C heat block.
2. Decapitate a G-CAMP2 mouse following CO_2 euthanasia (*see* **Note 1**). Cut the mandible bones with scissors and remove the lower jaw. Peel off the ridged upper palate tissue to expose the nasal cavity (Fig. 2a). Insert a surgical blade between the two front incisors to expose VNO. Lift the whole VNO from the nasal cavity by holding on the tail bone (Fig. 2b). Immediately transfer the VNO to oxygenated mACSF solution placed on ice.
3. Under dissection scope, separate the two VNOs by sliding the tip of a pair of #5 forceps gently along the wall of septal bone (Fig. 2c). Peel away the vomer bone encasing the VNO. Gently lift the VNO from the bone cavity (*see* **Note 2**).
4. Hold the posterior end of the VNO with a pair of forceps and gently submerge it into the melted agarose (Fig. 2d). Immediately place the tube on ice to solidify the agarose (*see* **Note 3**).
5. Proceed to sectioning immediately after the LMA solidifics. Supply cold oxygenated mACSF into the sectioning chamber and start cutting at 150–200 μm thickness per slice (Fig. 2d; *see* **Note 4**).
6. Collect and transfer the sectioned slices to mACSF incubation chamber. The slices are viable for 6–8 h in oxygenated mACSF at room temperature.

3.2 Imaging Chamber Setup

1. Place the VNO slice in the middle of the perfusion chamber and hold the slice down with a slice anchor (Warner Instruments, Hamden, CT). Oxygenated mACSF is delivered to the perfusion chamber through inlet port at ~1 mL/min and the liquid is drained through the suction needle at the opposite side of the perfusion chamber (*see* **Note 5**).
2. Fill a 30 mL syringe with Ringer's solution and clamp it to the syringe pump. Set the pump speed to 300–600 μl/min to

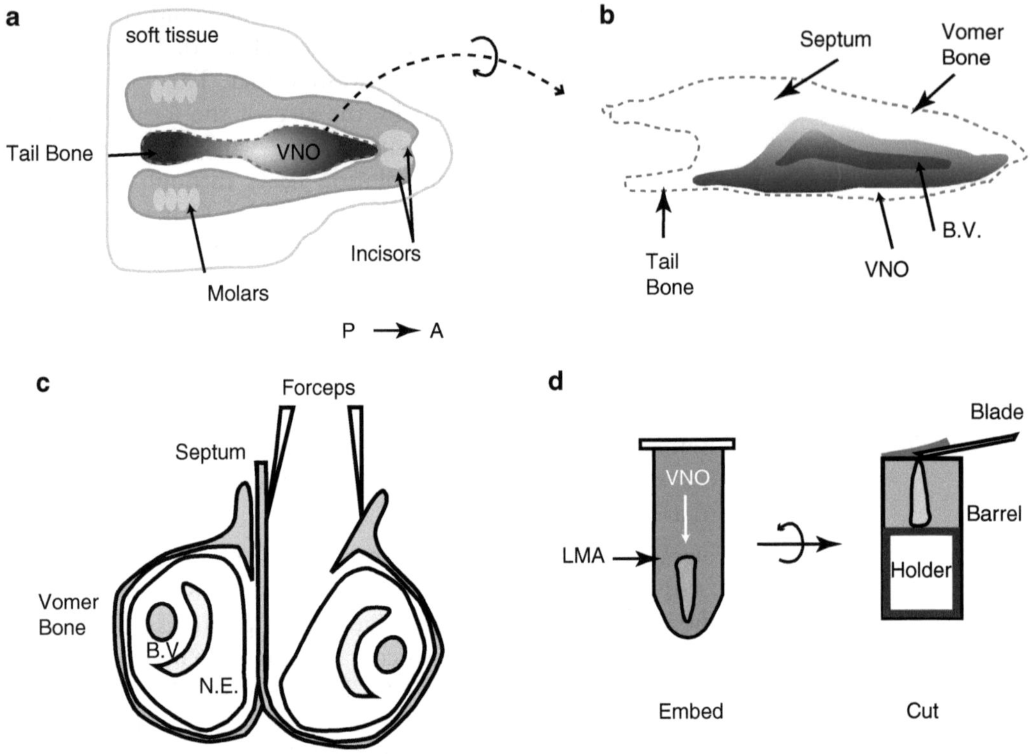

Fig. 2 Schematic illustration of VNO dissection process. (**a**) The anatomical location of VNO in the mouse head. The drawing illustrates the head of a mouse laid upside down, with the jaw removed and the palate peeled to expose the VNO. (**b**) A side view of the isolated VNO that is enclosed in the vomer bone. (**c**) A coronal view of the VNO and dissection process. One VNO is separated from the septum and the vomer bone can then be removed to extricate the neuroepithelium. (**d**) VNO is embedded into LMA. The embedded block is glued to the tissue holder for sectioning. The tissue holder is advanced at 180–200 μm per slice pushing the agarose block out of the metal barrel for sectioning. The cutting blade is positioned closely to the metal barrel. *B.V.* blood vessel. *N.E.* neuroepithelium

provide a continuous flow of Ringer's solution over the slice (Fig. 1c; *see* **Note 6**).

3. Connect the outlet of the Ringer's to the HPLC injection loop.
4. Adjust the perfusion tip under a 5× or a 10× lens so that the tip is about 1 mm away from the VNO slice.
5. Once the perfusion system is set up, test the device by measuring the sample delay time and duration using a solution containing a fluorescent dye. Load 0.1 % rhodamine 6G dissolved in Ringer's solution into the sample loop, switch the valve to inject position, and perform time-lapse imaging as would for pheromone stimuli (*see* **Note 7**).

3.3 Time-Lapse Imaging

1. Set the acquisition speed to one frame per second. Adjust the intensity of the light to minimize bleaching of the G-CaMP2 signals and photo damage to the cells.
2. Set the injection loop to "load" position. Load the injection loop with stimulus solution.
3. Start image acquisition.
4. Switch the injection valve from "load" to "injection" position at a specific time point (e.g., 5 s after start) for one set of experiment to obtain consistent time delay in all trials (*see* **Notes 6** and **8**).
5. End acquisition at desired time. Typically we acquire a 60-frame image stack (~60–70 s).
6. Wait for 4–10 min for the VSNs to recover before applying the next stimulus (*see* **Note 9**).

3.4 Data Analysis

1. Perform image registration of all the images acquired from one slice. We use a custom-written VBA script in AxioVision (Carl Zeiss, North America) to automate this process (*see* **Note 10**). All image frames within the same experiment are registered against a common chosen reference frame with elastic registration (*see* **Note 11**). This reference frame is chosen arbitrarily from the image stacks.
2. Perform image subtraction to identify the responding cells. We use custom-written macros in ImageJ v1.42 (http://rsb.info.nih.gov/ij/, NIH, Bethesda, MD) to automate this process. A minimal projection image is generated for each stack. Responding cells emerge after the minimal projection is subtracted from the raw stacks.
3. Identify the regions of interest (ROIs) from the subtracted stacks. Obtain the ROI coordinates using Multi-Measure PlugIn from ImageJ. Process all stacks for one experiment and save all ROI coordinates in an ROI master list.
4. Use the master list of ROI to measure cell responses from raw image stacks with custom-written macro and Multi-Measure PlugIn from ImageJ.
5. Data produced by Multi-Measure are exported into a spreadsheet for further analysis.

3.5 Alternative Approach

In addition to the mice discussed above, another mouse line could offer similar convenience and signal sensitivity for recording Ca^{2+} signals from VNO slices. Recently, a Cre-dependent reporter line that expresses a new version of G-CaMP, G-CaMP3, has been developed [26]. G-CaMP3 has been shown to have lower basal level of fluorescence, and therefore afford a better signal [24]. One could cross this reporter line with a line that expressed the Cre recombinase in the mouse olfactory systems [27, 28]. The resulting line is expected to express G-CaMP3 in the VNO. However, this alternative approach has not been tested. The G-CaMP3 expression is driven by the actin promoter. Whether it can drive G-CaMP3 expression to a level that allows slice imaging has yet to be confirmed.

4 Notes

1. Please follow the procedures approved by the Institutional Animal Care and Use Committee.
2. Extreme care should be taken in this step not to damage the neuroepithelium. Small bone fragments left on the tissue surface must be removed completely before embedding. Fragments left on the tissue may be caught by the cutting blade to pull the tissue out of the agarose block.
3. The embedding and cooling process should take less than 2 min to ensure the health of the tissue.
4. Adjust the advancing speed and vibration frequency so that the tissue is not compressed during sectioning and VNO does not fall off from the agarose.
5. The threads of the anchor should only press against the LMA part of the slice but not the VNO tissue.
6. Prior to the imaging experiment, air bubbles should be chased out to ensure smooth flow of the perfusion fluid.
7. Perform a test run using Ringer's solution as the stimulus. Readjust the perfusion setup if movement artifact is introduced during sample injection.
8. Perform a positive control run with mouse urine diluted at 1:100 in Ringer's solution. The typical maximal $\Delta F/F$ value of G-CaMP2 response to mouse urine is around 20–40 %. If needed, one can confirm the viability of the slice by delivering 10 mM KCl in Ringer's to stimulate the slices at the end of the experiments.
9. Wash the Hamilton syringe in Ringer's solution at least three times after loading one stimulus. Wash the sample loop with Ringer's solution at least three times between different stimuli. These steps prevent cross contamination among different samples.
10. Image registration can also be done in ImageJ.
11. Elastic registration, also known as nonlinear registration, is a category of image registration technique emphasizing the transformation of a target image non-rigidly to a reference image. Here we used the implementation from AxioVision.

Acknowledgments

This work is supported by funding from Stowers Institute and the NIH (NIDCD 008003) to C.R.Y.

References

1. Birch MC (1974) Pheromones. American Elsevier Pub. Co., New York
2. Wyatt TD (2003) Pheromones and animal behaviour: communication by smell and taste. Cambridge University Press, Cambridge, UK
3. Papes F, Logan DW, Stowers L (2010) The vomeronasal organ mediates interspecies defensive behaviors through detection of protein pheromone homologs. Cell 141:692–703
4. Leinders-Zufall T, Lane AP, Puche AC, Ma W, Novotny MV, Shipley MT, Zufall F (2000) Ultrasensitive pheromone detection by mammalian vomeronasal neurons. Nature 405: 792–796
5. Dulac C, Axel R (1995) A novel family of genes encoding putative pheromone receptors in mammals. Cell 83:195–206
6. Herrada G, Dulac C (1997) A novel family of putative pheromone receptors in mammals with a topographically organized and sexually dimorphic distribution. Cell 90:763–773
7. Matsunami H, Buck LB (1997) A multigene family encoding a diverse array of putative pheromone receptors in mammals. Cell 90: 775–784
8. Ryba NJ, Tirindelli R (1997) A new multigene family of putative pheromone receptors. Neuron 19:371–379
9. Pantages E, Dulac C (2000) A novel family of candidate pheromone receptors in mammals. Neuron 28:835–845
10. Zhang X, Rodriguez I, Mombaerts P, Firestein S (2004) Odorant and vomeronasal receptor genes in two mouse genome assemblies. Genomics 83:802–811
11. Liberles SD, Horowitz LF, Kuang D, Contos JJ, Wilson KL, Siltberg-Liberles J, Liberles DA, Buck LB (2009) Formyl peptide receptors are candidate chemosensory receptors in the vomeronasal organ. Proc Natl Acad Sci U S A 106:9842–9847
12. Riviere S, Challet L, Fluegge D, Spehr M, Rodriguez I (2009) Formyl peptide receptor-like proteins are a novel family of vomeronasal chemosensors. Nature 459:574–577
13. Yang H, Shi P, Zhang YP, Zhang J (2005) Composition and evolution of the V2r vomeronasal receptor gene repertoire in mice and rats. Genomics 86:306–315
14. Rodriguez I, Del Punta K, Rothman A, Ishii T, Mombaerts P (2002) Multiple new and isolated families within the mouse superfamily of V1r vomeronasal receptors. Nat Neurosci 5: 134–140
15. He J, Ma L, Kim S, Nakai J, Yu CR (2008) Encoding gender and individual information in the mouse vomeronasal organ. Science 320: 535–538
16. Holy TE, Dulac C, Meister M (2000) Responses of vomeronasal neurons to natural stimuli. Science 289:1569–1572
17. Chamero P, Marton TF, Logan DW, Flanagan K, Cruz JR, Saghatelian A, Cravatt BF, Stowers L (2007) Identification of protein pheromones that promote aggressive behaviour. Nature 450:899–902
18. Leinders-Zufall T, Ishii T, Mombaerts P, Zufall F, Boehm T (2009) Structural requirements for the activation of vomeronasal sensory neurons by MHC peptides. Nat Neurosci 12: 1551–1558
19. Kimoto H, Haga S, Sato K, Touhara K (2005) Sex-specific peptides from exocrine glands stimulate mouse vomeronasal sensory neurons. Nature 437:898–901
20. Leinders-Zufall T, Brennan P, Widmayer P, S PC, Maul-Pavicic A, Jager M, Li XH, Breer H, Zufall F, Boehm T (2004) MHC class I peptides as chemosensory signals in the vomeronasal organ. Science 306:1033–1037
21. Hendel T, Mank M, Schnell B, Griesbeck O, Borst A, Reiff DF (2008) Fluorescence changes of genetic calcium indicators and OGB-1 correlated with neural activity and calcium in vivo and in vitro. J Neurosci 28:7399–7411
22. Pologruto TA, Yasuda R, Svoboda K (2004) Monitoring neural activity and [Ca2+] with genetically encoded Ca2+ indicators. J Neurosci 24:9572–9579
23. Jayaraman V, Laurent G (2007) Evaluating a genetically encoded optical sensor of neural activity using electrophysiology in intact adult fruit flies. Front Neural Circuits 1:3
24. Tian L, Hires SA, Mao T, Huber D, Chiappe ME, Chalasani SH, Petreanu L, Akerboom J, McKinney SA, Schreiter ER, Bargmann CI, Jayaraman V, Svoboda K, Looger LL (2009) Imaging neural activity in worms, flies and

mice with improved GCaMP calcium indicators. Nat Methods 6:875–881

25. He J, Ma L, Kim S, Schwartz J, Santilli M, Wood C, Durnin MH, Yu CR (2010) Distinct signals conveyed by pheromone concentrations to the mouse vomeronasal organ. J Neurosci 30:7473–7483
26. Zariwala HA, Borghuis BG, Hoogland TM, Madisen L, Tian L, De Zeeuw CI, Zeng H, Looger LL, Svoboda K, Chen TW (2012) A Cre-dependent GCaMP3 reporter mouse for neuronal imaging in vivo. J Neurosci 32:3131–3141
27. Eggan K, Baldwin K, Tackett M, Osborne J, Gogos J, Chess A, Axel R, Jaenisch R (2004) Mice cloned from olfactory sensory neurons. Nature 428:44–49
28. Li J, Ishii T, Feinstein P, Mombaerts P (2004) Odorant receptor gene choice is reset by nuclear transfer from mouse olfactory sensory neurons. Nature 428:393–399

Chapter 16

The Electrovomeronasogram: Field Potential Recordings in the Mouse Vomeronasal Organ

Trese Leinders-Zufall and Frank Zufall

Abstract

Mammalian vomeronasal neurons (VSNs) located in the sensory epithelium of the vomeronasal organ (VNO) detect and transduce molecular cues emitted by other individuals and send this information to the olfactory forebrain. The initial steps in the detection of pheromones and other chemosignals by VSNs involve interaction of a ligand with a G protein-coupled receptor and downstream activation of the primary signal transduction cascade, which includes activation of ion channels located in microvilli and the dendritic tip of a VSN. The electrovomeronasogram (EVG) recording technique provides a sensitive means through which ligand-induced activation of populations of VSNs can be recorded from the epithelial surface using an intact, ex vivo preparation of the mouse VNO. We describe methodological aspects of this preparation and the EVG recording technique which, together with single-cell recordings, contributed significantly to our understanding of mammalian vomeronasal function, the identification of pheromonal ligands, and the analysis of mice with targeted deletions in specific signal transduction molecules such as Trpc2, Gαo, V1R, or V2R receptors.

Key words Field potential, Generator potential, Receptor potential, EOG, EVG, Multibarrel pipette, Agar bridge, Nonmetallic syringe needle, VNO dissection, TRP channel

1 Introduction

An important technique for measuring electrical activity of nasal sensory neuron populations is the local field potential recorded extracellularly from the surface of the nasal sensory epithelium in response to chemical stimulation. In the main olfactory epithelium, this technique is referred to as electroolfactogram (EOG) [1, 2] whereas it is called electrovomeronasogram (EVG) when signals from the sensory epithelium of the vomeronasal organ (VNO) in the accessory olfactory system are measured [3–5]. In both cases, these field potentials represent summated primary responses of sensory neuron populations induced by molecular stimuli. The size of the field potential was shown to be associated with a change in action potential firing frequency [6] as predicted for a generator (receptor) potential [7].

Kazushige Touhara (ed.), *Pheromone Signaling: Methods and Protocols*, Methods in Molecular Biology, vol. 1068,
DOI 10.1007/978-1-62703-619-1_16, © Springer Science+Business Media, LLC 2013

In this chapter we focus on EVG recordings from mouse VNO. Classical EOG and EVG recordings initially used vapor-phase stimulation methods, meaning that the odor stimulus was applied via an air puff. However, the mucous layer of the sensory epithelium is prone to dehydration. Therefore, a source of artifacts in these recordings can be caused by epithelial drying and thinning of the mucus [8] or a change in humidity of the airstream associated with an odorant stimulus [9]. We therefore developed submerged preparations for both mouse main olfactory epithelium [10–13] and VNO [12, 14–21], comparable to the "underwater" EOG first reported in snake, turtle, and fish [4, 22, 23]. These preparations are advantageous as they are independent of the volatility of a given chemosensory stimulus and allow for application of pharmacological agents without significantly altering the thickness of the aqueous layer covering the epithelium. Furthermore, liquid-phase stimulation mimics the natural environment for ligand access in the vomeronasal lumen. Under our conditions, the EVG time course (activation and deactivation), including its adaptational properties, can be precisely measured in a repeated manner [15, 21].

The EVG recording technique is ideal as a high-throughput method for identifying potential pheromonal ligands acting on populations of vomeronasal neurons (VSNs) by using precisely timed stimuli of known concentrations. Evaluation of complex mixtures together with vomeronasal axotomy established that EVG recordings reflect the summed neural activity of VSNs [4]. The first pheromonal ligands activating mouse VSNs were discovered using the submerged EVG technique [21] described in this protocol. Subsequently, VSN activity to these newly identified stimuli was confirmed using alternative recording methods with single-cell resolution, such as calcium imaging in acute vomeronasal tissue slices and electrophysiological action potential recordings from individual VSNs [21]. The EVG recording method proved useful in identifying vast ligand families, including peptide and protein pheromones, that stimulate mouse VSNs [14, 17, 24, 25]. The EVG recording method has also been instrumental in analyzing vomeronasal phenotypes of mice with targeted deletions in specific signaling molecules including seven transmembrane receptors, G proteins, and ion channels [14, 18, 20, 21, 24, 26]. For instance, mice harboring a null mutation of *Trpc2*, which encodes the canonical transient receptor potential channel type 2, show severely diminished EVG responses, a finding that indicated a central role of this cation channel in the detection of pheromones and other chemosignals in the VNO [20]. Using a similar approach, Gαo was shown to be essential for sensing peptide and protein pheromones by a subtype of VSNs [14]. Furthermore, the first experimental evidence that V1R receptors function as pheromone receptors was shown by EVG recordings [18]. Thus, although quantitative analysis of different mouse strains with specific vomeronasal gene deletions can be demanding, use of the EVG method has provided critical phenotypic information related to the function of such genes. However, in many cases subsequent analysis using methods that allow single-cell resolution, such as patch-clamp recording or calcium imaging, is required to verify results from EVG recordings. The same ex vivo mouse VNO preparation that we initially developed for EVG recordings has also been used successfully for confocal imaging of Ca^{2+} responses in single dendritic endings of VSNs [27].

2 Materials

2.1 Preparation of Solutions

For the preparation of all physiological solutions we use ultrapure water (>18.2 MΩ-cm resistivity at 25 °C, low ppt in divalent cations). There are various systems available in laboratories which filter pretreated water (reverse osmosis or demineralized water) to produce ultrapure water with <5 ppb total organic carbon (TOC) to reduce organic contaminants.

1. Prepare extracellular solution (S1: 120 mM NaCl, 25 mM $NaHCO_3$, 5 mM KCl, 5 mM *N*,*N*-Bis(2-hydroxyethyl)-2-aminoethanesulfonic acid (BES), 1 mM $MgSO_4$, 1 mM $CaCl_2$, 10 mM glucose) according to the table with ultrapure water. The pH will be ~7.3 after 10-min aeration with carbogen (95 % O_2/5 % CO_2), and the osmolarity 300 mOsm. If a higher osmolarity is required, it can be adjusted by adding more glucose (1 mM equals 1 mOsm). The solution is filtered twice using a 0.2 μm membrane filter to eliminate dust particles and possible bacterial contaminations. The solution is stored at 4 °C. Make sure that the solution is at room temperature and aerated for 10 min with carbogen (95 % O_2/5 % CO_2) before use.
2. Prepare extracellular 4-(2-hydroxyethyl)-1-piperazine-ethanesulfonic acid (HEPES)-based solution (S2: 145 mM NaCl, 5 mM KCl, 10 mM HEPES, 1 mM $MgCl_2$, 1 mM $CaCl_2$) with ultrapure water. The pH is adjusted to 7.3 using NaOH and the osmolarity to 300 mOsm using glucose (1 mM equals 1 mOsm; *see* also **item 1**). The solution is filtered twice using a 0.2 μm membrane filter to eliminate dust particles and possible bacterial contaminations. The solution is stored at 4 °C. Make sure that it is at room temperature before use in an experiment.

2.2 Preparation of Agar Bridges

The following items are needed to produce agar bridges for the EVG recordings: agar-agar (CAS # 9002-18-0), S2 solution (*see* Subheading 2.1), a heat source with small flame (Bunsen burner with pilot light tube or alcohol burner), polyethylene tubing

(1.193 × 1.574 × 0.215 mm), 1 cc syringe, pipette tip ("yellow," 200 μl), and glass scintillation vial.

1. Dissolve 4 % (w/v) agar-agar in S2 solution by heating and stirring the solution to approximately 60 °C.
2. Cut the polyethylene tubing in approximately 3.5 cm long pieces. Every piece of tubing will become one agar bridge and is exchanged daily. Therefore, we prepare enough for at least 2 weeks.
3. The polyethylene tubing will be bent near a small flame of a heat source (Bunsen burner with pilot flame) into an angle of 90° (Fig. 1a). Note that experienced agar bridge preparers do not precut the polyethylene tubing. They take the whole roll of tubing, bend the end of the tubing over the flame, make a cut for the total length (approximately 2.5 cm), and restart the procedure.
4. Take the syringe and fill it with the heated dissolved agar solution.
5. Take the pipette tip and cut as indicated in the figure (Fig. 1a). Now the pipette tip fits onto the syringe and can be used to fill the heated dissolved agar solution into the bent polyethylene tubing.
6. Fill the prepared polyethylene tubings with the agar solution and leave the tubing on a piece of paper to cool down. These filled tubings are the agar bridges.
7. Fill a glass scintillation vial with cool S2 solution and drop the agar bridges into this solution (Fig. 1a). This kind of agar bridge storage prevents the loss of water (evaporation) and keeps the amount of salt solution in the agar bridge constant.
8. Store the agar bridges at 4 °C until use.

2.3 Head Mounting Dissecting Dish

To enable secure positioning of the mouse head and later EVG recordings from the VNO, a special head mounting dish should be prepared. The following items are needed: a plastic Petri dish (square, 9–120 × 15 mm), Sylgard 184 silicone elastomer kit, and insect pins (needles).

1. Prepare the Sylgard elastomer in a 1:10 ratio of the two components (curing agent and elastomer base, respectively) by mixing the components thoroughly in the plastic Petri dish. Use plastic syringes to extract the two components from their containers and add them directly into the Petri dish. A volume of approximately 50 ml base and 5 ml curing agent fills a 90 mm square Petri dish perfectly, leaving enough latitude to later move and manipulate the preparation and the overflow of the perfusion solution.

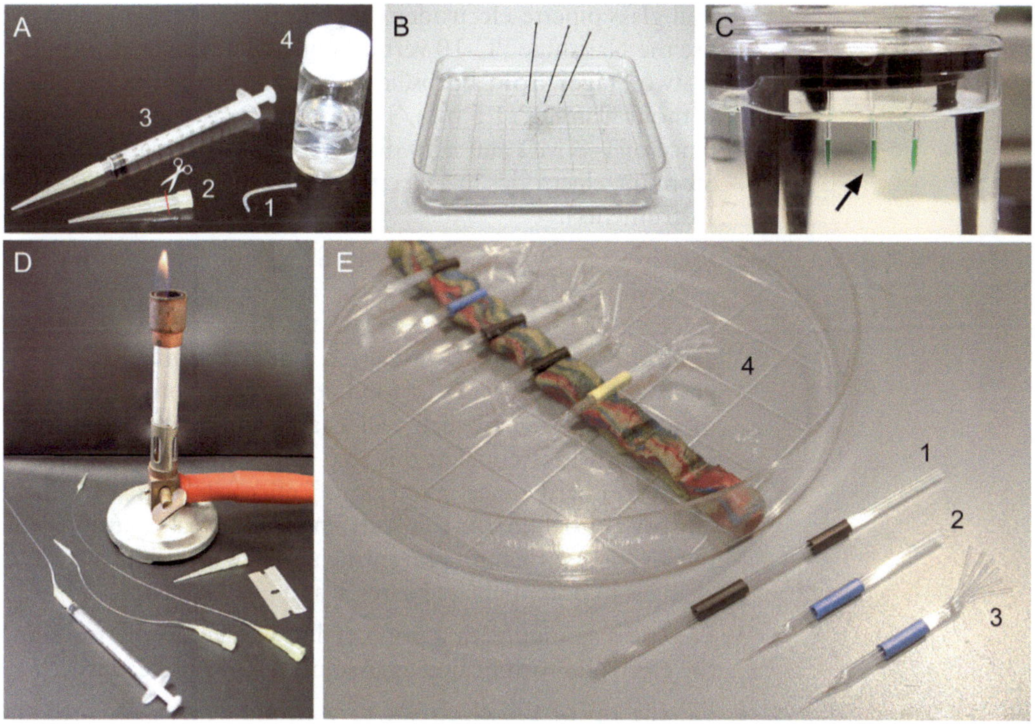

Fig. 1 (**a**) Tools for preparing and storing agar bridges: *1*, bent polyethylene tubing; *2*, "yellow" pipette tip (200 μl); *3*, syringe with pipette tip for filling the polyethylene tubing; *4*, scintillation vial for storing the agar bridges in S2 solution. To enable the yellow pipette tip to fit on the syringe, it has to be cut at the indicated position (see *red line* and a small scissor). (**b**) Head mounting dissection dish with three insect pins. The dissection dish contains hard Sylgard 184 elastomer enabling the secure positioning of the mouse head for the EVG recording. (**c**) Electrode storage jar containing S2 solution and three electrodes filled with colored agar solution (*green*). An *arrow* points out one of the electrodes. (**d**) To produce nonmetallic syringe needles a yellow pipette tip (200 μl) is heated over a pilot flame of a Bunsen burner and stretched to fabricate thin plastic tubings of preferred diameters. A single-edged razor blade cuts the needle at the desired length and the broad end of the yellow tip is cut similar as in (**a**) to fit the nonmetallic needle on a syringe. (**e**) Multibarrel stimulation pipettes are made from thin glass capillaries held together by shrinkage tubing (*1*, *black*). The glass capillaries are heated, turned, and pulled to produce a small-diameter tip using an automated puller (*2*). On the other end, the individual glass capillaries are heated and separated to enable tubing to connect to the individual capillaries (*3*). These multibarrel stimulation pipettes are stored in a big Petri dish until use

2. Let the liquid mixture cure to a flexible, transparent elastomer overnight.
3. Place three insect pins in the dish for later use (Fig. 1b; *see* Subheading 3.2, Dissection of Mouse Head and VNO). Insect pins are black-enameled stainless steel pins which offer extra protection against corrosion and also reduce glare.

2.4 Preparation of EVG Recording Electrodes

The following items are needed: patch-pipette puller, borosilicated glass, two electrode storage jars, S2 solution (*see* Subheading 2.1), agar-agar (CAS # 9002-18-0), food color, 1 cc syringe, and nonmetallic syringe needles (*see* Subheading 2.5).

1. Pull glass pipette electrodes using a programmable puller. We use the Narishige PC-10 vertical puller. The electrode is pulled in two stages from borosilicated thin-walled glass capillaries without filament (1.5 mm/1.12 mm OD/ID) and has a resistance of 1–3 MΩ and a tip diameter of approximately 1.5 μm. Store the electrodes in a electrode storage jar (World Precision Instruments).
2. Prepare 2 % (w/v) agar-agar in S2 solution by heating it to approximately 60 °C. When the agar is dissolved, add food color (McCormick Green). The tip of a colored agar-filled electrode can be easily visualized under the microscope. Be aware that not all types of food color can be used, since they may contain sugar which will change the osmolarity of the solution. It is essential at this step to have a agar solution that is not too hard, since it can rip off microvili or cilia from the sensory epithelium, or too soft, since it will leak solution onto the sensory epithelium preventing a solid contact for recording changes in potential (*see* **Notes 1** and **2**).
3. Prepare a syringe with a nonmetallic syringe needle for filling the electrode with heated agar solution (*see* Subheading 2.5).
4. Fill the tip of the electrode to a height of 0.5 cm with the heated agar solution (Fig. 1c). In this step it is essential to work fast and consistent, since the heated agar will cool and solidify relatively fast in the thin nonmetallic syringe needle. During this step it is very helpful to have another lab member hand and regain the electrodes after filling. The filled electrodes are put back into the electrode storage jar to cool down to room temperature.
5. Fill a second electrode storage jar with S2 solution (*see* subheading 2.1) to store the electrodes after visual inspection of the filling procedure.
6. Check visually under a microscope that the agar solution has reached the tip of the electrode without any air bubbles and that the agar solution did not creep up the outer glass wall. This latter point may cause false measurements of potential changes (*see* **Note 3**).
7. Store the agar-filled electrodes in the S2 solution-containing storage jar (Fig. 1c). Make sure that the electrode tips contact the solution to prevent dehydrating the agar in the electrodes. If the consistency of the agar is incorrect (low gel strength), the colored agar will leak out of the pipettes. In this case the weight/volume percentage of the agar solution has to be increased.
8. The electrodes are stored at 4 °C until use.

2.5 Preparation of Nonmetallic Syringe Needles

To fill a micropipette or glass-electrode without air bubbles is challenging if long, stainless steel hypodermic needles are used. Nonmetallic syringe needles can bend (therefore do not damage the glass capillaries) and can be bought from various companies. However, we produce our own nonmetallic needles from yellow pipette tips to reduce the cost of these items (Fig. 1d). We regularly exchange the needles due to the use of different filling solutions (*see* Subheading 2.6) or salt crystals clogging the needle. The following items are needed: pipette tips ("yellow," 200 µl), a heat source with small flame (Bunsen burner with pilot light tube or alcohol burner), and single-edge razor blade.

1. Heat up a plastic pipette tip approximately in the middle over a small flame of a heat source (pilot flame of Bunsen burner) holding both ends. Turn the plastic pipette tip slowly and notice the change in plastic coloring. Take care that the plastic does not catch fire. It is recommended to start turning the plastic pipette tip some distance from the flame and to slowly reduce the distance until one gets a sense of the optimal distance between the flame and pipette tip.
2. Move the pipette away from the heat source, the moment the plastic is changing its color and looks more liquified. Turn the pipette and hands 90° and move the hand containing the tip slowly in a constant velocity downward away from the hand holding the broad end of the yellow pipette tip. A thin plastic capillary is forming. The inner diameter of the plastic capillary can be controlled by the speed of the pull.
3. Blow air onto the plastic to harden it at the desired thickness and let it cool further down on a table.
4. Cut the plastic capillary in a 45° angle with a razor blade to the desired length. Make sure that the cut is not in a 90° angle since the plastic tubing will be squeezed together blocking the solution outflow.
5. To finalize the nonmetallic syringe needle, cut a small piece from the broad end of the plastic tip so that it fits on a 1 cc syringe (Fig. 1a, d).
6. Repeat the procedure to acquire the desired amount of nonmetallic syringe needles. The long and fine tip of this "needle" allows the filling of the glass capillaries (electrode and multibarrel stimulation pipettes; *see* Subheadings 2.4. and 2.6) very close to their narrow tips eliminating air bubble formation.

2.6 Preparation and Filling of Multibarrel Stimulation Pipette

Micropipette puller exists that can consistently and automatically produce multibarrel stimulation pipettes up to 7-barrels (PMP-107 puller, MicroData Instrument). Multibarrel pipettes (up to 15-barrels) can also be manufactured using a Narishige PE-22 puller. However, in this case the production of the pipettes depends on the aptitude of the preparer/investigator since the rotation of the pipette clamp (drill chuck) will rely on his/her skill. The flow of the stimulation solution from these pipettes can be precisely controlled by the amount of rotation in the glass pipette and the tip opening (*see* **step 6**). The following items are needed: puller, borosilicated glass, heat-shrink tubing (precut, 1 cm long), food color, 1 cc syringes, nonmetallic syringe needles (*see* Subheading 2.5), and a heat source with small flame (Bunsen burner with pilot light tube or alcohol burner). Here we describe the production and filling of 7-barrel stimulation pipettes.

1. Take seven borosilicated thin-walled glass capillaries without filament (1 mm/0.75 mm OD/ID) and fit onto either side heat-shrink tubing (Fig. 1e).
2. Heat the tubing by holding the glass capillaries on both ends and turning it over a small flame of a heat source (pilot flame of Bunsen burner). The tubing shrinks, providing a snug fit over the seven capillaries, and helps to clamp the barrels securely into the drill chuck of the puller without damaging the glass.
3. Fasten the seven capillaries into the pipette clamps of the puller (MicroData Instrument, PMP-107 puller) and use the puller's control and functions to pull the pipette in two stages. In case the Narishige Puller is used, the rotation of the drill chuck is being turned in the first heating/pulling step by hand. After the rotation, the power is stopped and air is blown onto the capillaries to cool the glass. The second pull is initialized and finalized by the puller without help.
4. Heat the multibarrel stimulation pipette (approximately between the end of the capillary and the shrinkage tubing) over a small flame of a heat source (pilot flame of Bunsen burner). Gently push down the thin glass capillaries using old forceps to separate the glass capillaries of the 7-barrel pipette and to enable plastic tubing to connect to these ends (Fig. 1e).
5. Store the pipettes in a dry space and cover them with a lid to prevent dust accumulation. We use big Petri dishes in combination with a strip of modelling clay (available from any children's store or supermarket) to hold the pipette securely in place (Fig. 1e).
6. Before filling the 7-barrel stimulation pipette, bevel its tip using a micropipette grinder (Narishige EG-400). The diameter and the angle of the stimulation pipette tip can be controlled by the microgrinder. The tip diameter can produce either (a) leakage of the stimulus solution causing adaptation of the sensory epithelium or (b)—if too narrow—a reduced flow rate of the stimulus, inevitably affecting the ligand concentration. In addition, the angle of the cut and the placement

of the stimulation pipette into the solution flow can cause backpropagation of the bath solution into the pipette also affecting the stimulus concentration. To prevent the accumulation of shavings in the tip of the stimulation pipette, it is essential that the grinder is equipped with a water-dripping device.

7. Secure the bevelled 7-barrel stimulation pipette on the edge of a glass beaker using modelling clay. The tip of the pipette should be covered by ultrapure water. Do not worry about the hygroscopic/capillary forces causing water to creep up into the glass pipettes. The ultrapure water boundary will be separated from the physiological backfilling solution (stimulus solution) by an air bubble. Later, the air bubble will be expelled by the air pressure from the Picospritzer (Parker Instruments) indicating when the stimulation solution arrived at the tip of the stimulation pipette.
8. Prepare stimulation solution and your control solution based on the extracellular S1 solution (*see* Subheading 2.1).
9. Fill a syringe with the solution and use for every syringe a separate nonmetallic syringe needle.
10. Test the output flow of the multibarrel stimulation pipette in the EVG setup containing a microscope and picospritzer (Parker) using a small Petri dish filled with S1 solution. The picospritzer is designed to rapidly and reproducibly eject small volumes from the stimulation pipette using air pressure. Air pressure is relayed via valves and tubings to the various ends of the 7-barrel stimulation pipette. The output flow of the individual barrels should be in the same direction and contain approximately the same amount of solution. The flow can be better visualized if the control and stimulation solution contain a small, but equal, amount of food color (1:1,000). The stimulus length is being controlled via the computer sending a TTL signal to the picospritzer.
11. Keep the completely prepared stimulation pipette in either a beaker or a Petri dish with S1 solution (*see* Subheading 2.1), until the tissue is prepared, and the stimulation pipette needs to be placed near the sensory epithelium for stimulation (*see* **Note 6**).

3 Methods

3.1 The Recording Setup

The recording setup consists of equipment for acquisition and recording of the electrical EVG potential changes and for visualizing and handling the specimen and the stimulation (*see* Subheading 2.6) and recording pipette (*see* Subheading 2.4) via micromanipulators which are placed on a vibration-free table. The table is enclosed by a Faraday cage to prevent electromagnetic disturbances. For the electrical recording of the signal, a recording electrode containing the glass pipette with agar in the tip and backfilled with S1 solution (*see* Subheading 2.1) is connected via an Ag/AgCl wire to a differential amplifier (DP-301, Warner Instruments). The second Ag/AgCl wire serving as the indifferent electrode is connected to an agar bridge (*see* Subheading 2.2) and placed near the preparation (*see* Subheading 3.2) using modelling clay. The amplifier output signals are directly transmitted to an analog-to-digital converter (Heka/Instrutech, ITC-16) that is connected to a computer and to an oscilloscope for real-time monitoring of the signal. Software such as Heka's Pulse/Pulsefit can be used to control the stimulation protocol and to record the potential changes. The stimulation/pulse protocol triggers the picospritzer (Parker) which rapidly ejects fluid from a particular glass capillary of the stimulation pipette using air pressure. The stereo microscope (SZX12, Olympus) should have a great zoom range and high magnification capabilities enabling to visualize not only the olfactory organs but also the electrode and stimulation pipette tip. Options like fluorescence illumination and detection could be incorporated to help visualize particular organ regions.

3.2 Dissection of Mouse Head and VNO

Please make sure that all animal experimental procedures are performed in accordance with the guidelines established by the animal welfare committees of the respective institutions.

1. Euthanize a mouse with a CO_2 overdose followed by decapitation.
2. Take the mouse head and eliminate all skin and fur to expose the skull and nasal bones. Start with a caudal to rostral cut along the sagittal suture to the tip of the nose using sharp Metzenbaum scissors. Remove the skin sideways and cut it off.
3. Detach the underjaw (mandible) from the head bone by cutting on both sides through the temporomandibular joint with sharp blunt scissors.
4. Cut off the top of the mouse incisors (front teeth) to ensure a stable position of the ventral part of the head on the surface of a cutting board (*see* Fig. 2a).
5. Using a single-edge razor blade, make a cut through the left nasal cavity and the whole mouse head. More precisely, when placing the head on the cutting board, the razor should be positioned with one of its sharp points next to the tip of the nose (Fig. 2a). To cut through the nasal cavity, the sharp surface of the razor blade is carefully placed in the same angle as the dorsal nose/head bones, parallel to the internasal suture (midline) and approximately 1/3 the distance of the nasal bone from the suture. Extreme care should be taken to move the blade in a 90° angle to the cutting board to prevent damage

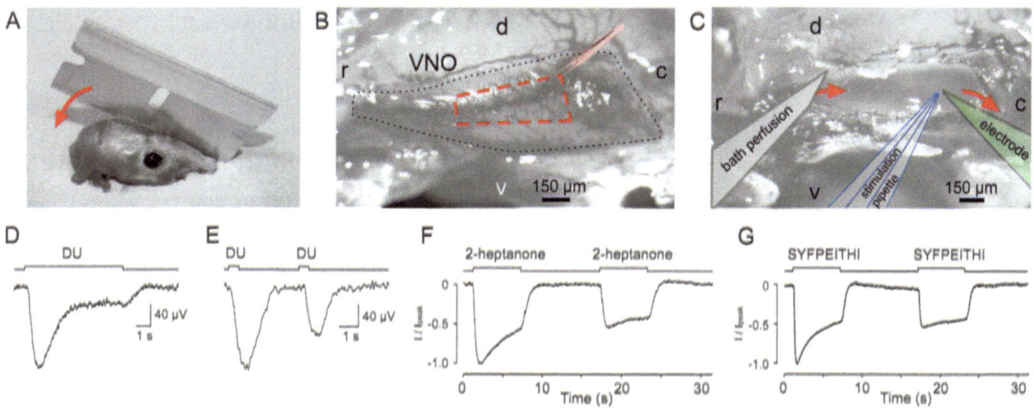

Fig. 2 (**a**) Photograph of a mouse head showing the placement of a single-edge razor blade for cutting through the nasal cavity and brain in a single stroke in order to prevent damage to the nasal cavity. *Arrow* indicates the cutting direction. (**b**) Sagittal view of the vomeronasal organ (VNO). The *red dashed box* indicates the outline of vomer bone that should be removed using forceps. *Thin black dashed line*, outline of the VNO. (**c**) View onto the ciliary surface of the VNO showing the placement of the perfusion input, the multibarrel stimulation pipette, and the *green* agar-filled electrode. *Red arrows* indicate the flow of the bath perfusion. *d* dorsal, *v* ventral, *r* rostral, *c* caudal. (**d**, **e**) Typical examples of EVG recordings from mouse VNO. The recorded potentials undergo time-dependent adaptation, evident as a decline of the response during sustained stimulation or as a peak response reduction during brief, repetitive stimulation. Stimulus, diluted urine (DU, 1:1,000). (**f**, **g**) Examples of stimulus-induced EVG recordings that were scaled to compare the adaptation time course. Responses were evoked by repetitive 6-s identical pulses of 2-heptanone (10 nM) or the major histocompatibility complex peptide ligand SYFPEITHI (0.3 pM). Interpulse interval was 10 s. (**d–g**, modified with permission from [15])

to the VNO which protrudes into the nasal cavity. Put pressure onto the angled top part of the razor blade and cut through the bones and brain tissue in one stroke (*see* arrow in Fig. 2a).

6. Pin the sagittally cut mouse head onto the premade mounting dissection dish (*see* Subheading 2.3) using three insect pins. Place the pins in three random brain areas.
7. Place the dish onto the recording stage under the stereo microscope (SZX12, Olympus) to perform more detailed dissection.
8. Take away any obscuring tissue overlaying the VNO so that there is a direct view onto the vomer bone. In case the cut was too shallow, the preparer has to cautiously cut away the leftover incisor. Its root can cover most of the VNO and may rip off the VNO with septum.
9. Using standard #5 Dumont forceps (0.10 × 0.06 mm) nibble a small rectangular window into the vomer bone (Fig. 2b, red dashed box). The window should be located in the middle of the vomer and not too far rostral or caudal. Care has to be taken not to extend the window too much dorsally: VNO axon bundles protrude underneath the dorsal part of the vomer bone along the septum and could be damaged. The cavenous

tissue of the VNO with its distinct blood vessel is now clearly visible through the window.

10. Put one of the legs of a Biologie #5 Dumont forceps (0.05 × 0.02 mm) from dorsal to ventral through the VNO lumen at the most caudal end of the window, pinch gently the forceps, and slowly tear the cavenous tissue from caudal to rostral, away from the sensory epithelial surface. Be meticulous in preventing the ciliary surfaces of the cavenous tissue and the sensory epithelium to touch each other. Cilia from both sensory and nonsensory epithelium could be lost due to their entanglement (velcro hook-and-loop closure effect) and therefore cause the absence of chemosensory responses.
11. Carefully clear the rest of the rostral and caudal vomer bones and cavenous tissue. Also check if the big blood vessel was eliminated which may block access to record from the sensory epithelium. Do not discard the ventral and dorsal vomer bone boundaries. They will service as a wall for the input and output perfusion system.
12. Perfuse, as quickly as possible, the sensory epithelium with oxygenated S1 solution to prevent dehydration and oxygen deprivation.

3.3 Perfusion of the Sensory Epithelium

1. Control the flow of the bath perfusion solution (S1 solution, *see* Subheading 2.1), e.g., with a pump or a gravity feed. The solution speed in a gravity feed system should be strictly controlled using a combination of solution volume (container) and gravity flow controllers [28]. Keep track of the perfusion solution volume to maintain relatively equal pressure on the tubing lines.
2. Install a 1-way stopcock directly at the source of the perfusate. This enables instantaneous stoppage of the solution flow, which is advantageous if more than one perfusate will be used during an experiment (e.g., in the presence of pharmaca or blockers). In addition, backpropagation of the flowing bath solution into the tubing of the other perfusion solution will be prevented.
3. Install a manifold, if more than one perfusate (S1 bath solution, *see* Subheading 2.1) will be used. The manifold output should be located within a very short distance from the VNO perfusion input to reduce solution exchange times.
4. Adjust the flow rate of the perfusate to approximately 100 μl/s. Always count the output flow and do not depend on the written information on the flow controllers.
5. Replace the insect pins in a way that ensures that the tip of the nose is located approximately 10–15° higher than the brain. This will help to modulate the perfusion height and speed in

the natural chamber composed by the boundaries of the nasal cavity and vomer bone.

6. Use an 18 G needle as the VNO perfusion input. Place the opening of the needle near the rostral VNO entry onto the sagittally cut mouse head so that the solution is flowing in an angle onto the tissue.
7. Make a small incision into the caudal region of the soft palate near the uvula. The excess of perfusion solution will exit the "chamber" constructed from the borders of the palate and nasal/brain bones. Sometimes it helps to put a small piece of tissue paper (Kimberly-Clark) at the height of the uvula which connects the "chamber" and the bottom of the head mounting dissection dish. The excess solution on the bottom of the dissection dish is also necessary to have electrical contact with the reference electrode.
8. Align the mounting dissection dish so that the sensory epithelium is centered under the microscope and secure the dish with modelling clay.
9. Place the vacuum suction (output of perfusate) near the wall of the Petri dish to prevent overflow. When using vacuum suction, it is necessary to precisely control the force of the suction to be able to fine-tune the fluid height in the Petri dish [28].

3.4 Recording and Analyzing EVG Data

1. Place the agar bridge with reference electrode near the dorsal mouse head (*see* **Note 4**).
2. Place the two micromanipulators containing the multibarrel stimulation pipette and the recording electrode near the stage.
3. Turn on the differential amplifier (DC mode, low-pass filter at 0.1 kHz, gain 100×).
4. Carefully lower the recording electrode near the surface of the sensory epithelium. The electrode should contact the bath solution. When the electrode comes in contact with the solution a straight baseline will appear in the oscilloscope window.
5. Gently lower the stimulation pipette so that its tip is near the tip of the recording electrode (Fig. 2c).
6. Check the amplifier gain and operation by using the 1 mV calibration signal of the amplifier by recording this signal into the computer using your acquisition software (Heka Pulse/Pulsefit).
7. Lower the recording electrode onto the surface of the sensory epithelium. This step has to be done very gently. At the same time the change in signal should be followed on the oscilloscope. A change in the baseline indicates that the agar-filled tip of the electrode connects to the tissue. The trick is to only connect to the surface containing the microvilli and the knobs. A strong dip

in the baseline indicates that the electrode went past the surface and is now at the level of the somata or worse at the vomer bone (glass electrode can break!). With practice, the investigator will notice the small changes in potential indicating that the electrode is near the epithelial surface (*see* **Notes 1** and **2**).

8. Lower the stimulation pipette near the electrode. Make sure that its tip is a little elevated and placed at a small distance from the electrode touching the sensory epithelium. Keep in mind that the solution will be pushed out of the glass capillary, but, due to the angle of the stimulation pipette, the solution will be directed either over the area of interest (the recording site) or before the area of interest which may cause a deflection from the epithelium surface over the area of interest. In both cases, the recording site may not be stimulated. To test the direction of the stimulation fluid, use control S1 solution (*see* Subheading 2.1) with food color by initiating a stimulation protocol via the computer via the external trigger of the picospritzer. The stimulus length of the test should not be more than 0.3 s (*see* **Notes 5** and **6**).
9. Start the experiment on the sensory epithelium by activating the stimulation protocols via the computer. Make sure to wait at least 4 min between stimulations to prevent adaptation of the sensory neurons (Fig. 2d–g) [11, 15, 21] (*see* **Notes 5** and **6**).
10. Software from Heka (Pulse/Pulsefit) and Wavemetrics (Igor) provide tools for measuring and analyzing the EVG signals. Parameters include response amplitude, latency, time-to-peak, and time constants of activation, adaptation/desensitization, and termination [11, 15, 21]. It is desirable to digitally filter the traces (low-pass 8-pole Bessel filter with corner frequency 1 kHz; digital filter frequency 60 Hz) (*see* **Note 7**).

4 Notes

These notes are provided to help identify some common problems in EVG recordings.

1. Check the tip of the recording electrode if a spontaneous loss of EVG responses occurred. Agar may have been lost or broken off from the electrode tip preventing electrical contact with the tissue. In some cases the agar is boiled too long increasing the gel strength and increasing the risk of losing agar on the surface of the sensory epithelium.
2. Check also the tip of the recording electrode, if no stable baseline can be obtained or no responses are measured. Colored agar solution may be leaking from the recording electrode. The gel strength was in this case too low.

3. Another obstacle in recording authentic EVGs is caused by a very thin film of agar on the outside of the recording electrode. In this case extreme hot and liquid agar solution leaked out of the tip of the glass capillary during the preparation of the electrode.
4. Chlorate the silver wires of the recording and reference electrodes regularly. The differential amplifier has a high tolerance to DC offsets, but cannot be adjusted indefinitely. Furthermore, a wrong placement of the reference electrode can cause a dramatic shift in this offset value.
5. Start to record and stimulate at the most caudal area of the sensory epithelium. In this way the more rostral sensory epithelium has not yet detected any stimulants due to the flow of the bath solution and should not be adapted to the ligand stimulation.
6. Extreme care should be taken with the multibarrel stimulation pipette. Any leakage from these pipettes can cause adaptation of the tissue and prevent the recording of meaningful data.
7. A chemosensory stimulus should be included in the experimental series that evokes broadly responses for a particular sensory neuron population to enable the detection of sudden changes in responsiveness at a particular electrode site.

Acknowledgements

This work was supported by grants from the Deutsche Forschungsgemeinschaft (to T.L.-Z. and F.Z.) and the Volkswagen Foundation (to T.L.-Z.). T.L.-Z. is a Lichtenberg professor of the Volkswagen Foundation.

References

1. Hosoya Y, Yoshida H (1937) Über die bioelektrischen Erscheinungen an der Riechschleimhaut. Jpn J Med Sci III Biophys 5:22–23
2. Ottoson D (1955) Analysis of the electrical activity of the olfactory epithelium. Acta Physiol Scand Suppl 35:1–83
3. Inouchi J, Wang D, Jiang XC, Kubie J, Halpern M (1993) Electrophysiological analysis of the nasal chemical senses in garter snakes. Brain Behav Evol 41:171–182
4. Taniguchi M, Wang D, Halpern M (1998) The characteristics of the electrovomeronasogram: its loss following vomeronasal axotomy in the garter snake. Chem Senses 23:653–659
5. Müller W (1971) Vergleichende elektrophysiologische untersuchungen an den sinnesepithelien des Jacobsonschen organs und der nase von amphibien (Rana), reptilien (Lacerta) und säugetieren (Mus). Z vergl Physiol 72: 370–385
6. Getchell TV, Shepherd GM (1978) Responses of olfactory receptor cells to step pulses of odour at different concentrations in the salamander. J Physiol 282:521–540
7. Davis H (1961) Some principles of sensory receptor action. Physiol Rev 41:391–416
8. Gesteland RC, Lettvin JY, Pitts WH (1965) Chemical transmission in the nose of the frog. J Physiol 181:525–559
9. Scott JW, Scott-Johnson PE (2002) The electroolfactogram: a review of its history and uses. Microsc Res Tech 58:152–160
10. Chen S, Lane AP, Bock R, Leinders-Zufall T, Zufall F (2000) Blocking adenylyl cyclase inhibits

olfactory generator currents induced by "IP(3)-odors". J Neurophysiol 84:575–580
11. Munger SD, Lane AP, Zhong H, Leinders-Zufall T, Yau KW, Zufall F, Reed RR (2001) Central role of the CNGA4 channel subunit in Ca^{2+}-calmodulin-dependent odor adaptation. Science 294:2172–2175
12. Spehr M, Spehr J, Ukhanov K, Kelliher KR, Leinders-Zufall T, Zufall F (2006) Parallel processing of social signals by the mammalian main and accessory olfactory systems. Cell Mol Life Sci 63:1476–1484
13. Spehr M, Kelliher KR, Li XH, Boehm T, Leinders-Zufall T, Zufall F (2006) Essential role of the main olfactory system in social recognition of major histocompatibility complex peptide ligands. J Neurosci 26:1961–1970
14. Chamero P, Katsoulidou V, Hendrix P, Bufe B, Roberts R, Matsunami H, Abramowitz J, Birnbaumer L, Zufall F, Leinders-Zufall T (2011) G protein Gαo is essential for vomeronasal function and aggressive behavior in mice. Proc Natl Acad Sci U S A 108:12898–12903
15. Spehr J, Hagendorf S, Weiss J, Spehr M, Leinders-Zufall T, Zufall F (2009) Ca^{2+}-calmodulin feedback mediates sensory adaptation and inhibits pheromone-sensitive ion channels in the vomeronasal organ. J Neurosci 29:2125–2135
16. Zufall F, Ukhanov K, Lucas P, Liman ER, Leinders-Zufall T (2005) Neurobiology of TRPC2: from gene to behavior. Pflugers Arch 451:61–71
17. Leinders-Zufall T, Brennan P, Widmayer P, Chandramani SP, Maul-Pavicic A, Jager M, Li XH, Breer H, Zufall F, Boehm T (2004) MHC class I peptides as chemosensory signals in the vomeronasal organ. Science 306:1033–1037
18. Del Punta K, Leinders-Zufall T, Rodriguez I, Jukam D, Wysocki CJ, Ogawa S, Zufall F, Mombaerts P (2002) Deficient pheromone responses in mice lacking a cluster of vomeronasal receptor genes. Nature 419:70–74
19. Zufall F, Kelliher KR, Leinders-Zufall T (2002) Pheromone detection by mammalian vomeronasal neurons. Microsc Res Tech 58:251–260
20. Leypold BG, Yu CR, Leinders-Zufall T, Kim MM, Zufall F, Axel R (2002) Altered sexual and social behaviors in trp2 mutant mice. Proc Natl Acad Sci U S A 99:6376–6381
21. Leinders-Zufall T, Lane AP, Puche AC, Ma W, Novotny MV, Shipley MT, Zufall F (2000) Ultrasensitive pheromone detection by mammalian vomeronasal neurons. Nature 405:792–796
22. Tucker D, Shibuya T (1965) A physiologic and pharmacologic study of olfactory receptors. Cold Spring Harb Symp Quant Biol 30:207–215
23. Silver WL, Caprio J, Blackwell JF, Tucker D (1976) The underwater electro-olfactogram: a tool for the study of the sense of smell of marine fishes. Experientia 32:1216–1217
24. Haga S, Hattori T, Sato T, Sato K, Matsuda S, Kobayakawa R, Sakano H, Yoshihara Y, Kikusui T, Touhara K (2010) The male mouse pheromone ESP1 enhances female sexual receptive behaviour through a specific vomeronasal receptor. Nature 466:118–122
25. Kimoto H, Sato K, Nodari F, Haga S, Holy TE, Touhara K (2007) Sex- and strain-specific expression and vomeronasal activity of mouse ESP family peptides. Curr Biol 17:1879–1884
26. Leinders-Zufall T, Ishii T, Mombaerts P, Zufall F, Boehm T (2009) Structural requirements for the activation of vomeronasal sensory neurons by MHC peptides. Nat Neurosci 12:1551–1558
27. Riviere S, Challet L, Fluegge D, Spehr M, Rodriguez I (2009) Formyl peptide receptor-like proteins are a novel family of vomeronasal chemosensors. Nature 459:574–577
28. Leinders-Zufall T (1998) Technique for setting up the perfusion of a recording chamber for physiological studies with cultured and acutely dissociated cells. http://wwwwarneronlinecom/pdf/whitepapers/perfusion_strategiespdf

Chapter 17

Electrical Recordings from the Accessory Olfactory Bulb in VNO-AOB Ex Vivo Preparations

Julian P. Meeks and Timothy E. Holy

Abstract

Electrical recordings from individual accessory olfactory bulb neurons allow exploration of the functional properties of this important pheromonal processing circuit. Several approaches to performing such recordings have been used. Here, we describe ex vivo methods that we have found useful for recording from accessory olfactory bulb neurons using simple extracellular glass electrodes.

Key words Accessory olfactory bulb, Vomeronasal organ, Vomeronasal nerve, Ex vivo, Microelectrode, Extracellular recording, Tissue perfusion, Electrophysiology

1 Introduction

Electrical recordings from individual accessory olfactory bulb (AOB) neurons have been made in awake mice [1], in anesthetized mice [2, 3], and in an ex vivo experimental preparation that maintains functional inputs from the vomeronasal organ [4]. The ex vivo preparation takes advantage of the fact that the axons of vomeronasal sensory neurons project medially through the ipsilateral septal tissue, and do not cross the midline before ramifying in the AOB glomerular layer. Although the experimental dissection is quite involved, this procedure has many advantages in experimental contexts in which active odor sampling and/or centrifugal connectivity are not required.

Because the vomeronasal synaptic inputs remain functional, one can perform experiments linking peripheral sensory stimulation to AOB neural activity. Because the AOB is self-contained inside an active tissue perfusion chamber, one can use pharmacology to help elucidate mechanisms of AOB circuit function. We have used single-glass electrode recordings to monitor the activity of individual neurons at various depths in the AOB, but this configuration is also compatible with more advanced electrophysiological

Kazushige Touhara (ed.), *Pheromone Signaling: Methods and Protocols*, Methods in Molecular Biology, vol. 1068,
DOI 10.1007/978-1-62703-619-1_17, © Springer Science+Business Media, LLC 2013

techniques, including guided or blind patch-clamping and multi-electrode array recording.

With practice, one should expect to produce successful ex vivo preparations in approximately 80 % of attempts, and 2 targeted extracellular glass electrode recordings from each preparation per hour. The major cell types in the first 300 μm along the depth of the AOB remain mostly devoid of cell degradation markers for up to 8 h, but a potential limitation of this preparation is that the deepest granule cell layers begin to show signs of cellular distress at 4 h post dissection.

The major steps necessary to perform targeted extracellular recordings from ex vivo preparations are.

1. Initial fast dissection to isolate one hemisphere of the mouse snout and rostral cranium.
2. Secondary detailed dissection to expose vomeronasal axons to the superfusion solution.
3. Insertion of a thin polyimide cannula into the vomeronasal organ lumen.
4. Insertion of microelectrodes into the AOB.

2 Materials

2.1 Reagents

1. Chilled dissection artificial cerebrospinal fluid (aCSF): 125 mM NaCl, 2.5 mM KCl, 2 mM $CaCl_2$, 10 mM $MgCl_2$, 25 mM $NaHCO_3$, 1.25 mM NaH_2PO_4, 0.4 mM Na-ascorbate, 2 mM Na-pyruvate, 3 mM myo-inositol, 25 mM glucose.
2. Superfusion aCSF: 125 mM NaCl, 2.5 mM KCl, 2 mM $CaCl_2$, 1 mM $MgCl_2$, 25 mM $NaHCO_3$, 1.25 mM NaH_2PO_4, 0.4 mM Na-ascorbate, 2 mM Na-pyruvate, 3 mM myo-inositol, 25 mM glucose.
3. Stimulating Ringer's saline: 115 mM NaCl, 5 mM KCl, 2 mM $CaCl_2$, 2 mM $MgCl_2$, 25 mM $NaHCO_3$, 10 mM HEPES, 10 mM glucose.

2.2 Dissecting Tools

Decapitation scissors, sharp curved fine dissecting scissors, #11 scalpel blade and handle, two pairs of Adson Forceps (Fine Science Tools, Foster City, CA), standard carbon steel safety razor blade, two pairs of #5 Dumont forceps (Fine Science Tools), two pairs of #3 Dumont forceps (Fine Science Tools).

2.3 Equipment and Supplies

1. 60 mm plastic petri dishes.
2. Small plastic plank, approximately 1 cm × 0.6 cm × 1 mm.
3. Dow Corning® high vacuum grease.
4. Perfusion dish or chamber (*see* Fig. 1).

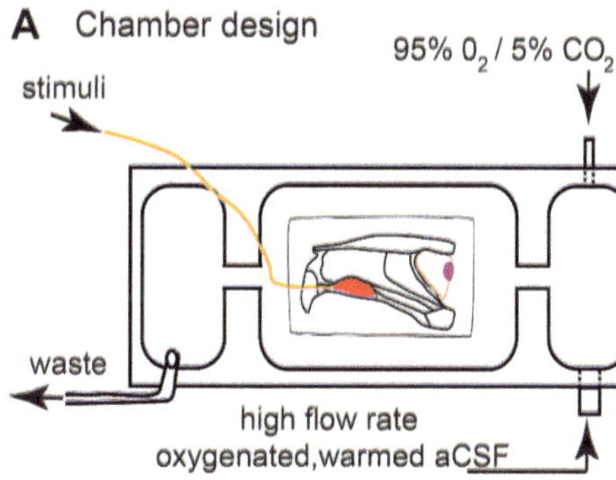

Fig. 1 Schematic of ex vivo tissue chamber. Warmed, oxygenated aCSF flowing at 5–8 mL/min enters a secondary oxygenation area prior to reaching the tissue. Aiming the main chamber inlet stream directly at the surface of the AOB improves tissue health over time. Note that both inlet and suction may require electrical grounding

5. Peristaltic or gravity-fed perfusion apparatus and suction (Gilson® Minipuls 3 or an equivalent).
6. 3 M™ Vetbond™ tissue glue.
7. Stereomicroscope.
8. Long (3–4 in.), thin polyimide cannula (0.0056″ ID, 0.001″ wall thickness) attached to a pressurized fast perfusion device (Automate Scientific® ValveLink® or similar).
9. Borosilicate capillary glass, pulled to diameter and impedance specified below.
10. Fine polyimide electrode-filling syringe (World Precision Instruments™ MicroFil™ or an equivalent).
11. 0.22 μm syringe filter (EMD Millipore® Millex or an equivalent).
12. Electrode holder, headstage, extracellular amplifier, and analog/digital recording and storage devices.
13. Oscilloscope attached to the voltage output of the microelectrode amplifier.
14. Optional audio device attached to voltage output of the microelectrode amplifier.
15. 4-D microelectrode positioning device, with the penetration axis controlled with fine motion, preferably motorized (e.g., Siskiyou Corporation MX1641).

3 Methods

3.1 Fast Initial Dissection (See Note 1)

1. Deeply anesthetize animal using inhaled isoflurane in an induction chamber.
2. Decapitate, and immediately place the tissue into a plastic petri dish filled with chilled dissection aCSF. Fully immerse the tissue for 10–20 s (*see* Fig. 2A).
3. Grasp the head by the caudal/dorsal scalp. Remove lower jaw and tongue using curved scissors.
4. Using Adson forceps, grasp the scalp at the caudal edge, and then peel by pulling rostrally. At nostrils, remove the scalp tissue with a single cut (*see* Fig. 2B).
5. Using Adson forceps, carefully remove the skull, starting at the caudal edge overlying the cerebellum or brain stem. Stop at the rhinal sinus (the dense bone anterior to bregma; *see* Fig. 2C).
6. Make a single bilateral coronal cut approximately 5 mm caudal to the rostral sinus (through the frontal pole of neocortex). Remove the brain caudal to the cut using a forceps or a curved scoop.
7. Using sharp scissors, cut through the remaining cranial bones at the same location of the previous caudal cut (*see* Fig. 2D1). Place the tissue into a 60 mm petri dish filled with chilled dissection aCSF for 5–10 s.

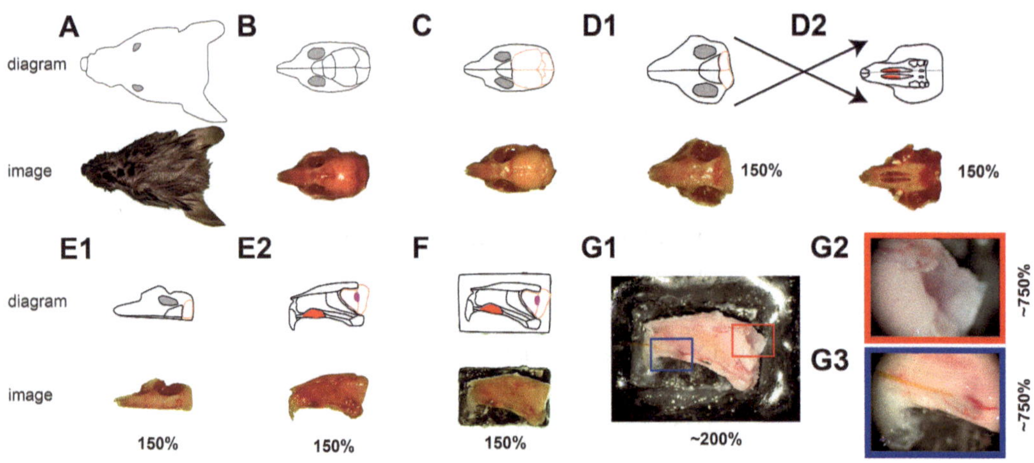

Fig. 2 Overview of the ex vivo dissection. Images taken immediately after Step 1.2 (**A**), Step 1.4 (**B**), Step 1.5 (**C**), Step 1.7 (**D1**), Step 1.8 (**D2**), Step 1.15 (**E**), Step 1.16 (**F**), and Step 3.5 (**G1–G3**). (**G1**) Ex vivo preparation with cannula inserted into VNO. (**G2**) Magnified view of the accessory olfactory bulb (*red boxed* region in **G1**). (**G3**) Magnified view of the polyimide cannula (*orange/brown line* entering from the *left*) placed inside the vomeronasal organ (*blue boxed* region in **G1**)

8. Grasping the remaining tissue from the anterior side near the eye sockets, turn the tissue ventral side up (*see* **Note 2**). Then, using the #11 scalpel, detach the soft palate at the attachment point to the incisors. Grasp the cut end of the palate with Adson forceps, and peel it away with a caudal pull (*see* Fig. 2D2).
9. Place the sharp tip of the #11 scalpel blade into the small notch between the molars on the left hemisphere (caudal to the vomeronasal organ and the palatine foramen). Crack the bone by rotating the blade handle.
10. Make a careful caudal-rostral cut through the anterior-most portion of the palatine foramen, guiding the blade between the nasal cavity rostral to the left VNO to the area between the two incisors.
11. Grasping the tissue from the ventral side around the maxillary bones, flip the tissue dorsal side up. Using sharp scissors cut the right zygomatic bones and thin portion of the maxilla near its intersection with the lateral snout. Remove any attached soft tissue with Adson forceps.
12. Grasp the tissue firmly with fingers at the anterior portion of the tissue. Orient the safety razor blade parallel to the midline behind the caudal end of the tissue. Place the safety razor blade just left of the midline (<1 mm) and touching the frontal cortical lobe and both dorsal and ventral bones.
13. Gently rock the razor blade along the dorsal/ventral axis, using the rocking motion to slowly cut through the bones just left of the midline. Continue using this rocking motion to cut through the bone until approximately 5 mm rostral to the cribriform plate.
14. Grasp both hemispheres near the caudal-most region of tissue with two pairs of Adson forceps or the thumb and index finger of both hands. Gently separate the hemispheres. At the anterior attachment point, make a small cut using sharp scissors to separate the two hemispheres.
15. Retain the right hemisphere, which will have both vomeronasal organs but just one hemisphere of the olfactory bulb, placing it immediately into chilled dissection aCSF (*see* Fig. 2E).
16. Place the plastic plank on a flat piece of paper towel or filter paper. Apply a small amount (~5–10 μL) of Vetbond™ tissue glue on one half of the plank. Grasp the right hemisphere with Adson forceps via the dorsal and ventral bones. Transfer the tissue to the paper towel or the filter paper a few centimeters from the plank to wick away excess fluid, and then place the tissue on the plank with the rostral tissue facing the short end of the plank (*see* Fig. 2F).

3.2 Secondary Dissection (See Note 3)

1. Coat a portion of the perfusion dish or chamber by a thin (1 mm) layer of Dow Corning® silicone vacuum grease. Press the flat end of the plastic plank and tissue firmly down into the vacuum grease.
2. Initiate superfusion of the preparation. We use a peristaltic pump from Gilson™ with a measured flow rate of 5–8 mL/min. We additionally target the perfusion outlet towards the AOB surface to increase convective exchange between the tissue and perfusate.
3. While using a pair of steel #3 forceps to hold the plank in place, use the other pair of #5 forceps, with the tips brought together, to carefully separate the frontal cortical tissue away from the olfactory bulb, but leaving it attached at the ventral edge (including the lateral olfactory tract). A dense "net" of blood vessels may be visible at higher magnification. It is helpful for electrical recordings to remove these vessels, but it is a delicate task. To accomplish this, one can, at this stage, use a sharp pair of #5 forceps to grasp the net of vessels at the anterior/lateral edge of the AOB, and then carefully pull the net of vessels caudally away from the AOB, and then ventrally and medially to peel it away. If you have already removed the frontal cortical tissue, this is often quite difficult. After peeling the vessels from the AOB, remove the frontal cortex from the preparation using forceps.
4. Again while using a pair of #3 forceps to hold the plank in place, detach the contralateral VNO with #5 forceps. This can be accomplished by stabbing through the vomeronasal bone and iteratively peeling it away in chunks. Alternatively, one can remove the entire contralateral VNO by placing a single tip of #5 forceps between the two vomeronasal organs at the caudal/dorsal edge, and then gently running the forceps anteriorly until they begin separating. At this point the entire VNO, including the bone, can be removed by grasping the dorsal/caudal edge with #5 or #3 forceps.
5. The septal cartilage will appear opaque white, and covers the vomeronasal nerves. This must be removed in order to expose the vomeronasal nerves to superfusion. Using a pair of #3 forceps, grasp the caudal-most portion of the septal cartilage and gently pull it slightly away from the preparation. Do not attempt to remove the septal cartilage quickly. Instead, use a second pair of #3 forceps (in the other hand) to gently separate the cartilage from the septal tissue along the anterior and dorsal edges until reaching the rostral connective tissue. Here, carefully sever the connections using #5 or #3 forceps. If this process is not done carefully, the septal tissue may become strained, and can detach from the VNO, severing the vomeronasal nerves.

6. The septal bone will now be visible just rostral to the cribriform plate. It is a transparent bone with blood vessels visible in its interior. To fully expose the vomeronasal nerves, this must also be removed. To do so, use a pair of #3 forceps in the left hand and a pair of #5 forceps in the right. Very carefully, approaching from an angle nearly parallel to the septal tissue, grasp the septal bone near the caudal/dorsal edge. While gently pulling the caudal/dorsal edge away from the tissue, take the #5 forceps, with tips together, and perforate the septal bone just rostral to the cribriform plate. This will assist in forming a linear crack in the bone. Once the bone cracks, it can be removed easily using the forceps in the left hand (*see* **Note 4**).

3.3 Insertion of Polyimide Cannula into Vomeronasal Organ

1. Before inserting the polyimide cannula into the VNO, it is important to inspect the VNO at its most rostral end. Note that the VNO is connected to the nasal cavity through a very thin cartilaginous tube. It is best to remove the cartilage tube just as it touches the rostral-most VNO. Using #5 forceps, sever the cartilage at the VNO junction by pinching the two forceps points together. Keep the hand forceps pinched at this junction, and then use the forceps in the other hand to pull the cartilage tube away from the VNO.
2. Bring the polyimide cannula within 1 cm of the ex vivo preparation. Position the cannula so that it will have at least 1 cm of extra slack (i.e., such that it can reach many areas around the VNO, and will not be stretched to do so). It is helpful to increase magnification at this stage if possible to focus on the anterior VNO.
3. Initiate Ringer's flow through the cannula (with a flow rate between 0.1 and 0.4 mL/min). Submerge the cannula using a pair of #3 forceps. Gently grasp the cannula approximately 2 mm from its tip, and bring it near the opening to the VNO at the rostral end of the snout. It is common for the effective flow rate out of the cannula to decrease while the tube is being held in the forceps.
4. Bring the cannula opening near the VNO opening. Use the Ringer's flow out of the cannula to help align the cannula to the VNO opening. When the cannula flow is parallel to the VNO opening, you will observe the VNO "inflating," which causes the blood in the blood vessels supplying the VNO to empty. Upon observing this, slowly move the cannula into the VNO, stopping and retrying if your VNO "deflates" due to movement of the cannula.
5. Stop advancing the cannula when it is half way into the VNO lumen. When the cannula is released, it should relax to the lateral portion of the VNO, containing the pump organ.

This may disrupt the pump organ, but minimally disrupts the VNO epithelium. Check to ensure that the flow rate through the cannula is between 0.1 and 0.4 mL/min and that it is stable.

3.4 Single-Glass Electrode Recordings

1. Prepare 4–10 borosilicate glass electrodes using a microelectrode puller. Tip diameter should be no less than 1 μm, and impedance should be 3–8 MΩ. Taper and smoothness of the microelectrode tip are matters of personal preference.
2. At the time of recording, backfill the electrode with syringe-filtered aCSF, removing any bubbles with gentle finger flicks.
3. Mount the electrode on the holder, and then use the micropositioner to bring the electrode within 100 μm of the AOB surface, using the dissection microscope.
4. Using slow advances of the fine (penetrating) direction of the micropositioner (no more than 5–10 μm per step) advance the electrode into the AOB. Upon encountering the AOB surface, the pitch (when listening via an acoustic monitor) will suddenly shift from high to low. Use this audible transition to mark your positioner to "zero depth."
5. Advance the electrode into the AOB tissue, to the desired depth, going no faster than 1 μm/s. Steps of 5 μm are convenient, but best results require one to wait at least several seconds after each step.
6. The following recording environments are typically encountered when advancing electrodes along the AOB transverse axis (*see* **Note 5**).
   (a) In the initial 80 μm of depth, small-amplitude, negative-polarity spikes are often encountered, which may correspond to action potentials in periglomerular dendrites or to backpropagating action potentials in mitral cell dendrites. Also, it is common when using smooth-tipped microelectrodes to occasionally encounter a large transition from high-frequency, low-amplitude noise to extremely large-amplitude (millivolts RMS), low-frequency oscillations. These are likely the result of contacting blood vessels, which can clog the tip and cause large changes in tip impedance. Retracting the tip 20–50 μm will often reverse this change. In some cases, a new electrode penetration origin will be necessary to bypass the blood vessel.
   (b) From 80 μm of depth to 250 μm in depth, larger amplitude spikes can be isolated, corresponding to the proximal dendrites and somata of cell types in the external cellular layer. Extremely large amplitude spikes (sometimes greater than 1 mV) that are positive in polarity can be encountered,

likely reflecting a switch in recording microenvironment to one in which the large capacitive currents of mitral cell membranes dominate ionic currents. These positive-polarity waveforms are often multi-peaked, reflecting the propagation of action potentials through the axonal, somatic, and dendritic regions of these larger cells. This is especially the case when using higher impedance and smoothed electrode tips.

(c) Recordings from depths deeper than 250 μm in depth require extra patience. The time necessary to properly penetrate to this depth, along with the practical likelihood that at this depth the thicker portion of the microelectrode taper may be entering tissue and exerting some downward forces, means that advancement must be done quite slowly. Because the granule cells located deep to the lateral olfactory tract are small in diameter and often do not respond strongly to stimuli, encountering spikes at these depths is rare. An alternative strategy for recording from these neurons is to advance the electrode from a more extreme medial or dorsal-medial angle. This can allow some granule cells to be encountered within 250 μm from the entry point.

7. Post hoc spike sorting is sometimes necessary, as even smooth, 4–8 MΩ pipettes are capable of detecting spikes from multiple nearby cells. Principal component analysis of the spike waveforms, along with manual, k means, or Gaussian mixture model segmentation, is typically able to perform this function with high confidence.

4 Notes

1. The goal for completing the initial dissection should be approximately 5–8 min. Expect to practice this dissection dozens of times before being able to complete the initial dissection in this amount of time.
2. Recall that when the tissue is facing ventral-up the left hemisphere will be on the right side of the midline.
3. The design of the tissue chamber is an important consideration for any researcher intending to perform AOB recordings. The ex vivo tissue chamber described in Meeks and Holy [4] was custom-built to allow various electrode penetration angles while keeping the chamber volume minimized. Keep in mind that this chamber also contained an aCSF oxygenation region just before the tissue, which can generate electrical artifacts.

4. If a clear break of the septal bone does not occur on the initial perforation, repeat one time. If the bone cracks in an unintended place, one must remove the bone by peeling a small piece at a time. There is a large risk of puncturing the septal tissue at this stage, so proceed with great caution.
5. During electrode penetrations, it is helpful to periodically stimulate the VNO with a positive control stimulus in place of the standard Ringer's solution. For example, a 50 mM KCl Ringer's solution (substituting KCl for NaCl on an equimolar basis) will depolarize vomeronasal neurons and activate widespread AOB activity. This has the effect of increasing spontaneous firing rates in the AOB, which can improve the rate of identifying single neural units.

References

1. Luo M, Fee MS, Katz LC (2003) Encoding pheromonal signals in the accessory olfactory bulb of behaving mice. Science 299:1196
2. Hendrickson RC, Krauthamer S, Essenberg JM, Holy TE (2008) Inhibition shapes sex selectivity in the mouse accessory olfactory bulb. J Neurosci 28:12523
3. Ben-Shaul Y, Katz LC, Mooney R, Dulac C (2010) In vivo vomeronasal stimulation reveals sensory encoding of conspecific and allospecific cues by the mouse accessory olfactory bulb. Proc Natl Acad Sci U S A 107:5172
4. Meeks JP, Holy TE (2009) An ex vivo preparation of the intact mouse vomeronasal organ and accessory olfactory bulb. J Neurosci Methods 177:440

Chapter 18

Pheromone-Induced Expression of Immediate Early Genes in the Mouse Vomeronasal Sensory System

Sachiko Haga-Yamanaka and Kazushige Touhara

Abstract

Immediate early genes (IEGs) are powerful tools for visualizing activated neurons and extended circuits that are stimulated by sensory input. Several kinds of IEGs (e.g., *c-fos*, *egr-1*) have been utilized for detecting activated receptor neurons in the pheromone sensory organ called the vomeronasal organ (VNO), as well as for mapping the neurons within the central nervous system (CNS) excited by pheromones.

In this chapter, we describe the procedure for the detection of pheromone-induced neural activation in the VNO and CNS using the c-Fos immunostaining technique.

Key words Vomeronasal organ (VNO), Accessory olfactory bulb (AOB), Bed nucleus of the stria terminalis (BST), Medial amygdala (Me), Posteromedial cortical amygdaloid nucleus (PMCo), Medial preoptic area (MPA), Ventromedial hypothalamic nucleus (VMH), Immediate early genes (IEGs), c-Fos

1 Introduction

In the vast majority of terrestrial animals, pheromone signals are detected and processed in the vomeronasal system. This sensory system consists of a sensory organ called the vomeronasal organ (VNO) where the pheromones are received by their corresponding receptor neurons, the accessory olfactory bulb (AOB) where the axons of the VNO neurons project, and a set of nuclei in the central nervous system (CNS) where the information is transferred by the AOB and is further processed and integrated. The nuclei in the CNS involved in pheromone signal processing include those projected by the AOB, such as the bed nucleus of the stria terminalis (*BST*), the medial amygdala (*Me*), and the posteromedial cortical amygdaloid nucleus (*PMCo*), whereas these nuclei further project to the higher brain areas in the hypothalamus such as the medial preoptic area (*MPA*) and the ventromedial hypothalamic nucleus (*VMH*) [1–3].

Kazushige Touhara (ed.), *Pheromone Signaling: Methods and Protocols*, Methods in Molecular Biology, vol. 1068, DOI 10.1007/978-1-62703-619-1_18, © Springer Science+Business Media, LLC 2013

To those researchers who seek to understand the modes of action of pheromones, the immediate early gene (IEG)-based activity mapping strategy has provided valuable information about the receptor neurons, as well as the neural pathways through which pheromone signals are processed. Among various kinds of IEGs, c-Fos has become the most widely used functional marker to reveal activated neurons for several reasons: (1) it is expressed at low levels in the VNO, AOB, and CNS under basal conditions; (2) it is stereotypically induced in response to several extracellular signals including pheromones and neurotransmitters; (3) the response is transient; (4) its expression can be detected relatively easily; and (5) its immunostaining can be readily combined with the detection of various other molecules including sensory receptor proteins, neuroanatomical marker proteins, retrograde tracers, and other activity markers.

Indeed, by monitoring the c-Fos-inducing activity during purification steps, a male mouse pheromone, the Exocrine gland-Secreting Peptide 1 (ESP1), was identified from the lachrymal grand [4, 5]. By analyzing the co-localization of the pheromone-inducing c-Fos and the pheromone receptors, the receptor of ESP1, V2Rp5, was identified out of more than 300 types of pheromone receptors [6]. By using a similar technique, it was also revealed that pheromone receptors detect not only intraspecific cues but also interspecific signals such as predators [7]. Moreover, by investigating c-Fos expression patterns, neural circuitries involved in processing distinct pheromone signals have been successfully investigated [8–10].

Here we describe the basic procedures for detecting c-Fos-immunoreactive neurons in different areas of the mouse nervous system (VNO, AOB, and CNS) stimulated by a pheromone input. As for the detailed protocols for double-staining with other markers, readers are directed to other recent publications [6, 7, 10].

2 Materials

2.1 Pheromone Stimulation

1. Mice: 10-week-old female mice single-housed for a week prior to the experiment without bedding change.
2. Pheromone: 10 μg ESP1 in 20 mg cotton swab (ESP1 was dissolved in 100 μL phosphate-buffered saline (PBS), absorbed by cotton swab and dried).
3. Control stimulation: 20 mg cotton swab soaked with 100 μL PBS and dried.
4. Anesthesia: Sodium pentobarbital (150 μg/g mouse), 1 mL syringe, 27G injection needle.

5. Dissection tools: 70 % ethanol (EtOH), decapitation scissors, surgical scissors, curved dissecting forceps, Dumont #7 forceps, petri dish, chilled 4 % paraformaldehyde (PFA) in PBS.
6. Perfusion and fixation: 10 mL syringe with chilled PBS, 20 mL syringe with chilled 4 % PFA/PBS, 26G injection needle.

2.2 Specimen Preparation

1. Postfixation: 4 % PFA in PBS, glass vials (20 mL, with lid).
2. Decalcification: 500 mM ethylenediaminetetraacetic acid (EDTA).
3. Freezing protection: 15 and 30 % sucrose in PBS.
4. Embedding: Tissue-Tek® O.C.T. (Optimal Cutting Temperature) Compound (Sakura), Tissue-Tek® Cryomold (Sakura), curved dissecting forceps.
5. Slicing the VNO and AOB: Research cryostat (Leica, CM3050S), MAS coated glass slide (Matsunami), hair drier, slide glass case.
6. Slicing the brain: Sliding microtome (Yamato, REM-710), electro-freezing component (Yamato, MC-802A), 24-well cell culture plate (Corning), ink brush, PBS.

2.3 Immunostaining

1. Staining jars, racks for the slide glass.
2. Agglutination plate (TOMY, T-2).
3. Ink brush.
4. Super Pap pen liquid blocker.
5. Fixation: 4 % PFA in PBS.
6. Rinse: Tris-buffered saline (TBS; 100 mM Tris–HCl, 150 mM NaCl, pH 7.5) or Tris-buffered saline with Triton X-100 (TBST; 100 mM Tris–HCl, 150 mM NaCl, 0.1 % Triton X-100, pH 7.5).
7. Peroxidase inactivation: 1, 0.1, or 0.3 % hydrogen peroxide (H_2O_2) in PBS.
8. Moisture chamber.
9. Blocking solution: 3 % bovine serum albumin (BSA) in TBST.
10. Diluent solution: 3 % BSA in TBST.
11. Primary antibody: Rabbit anti-c-Fos antibody (Calbiochem, PC38, Ab-5).
12. Secondary antibody: Biotinylated goat anti-rabbit IgG antibody (Vector Laboratories, BA-1000).
13. ABC solution: Avidin and biotin (Vector Laboratories, VECTASTAIN® ABC system, PK-6100) in PBS.
14. Detection: Diaminobenzidine (DAB; SIGMAFAST™ 3,3′-diaminobenzidine tablets, D4293), 0.6 % (w/v) ammonium nickel (II) sulfate.

15. Dehydration: H_2O, EtOH, xylene.
16. Mounting: Mounting medium (Mount-Quick), cover glass (Matsunami glass, Neo).

2.4 Image Acquisition

1. Upright microscope (Olympus, BX 60).
2. Digital camera (Olympus, DP10).

3 Methods

3.1 Pheromone Stimulation and Intracardial Perfusion

1. Subject a mouse to either the pheromone (ESP1) or the control stimulation between 22:00 and 24:00 (light period: 9:00–21:00). Introduce a piece of cotton containing either ESP1 or PBS into the mouse cage. Stimulate the mouse for 90 min (*see* **Note 1**).
2. Anesthetize the mouse with intraperitoneal injection of pentobarbital using 1 mL syringe with 27G injection needle (*see* **Note 2**).
3. Place the mouse on the surface of a styrofoam plate with abdomen facing up. Using the small needles, secure the four paws to the surface spreading them as wide as possible.
4. Wet the abdomen with 70 % EtOH.
5. Grab skin with curved dissecting forceps at the level of the diaphragm, and cut to expose the liver. Cut laterally and then up, cutting through the ribs. Lift flap and continue cutting until the heart is easy to access.
6. Place 26G injection needle connected to 10 mL syringe with PBS into the left ventricle and immediately cut the right atrium.
7. Take 5 min to slowly inject 5 mL PBS.
8. Change the 10 mL syringe to a 20 mL syringe containing 4 % PFA/PBS.
9. Spend 15 min to slowly inject 15 mL 4 % PFA/PBS (*see* **Note 3**).
10. Remove the small needles securing the mouse paws.
11. Decapitate and cut the skin from the base of head to the tip of nose.
12. Peel the skin, expose the entire skull, and remove eyes using curved dissecting forceps.
13. Remove lower jaw and tongue using surgical scissors.
14. Grasp the lateral side of nose and peel the dorsal and lateral skull using Dumont #7 forceps starting at the caudal edge of the brain stem to the rostral tip of the olfactory bulbs (*see* **Note 4**).

15. Hold the lateral side of the nose with dorsal side down and carefully take out the brain with the olfactory bulbs from the skull using curved dissecting forceps. Carefully drop the brain with the olfactory bulb into a petri dish filled with chilled 4 % PFA/PBS.
16. Cut the bone where the tip of the olfactory bulb existed and place the nasal tissue into a petri dish filled with chilled 4 % PFA/PBS.

3.2 Specimen Preparation (Nose)

1. Incubate the nose in a vial with 4 % PFA/PBS for 3 h at 4 °C (*see* **Note 5**).
2. Incubate the nose in a vial with 500 mM EDTA/PBS for two nights at 4 °C (*see* **Note 6**).
3. Incubate the nose in a vial with 15 % sucrose/PBS for 3 h at 4 °C.
4. Incubate the nose in a vial with 30 % sucrose/PBS overnight at 4 °C.
5. Embed the nose in O.C.T. compound in a cryomold (*see* **Note 7**) and freeze. Embedded samples can be stored at –80 °C.
6. Cut 14 μm thick coronal sections using the cryostat, collect sections containing the VNO on slides, and dry the sections with hair drier (*see* **Note 8**). Sections can be stored in a slide glass storage case at –80 °C.

3.3 Specimen Preparation (Brain)

1. Incubate the brain in a vial with 4 % PFA/PBS for 3 h at 4 °C.
2. Incubate the brain in a vial with 15 % sucrose/PBS for 3 h at 4 °C.
3. Incubate the brain in a vial with 30 % sucrose/PBS overnight at 4 °C.
4. Cut the brain at about 2.8 mm anterior from the bregma in order to separate the olfactory bulb from the brain.
5. Embed the olfactory bulb in the O.C.T. compound in a cryomold (*see* **Note** 7) and freeze. Embedded samples can be stored at –80 °C.
6. Cut 25 μm thick parasagittal sections of the olfactory bulb using a cryostat, collect sections containing AOB on slides, and dry the sections with hair drier (*see* **Note 8**). Sections can be stored in a slide glass storage case at –80 °C.
7. Cut 50 μm thick coronal sections of the brain using a sliding microtome. Sections are collected in PBS in a 24-well plate using an ink brush.

3.4 Immunostaining (for VNO and AOB Specimens)

1. Seal the edge of slides with the liquid blocker and place them in a staining rack (Fig. 1a).
2. Put the staining rack in a staining jar containing TBST and rinse the slides in TBST for 5 min (Fig. 1a).

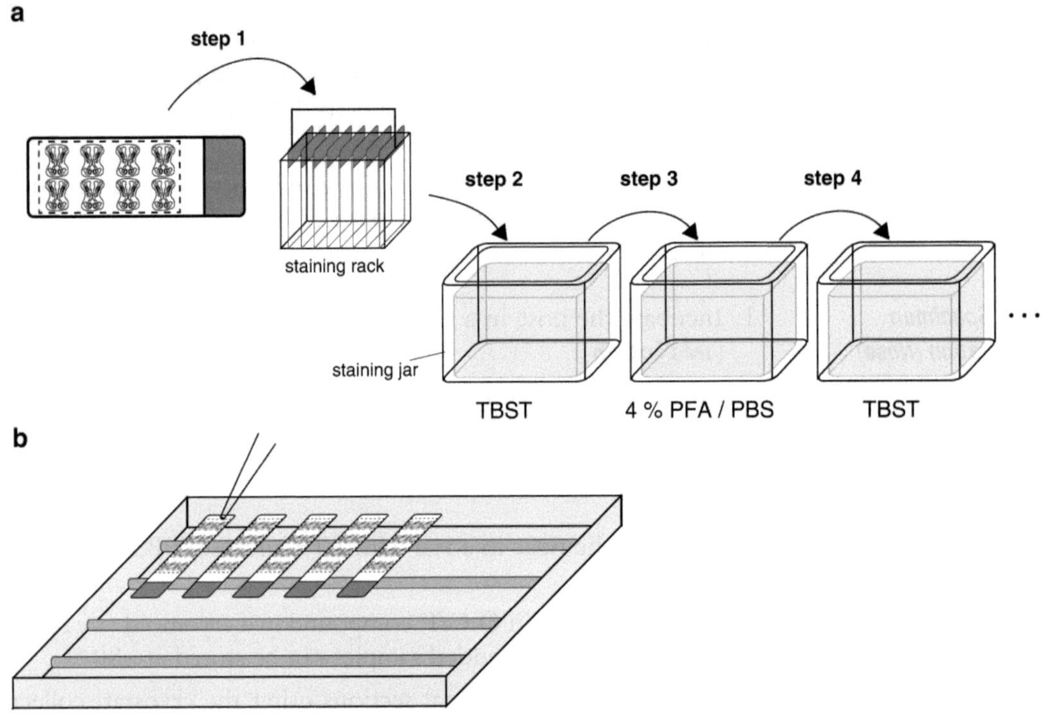

Fig. 1 Schematic illustration of the immunostaining procedures for the VNO and AOB. (**a**) Seal the edge of slides with the liquid blocker (*dashed line*) and place them in a staining rack (*see* **step 1**). Transfer the staining rack between staining jars containing different solutions. (**b**) Place the slides with the sections on the top in a moisture chamber flatly and gently apply 200 μL of each solution

3. Put the staining rack in a staining jar containing 4 % PFA/PBS and incubate the slides at 4 °C for 20 min (Fig. 1a).
4. Put the staining rack in a staining jar containing TBST and rinse the slides in TBST for 5 min.
5. Put the staining rack in a staining jar containing H_2O_2/PBS (1 % for nose, 0.1 % for AOB) and incubate the slides at room temperature (RT) for 30 min (*see* **Note 9**).
6. Put the staining rack in a staining jar containing TBST and rinse the slides in TBST for 5 min with agitation (repeat three times).
7. Place the slides with the sections on the top in a moisture chamber flatly and apply 200 μL blocking solution to each slide. Incubate the slides at 25 °C for 45 min (Fig. 1b).
8. Drain off the blocking solution, place the slides back into the chamber, and apply 200 μL diluent solution containing the primary antibody (1/5,000). Incubate the slides in a moisture chamber at 4 °C for three nights (Fig. 1b).
9. Drain off the antibody solution and place the slides in a staining rack. Put the staining rack in a staining jar containing TBST

and rinse the slides in TBST for 5 min with agitation (repeat three times).

10. Place the slides with the sections on the top in a moisture chamber flatly and apply 200 μL TBST containing the secondary antibody (1/200). Incubate the slides at 25 °C for 1 h (Fig. 1b).
11. Prepare the ABC solution of PBS containing avidin (1/100 dilution) and biotin (1/100 dilution) at 15 min before **step 12** (*see* **Note 10**).
12. Drain off the antibody solution and place the slides in a staining rack. Put the staining rack in a staining jar containing TBS and rinse the slides in TBS for 5 min with agitation (repeat three times).
13. Place the slides with the sections on the top in a moisture chamber flatly and apply 200 μL ABC solution. Incubate the slides at 25 °C for 30 min (Fig. 1b).
14. Drain off the ABC solution and place the slides in a staining rack. Put the staining rack in a staining jar containing TBS and rinse the slides in TBS for 5 min with agitation (repeat three times).
15. Place the slides with the sections on the top in a moisture chamber flatly and apply DAB solution. Incubate the slides for 5–10 min (until brown signals appear) (Fig. 1b).
16. Drain off the DAB solution and place the slides in a staining rack. Put the staining rack in a staining jar containing TBS and rinse the slides in TBS for 5 min with agitation (repeat three times).
17. Put the staining rack in a staining jar containing EtOH/H_2O. Dehydrate the samples in a series of 80, 95, and 100 % EtOH (two times each) and 100 % xylene (two times) at RT for 2 min each.
18. Mount the slides with 50 μL of mounting medium and put a cover glass on the top.
19. Observe the specimens under the microscope and take images with a digital camera (Fig. 2).

3.5 Immunostaining (for Brain)

1. Transfer brain sections with the *BST*, *MEA*, *PMCo*, *MPA*, and *VMH* (Fig. 3) to the wells of the agglutination plate containing 500 μL PBS, at most two sections per well, using ink brush and rinse the sections for 10 min (Fig. 4).
2. Transfer the sections to the wells containing 500 μL 0.3 % H_2O_2/PBS and incubate them at RT for 30 min (Fig. 4 and *see* **Note 11**).
3. Transfer the sections to the wells containing PBS and rinse them for 10 min (two times).

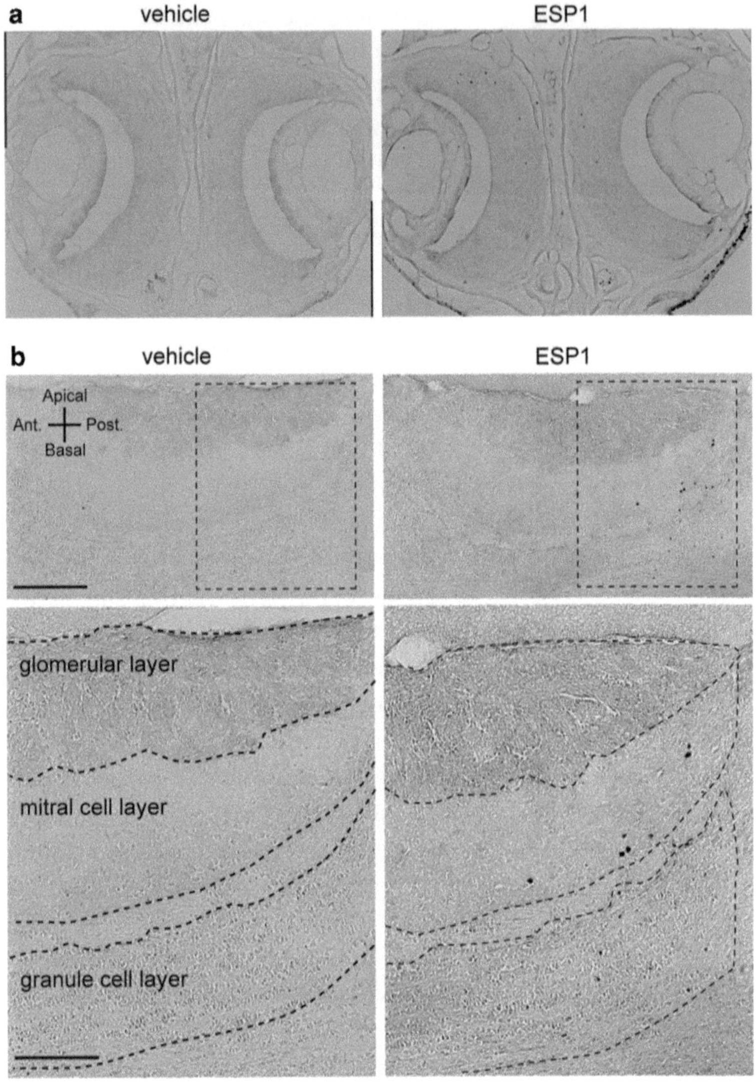

Fig. 2 ESP1 induces c-Fos expression in the VNO and AOB. (**a**) Representative images of anti-c-Fos immunostaining in the VNO stimulated by PBS (*left*) or ESP1 (*right*). (**b**) Representative images of anti-c-Fos immunostaining in the AOB stimulated by PBS (*left*) or ESP1 (*right*). *Lower panels* show the enlarged view of the *rectangles* in the *upper panels*. The scale bars, 200 μm (*upper panel*) and 100 μm (*lower panel*)

4. Transfer the sections to the wells containing 200 μL blocking solution and incubate them at RT for 30 min.
5. Transfer the sections to the wells containing 150 μL diluent solution with the primary antibody (1/30,000). Seal the wells with the tape and incubate the sections at 4 °C overnight (*see* **Note 12**).
6. Transfer the sections to the wells containing 500 μL PBS and rinse them for 10 min (three times) (*see* **Note 12**).

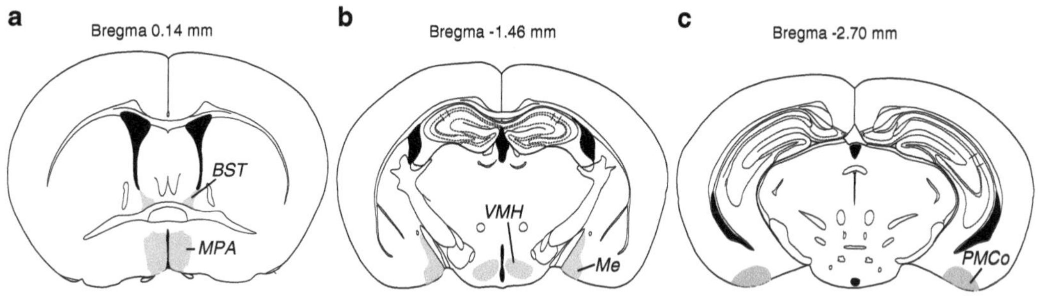

Fig. 3 Schematic illustration of the nuclei involved in pheromone information processing. Representative brain sections containing the *BST* and *MPA* (**a**), *VMH* and *Me* (**b**), and *PMCo* (**c**). Relative distance from the bregma is shown at the top of each image. The illustrations are modified from [11]

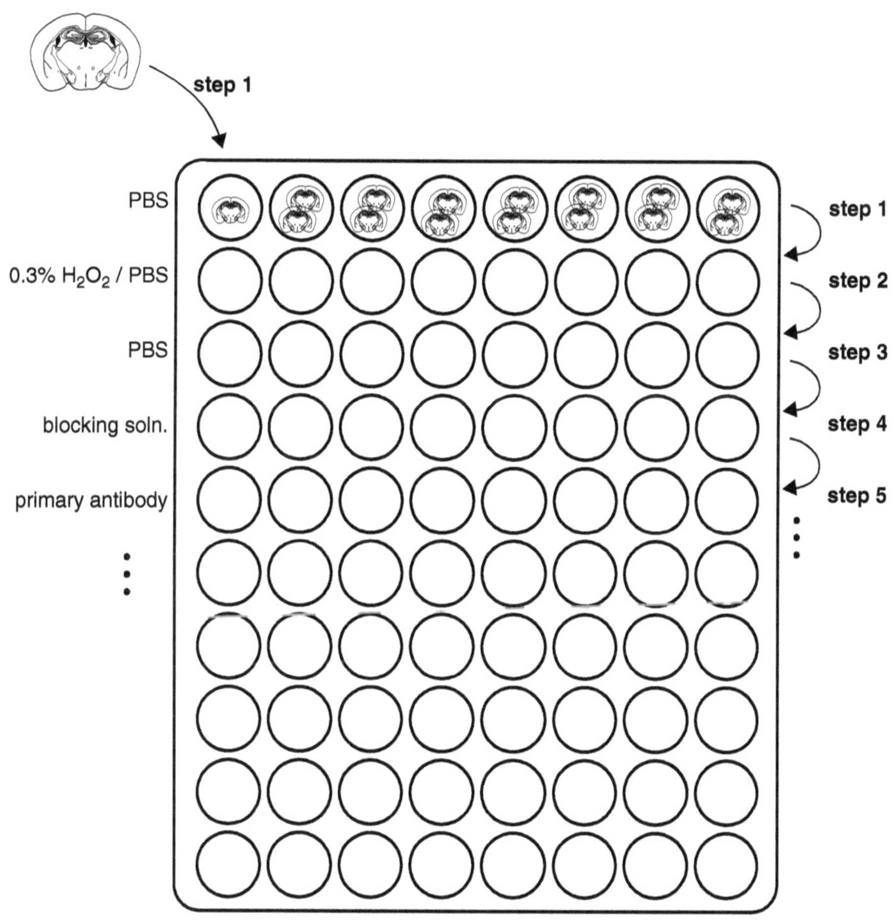

Fig. 4 Schematic illustration of the immunostaining procedures for brain. Put the brain sections into the wells of the agglutination plate, at most two sections per well, using an ink brush. Transfer the sections to the wells containing each solution

7. Transfer the sections to the wells containing 150 μL TBST with the secondary antibody (1/2,000) and incubate them at 25 °C for 1 h (*see* **Note 12**).
8. Prepare the ABC solution of PBS containing avidin (1/200 dilution) and biotin (1/200 dilution) just before **step 9** (*see* **Note 10**).
9. Transfer the sections to the wells containing 500 μL PBS and rinse them for 10 min (three times) (*see* **Note 12**).
10. Transfer the sections to the wells containing 200 μL ABC solution and incubate them at 25 °C for 30 min (*see* **Note 12**).
11. Transfer the sections to the wells containing 500 μL PBS and rinse them for 10 min (three times) (*see* **Note 12**).
12. Prepare DAB–Ni solution during **step 11**.
13. Transfer the sections to the wells containing 500 μL DAB–Ni solution and incubate them for 5–10 min (until black signals appear) (*see* **Notes 12** and **13**).
14. Transfer the sections to the wells containing 500 μL PBS and wash out DAB solution (*see* **Note 12**).
15. Transfer the sections to the wells containing 500 μL PBS and rinse them for 10 min (two times).
16. Affix the sections to slide glasses using an ink brush and let them dry. Place slides in a staining rack.
17. Put the staining rack in a staining jar containing EtOH/H_2O. Dehydrate the samples in a series of 80, 95, and 100 % EtOH (two times each) and 100 % xylene (two times) at RT for 2 min each.
18. Mount the slides with 50 μL of mounting medium and put a cover glass on the top.
19. Observe the specimens under the microscope and take images with a digital camera (Fig. 5).

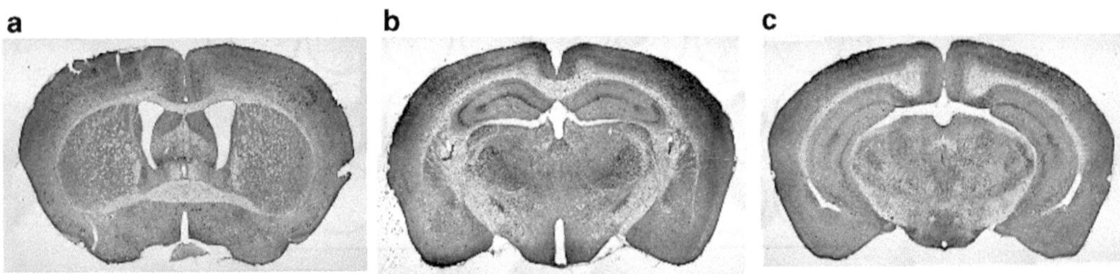

Fig. 5 ESP1-induced c-Fos expression in the brain. Representative images of anti-c-Fos immunostaining in the brain stimulated by ESP1. Each section corresponds to Fig. 3

4 Notes

1. Mice are stimulated in the dark or under the red light in a quiet environment. In order to reduce the neural activation for background environment, we house mice individually in mouse isolators and stimulate them in the isolator without moving the cages before stimulation.
2. To ensure that the mouse is properly sedated, pinch the toes to judge its level of response to painful stimulus.
3. Mouse should be quite stiff at the end of this step.
4. Be careful not to damage the brain tissue with forceps. After removing the skull, it may also be necessary to remove dura mater using Dumont #7 forceps.
5. In order to remove the air in the nostril cavities, fill the nostril cavities with 4 % PFA/PBS using a pipette.
6. After incubation for two nights, the bone becomes soft; or if not, refresh the EDTA solution and keep incubating the nose for one more night.
7. Briefly absorb the moisture of the tissue surface with a piece of Kimwipe and soak the tissue into the O.C.T. compound in a petri dish. Fill the nostril cavities with O.C.T. compound using a pipette in the petri dish, and then transfer the sample into a cryomold filled with O.C.T. compound.
8. Use cold air mode to dry slides.
9. Remove bubbles generated by the peroxidase reaction from the surface of specimens by moving the slide rack up and down slowly several times.
10. Prepare ABC solution 30 min prior to the application (follow the user instruction of the kit).
11. Remove bubbles generated by the peroxidase reaction from the surface of specimens by gently touching with an ink brush.
12. Rinse the ink brush thoroughly to avoid contamination in later steps.
13. Keep the reaction time consistent among all the specimens using a timer.

References

1. Wyatt TD (2003) Pheromones and animal behaviour. Cambridge University Press, Cambridge
2. Dulac C, Torello AT (2003) Molecular detection of pheromone signals in mammals: from genes to behaviour. Nat Rev Neurosci 4: 551–562
3. Dulac C, Wagner S (2006) Genetic analysis of brain circuits underlying pheromone signaling. Annu Rev Genet 40:449–467
4. Kimoto H, Touhara K (2005) Induction of c-Fos expression in mouse vomeronasal neurons by sex-specific non-volatile pheromone(s). Chem Senses 30(Suppl 1):i146–i147

5. Kimoto H, Haga S, Sato K, Touhara K (2005) Sex-specific peptides from exocrine glands stimulate mouse vomeronasal sensory neurons. Nature 437:898–901
6. Haga S, Hattori T, Sato T, Sato K, Matsuda S, Kobayakawa R, Sakano H, Yoshihara Y, Kikusui T, Touhara K (2010) The male mouse pheromone ESP1 enhances female sexual receptive behaviour through a specific vomeronasal receptor. Nature 466:118–122
7. Isogai Y, Si S, Pont-Lezica L, Tan T, Kapoor V, Murthy VN, Dulac C (2011) Molecular organization of vomeronasal chemoreception. Nature 478:241–245
8. Halem HA, Cherry JA, Baum MJ (1999) Vomeronasal neuroepithelium and forebrain Fos responses to male pheromones in male and female mice. J Neurobiol 39:249–263
9. Halem HA, Baum MJ, Cherry JA (2001) Sex difference and steroid modulation of pheromone-induced immediate early genes in the two zones of the mouse accessory olfactory system. J Neurosci 21:2474–2480
10. Choi GB, Dong HW, Murphy AJ, Valenzuela DM, Yancopoulos GD, Swanson LW, Anderson DJ (2005) Lhx6 delineates a pathway mediating innate reproductive behaviors from the amygdala to the hypothalamus. Neuron 46:647–660
11. Franklin KBJ, Paxinos G (2007) The mouse brain in stereotaxic coordinates, 3rd edn. Academic, San Diego

Part IV

Protocols in Analyzing Behavioral and Endocrine Changes Elicited by Pheromones

Chapter 19

Insect Pheromone Behavior: Fruit Fly

Daisuke Yamamoto, Soh Kohatsu, and Masayuki Koganezawa

Abstract

The amenability to genetics of *Drosophila melanogaster* has made this organism one of the best-suited models for studying the neurobiology of pheromone-guided behavior. Single-male assays use the minigene encoding the thermosensitive channel *dTrpA1* to activate neurons expressing *fruitless* (*fru*), a major courtship regulator gene, and thereby induce most of the elementary courtship acts in a solitary male exposed to temperature increase. Tethered male assays allow Ca^{2+}-imaging of neuronal activities of a male fly displaying courtship behavior on a treadmill when stimulated with a female or pheromones. Here we describe technical details of these assays.

Key words *Drosophila*, Thermogenetics, Ca^{2+}-imaging, MARCM, Courtship, *fruitless*

1 Introduction

The explosive development of optogenetic and thermogenetic tools has opened a new era in behavioral neurobiology [1–3]. These new tools are most powerful in resolving the neural circuitry for behavior at the single-cell level, when combined with other sophisticated genetic techniques available in the model organism *Drosophila melanogaster* [4]. What is crucial in these studies is our ability to rigorously identify the neurons whose activities are monitored and manipulated in an effort to capture neural correlates of specific behavior [5–13]. Refined versions of the GAL4-UAS system, for example, allow us to restrict the target of study to even a single neuron, which can be selectively labeled, activated, inactivated, or activity-monitored in an animal on the move [14]. The *M*osaic *A*nalysis with a *R*epressible *C*ell *M*arker (MARCM) technique [15] has been shown to be particularly useful in specifying central neurons for the generation of male courtship behavior [16, 17]. Two critical components in MARCM, in addition to GAL4- and UAS-fused transgenes, are the GAL4 repressor GAL80 and the Flp-FRT system that enables one to flip out GAL80 sporadically from the chromosome in a dividing cell [15]. As a result, one of

Kazushige Touhara (ed.), *Pheromone Signaling: Methods and Protocols*, Methods in Molecular Biology, vol. 1068,
DOI 10.1007/978-1-62703-619-1_19, © Springer Science+Business Media, LLC 2013

the two daughter cells loses GAL80, subsequently forming a neuronal clone with GAL4 activity when flipping occurs in a neuroblast that remains proliferative, or forming a single neuron with GAL4 activity when it happens in a ganglion mother cell that undergoes the final division. In the entire brain, only these GAL4-expressing neurons can be labeled, manipulated, or monitored for activity recording, since the rest of the brain is not affected by GAL4-responsive transgenes. For labeling neurons, a variety of markers are available [4], including the widely used mCD8::GFP [15]. The protein most favored as a neural activator is dTrpA1, a warmth-sensitive ion channel that excites the expressing cell when the ambient temperature is elevated beyond a certain level, e.g., from the permissive temperature 20 °C to the restrictive temperature 30 °C [18]. The other option often employed to activate neurons is the use of channelrhodopsin, a light-sensitive channel that makes the expressing neuron excitable upon illumination [19]. In inactivating neurons, reagents that reduce electrical excitability or block synaptic outputs are typically used [4]. Halorhodopsin is a light-sensitive Cl-transporter that hyperpolarizes the membrane potential of the expressing neuron and reduces excitability [20]. Conditional and reversible synaptic blockage is attained with a temperature-sensitive dominant-negative form of the Dynamin-coding gene *shibire* (shi^{ts}) by raising the temperature beyond 29 °C [1] while the Tetanus toxin light chain (TNT) constitutively blocks the transmission [21]. To visualize neural activities, changes in the intracellular concentration of various signal molecules associated with excitation are measured based on the *F*luorescence *R*esonance *E*nergy *T*ransfer (FRET; [4]). Among those, the FRET-based Ca^{2+} sensor yellow cameleon [22] is widely used for the detection of neural activities in animals displaying behavior under conditions in which the movement of the recording site (e.g., the brain) is minimized. In this chapter, we describe a few technical tips that are critical for the effective application of these neurogenetic tools to the analysis of cellular mechanisms of behavior, based on our recent success in elucidating the neural circuitry for male courtship [17].

2 Materials

2.1 Generation of Flies with Desired Genotypes

Depending on the behavior and cellular types to be analyzed, different GAL4 lines are required. Here we use an enhancer-trap line, fru^{NP21}, which expresses GAL4 in approximately 80 % of anti-Fru antibody-positive neurons, which are central to the production of male courtship behavior [16, 23].

1. Obtain fly lines harboring either the GAL4 driver fru^{NP21}, *UAS-dTrpA1, UAS-TNT, UAS-IMP-TNT, UAS-yellow cameleon 2.1, UAS-Channelrhodopsin, UAS-Halorhodopsin,* and

UAS-shi^{ts} or any other *UAS*-driven transgenes of interest from stock centers or relevant laboratories (*see* **Note 1**).

2. For MARCM, obtain an additional fly line carrying *hs-flp, FRT G13 tub-Gal80/FRT G13 UAS-mCD8::GFP* from a stock center.
3. Construct flies carrying all the genetic elements necessary for addressing specific problems, e.g., (1) *hs-flp; FRTG13 tub-Gal80/FRTG13 UAS-mCD8::GFP; fru*^{NP21}*/UAS-dTrpA1* for activating a *fru*-expressing neuronal clone, (2) *hs-flp; FRTG13 tub-Gal80/FRTG13 UAS-mCD8::GFP; fru*^{NP21}*/UAS-shi*^{ts} for inactivating a *fru*-expressing neuronal clone, or (3) *hs-flp; FRTG13 tub-Gal80/FRTG13 UAS-mCD8::GFP;fruNP21UAS-Yellow Cameleon 2.1/UAS-Yellow Cameleon* for Ca^{2+}-imaging of a *fru*-expressing neuronal clone.
4. Isolate newly emerged female and male flies and keep singly in small vials with cornmeal-yeast medium for several days before use. Transfer the flies to new vials at least every 3 days.
5. For MARCM analysis to determine the brain foci for elementary courtship acts, induce clones expressing *dTrpA1* (or *shi*^{ts}) in the brain during development. For this purpose, collect embryos within 24 h of egg laying and subject them to heat shock at 37 °C for 60 min.

2.2 Single-Male Assays to Induce Courtship Behavior in a Solitary Male

1. The observation chambers are constructed by cutting circular holes of 10 mm in diameter and 3 mm in height through a copper plate, and covering them on either side with a glass slip.
2. A thermal cycler (e.g., Gene Amp® PCR system 9700; Perkin Elmer) is used to control the temperature in the observation chamber; the chamber plate is directly mounted on the aluminum block of the thermal cycler.
3. A dual LED illuminator, type W-E (Micronet), is used as a light source to illuminate the chamber.
4. A WAT-221S (Watec) CCD camera equipped with a Micro-NIKKOR lens (35 mm, 1:28; Nikon) is used to take movies of fly behavior.
5. ArcSoft ShowBiz DVD software is used to control the movie recordings.

2.3 Tethered Male Preparation

2.3.1 Fly Preparation

1. Fly holder. Punch a hole of approximately 8 mm in diameter in a plastic coverslip. Seal it on one side with a piece of polyethylene film using UV glue (Fig. 1a) [24].
2. UV curing glue and UV pen light.
3. Syringe needle (24–27 G: Terumo).
4. Tungsten needle of 0.15 mm in diameter, which is attached to a holder.

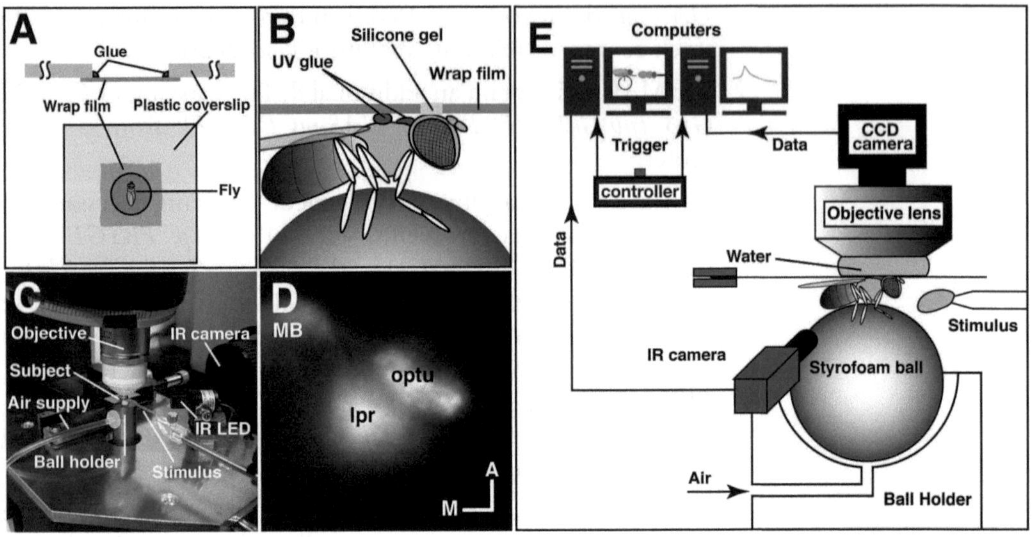

Fig. 1 The experimental setup for tethered male preparation. (**a**) The coverslip with wrap film to cover the head opening for recording. (**b**) Lateral view of a fly attached to the wrap film. (**c**) Arrangement of instruments on the microscopic stage. (**d**) Yellow cameleon-derived YFP fluorescence detected in the mushroom body (MB), lateral protocerebrum (lpr), and optic tubercle (optu). *A* anterior, *M* medial. (**e**) Diagram of the entire tethered mate system

5. Silicone elastomer (Kwik-Sil; World Precision Instruments).
6. Fly saline: 130 mM NaCl, 5 mM KCl, 2 mM $MgCl_2$, 2 mM $CaCl_2$, 36 mM sucrose, 5 mM HEPES–NaOH (pH 7.3) [25].
7. Cool anesthetization stage made of a brass plate of 50 × 50 × 5 mm.
8. Pulled glass capillaries.

2.3.2 Recording Setup

The recording system is a composite of two separable elements. The first element is used to acquire fluorescent signals from the brain. The second element serves to monitor behavior by video-recording lateral views of the subject male (Fig. 1c, e). An electrical trigger signal synchronizes the operation of these two elements. The subject fly is stimulated by touching his foreleg with a stimulant fly's body attached to a holder mounted on the arm of a manual micromanipulator.

1. Upright fluorescent microscopy.
2. W Plan-apochromat, ×40, NA = 1.0 (Carl Zeiss), a water immersion objective lens with high NA values and a long working distance (*see* **Note 2**).
3. C7780-20 and Aquacosmos (Hamamatsu Photonics), a CCD camera and the software to control it.
4. HBO-100 (Carl Zeiss), a high-pressure mercury lamp.

5. ND Filters. For the preparation of excitation light and separation of emission light, a 440/20 band-path filter, a 455 dichroic filter, and a 460 nm long-path filter are used. Emission light is further divided by a beam-splitting prism embedded in the CCD camera (460–490 nm for CFP and 490–570 nm for YFP signals), and then projected upon separate CCD chips (*see* **Note 3**).
6. STC-H400 (Sensor Technology Co., Ltd.), an infrared CCD camera.
7. 12× zoom system (Navitar), a macro zoom lens.
8. VTG-55D (FOR-A), an analog video timer.
9. DVMC-DA2 (Sony), an A/D video converter.
10. Styrofoam balls of 4–5 mm in diameter.
11. Styrofoam ball holder. A brass base equipped with a hemispherical pocket of 6 mm in diameter and an associated air inlet of 1 mm in diameter at the bottom of the pocket. The holder is attached to the microscope stage with appropriate adaptors for fine adjustment of the position.
12. Air source. A compressor produces an airstream, which, after being decompressed, passes through cotton, activated charcoal, and water in sequence for the filtration of contaminants, and flows into the ball holder through a needle valve. Adjust the airflow rate so as to float the ball stably (~1 ml/s).
13. Manual micromanipulator for handling the stimulus fly's body.

2.3.3 Data Analysis

1. ImageJ [17] (http://rsbweb.nih.gov/ij/ and Excel, Microsoft) as a software environment for image processing (*see* **Note 4**) [26].

2.4 Immunohistochemistry of Brain Neurons

1. Purchase reagents for immunostaining: Rabbit polyclonal anti-GFP antibody (Molecular Probes, Eugene, OR) and mouse monoclonal nc82 as primary antibodies, and Alexa548-conjugated goat anti-mouse IgG (Invitrogen, Carlsbad, CA) and Alexa488-conjugated goat anti-mouse IgG (Invitrogen) as secondary antibodies.
2. Prepare solutions for dissection of tissues (PBS), fixation (4 % paraformaldehyde in PBS), washing (PBT = 0.4 % TritonX in PBS), blocking (PBTN = 10 % normal goat serum in PBT), and staining (anti-GFP at 1:1,000 dilution, nc82 at 1:20, anti-rabbit IgG at 1:200, and anti-mouse IgG at 1:200, all in PBTN). Purchase VECTORSHIELD (Vector Laboratories) as a clearing and mounting reagent.
3. The LSM510META confocal microscopy (Carl Zeiss) yields a series of optical sections which are used to obtain reconstituted images with the aid of the ImageJ software.

3 Methods

3.1 Single-Male Assays

1. Introduce a single male with *dTrpA1* (or *shi*^{ts})-expression by aspiration into a chamber placed on a thermal cycler.
2. Start video-recordings of the fly's behavior.
3. Maintain the temperature in the chamber at 22 °C for the initial 5 min and then increase up to 35 °C by a series of 1 °C temperature steps of 5-min duration. The fly commences to exhibit elements of courtship behavior when the temperature exceeds certain threshold levels.
4. Estimate the locomotor speed of the fly based on automatic measurement using the program Move-tr/2D version 7.2 (Library Co., Tokyo, Japan).
5. Calculate the event frequency as the proportion of time the male spent performing each courtship action, e.g., tapping, wing extension, licking, or attempted copulation.
6. Construct perievent histograms in which the midpoint of every tapping event is taken as reference time zero. For a 6-s period starting 3 s before and ending 3 s after the midpoint of every tapping event, fly behavior is analyzed every 0.1 s. This analysis yields 61 data points associated with a single tapping event that serves as the time-zero reference, defining a single data set. The perievent histogram is regarded as the probability distribution of each courtship element with time. For example, if licking was observed at the 0.1-s period between 2.0 and 2.1 s before time zero in 50 out of 100 data sets, then the probability of observing licking at this time point is estimated to be 50/100 = 0.5.
7. In MARCM analysis of the courtship circuitry, classify the assayed flies into two groups: courters, who display unilateral wing extension or any other appropriate act, and noncourters, who do not. Both groups are subject to immunohistochemistry to identify the neurons expressing *dTrpA1* (or *shi*^{ts}).
8. For the observation of neurons, fix the dissected brain with 4 % paraformaldehyde for 60 min and incubate it in 10 % normal goat serum for 12 h at room temperature (25 °C). Thereafter, the tissues are reacted with primary antibodies, i.e., the anti-GFP antibody (rabbit polyclonal: A6455; Molecular Probes, Eugene, OR) and nc82 (mouse monoclonal; Developmental Studies Hybridoma Bank) at dilutions of 1:1,000 and 1:10, respectively, for 72–96 h at 4 °C. After washing, the brain is treated with Alexa Fluor 488 goat anti-rabbit IgG (Invitrogen A11034) or Alexa Fluor 546 goat anti-mouse IgG (Invitrogen A11030) at a 1:500 dilution.
9. Observe the stained brain with the LSM 510 META confocal microscope.

3.2 The Tethered Male Preparation

1. Anesthetize a male fly by placing it for ~20 s in a glass tube that has been cooled by ice (*see* **Note 5** and **6**).
2. After mounting the anesthetized fly on an ice-cooled preparation stage, fill the gap between the head and thorax of the fly with UV glue to restrict head movements (*see* **Note 7**).
3. Put a small amount of UV glue on the notum, which is then fixed to a polyethylene film placed on the fly holder. The head cuticle flanked by the base of the antennae and ocelli must be rigidly attached to the film, yet the aristae on the antennae should not be in contact with the film.
4. Hold the fly ventral side down in a humid chamber and let him grasp a Styrofoam ball that prevents him from flying. Prepare a dozen flies at a time according to **steps 1–3** above.
5. Keep the flies in the humid chamber for approximately 2 h to let them recover from manipulation.
6. Make a square opening in the wrap film covering the head by cutting it with the edge of a syringe needle, and cover the opening with a drop of saline (~100 μl). Be sure that saline does not spill over the film opening onto the reverse side. At this step, it is important that the edge of the cut film is not sealed against the head, but the head is in tight contact with the film.
7. Open a window on the head capsule by cutting it with a syringe needle so as to make the brain observable in situ by microscopy. Remove the fat bodies with fine forceps and gently set the air sacs aside to expose the brain, particularly its lateral protocerebral region (*see* **Note 8**). Special attention must be paid not to damage the air sacs.
8. Mix the two components of silicone elastomer, which will be used to seal the head opening. Drain the saline covering the dorsal surface of the brain using a micropipette and twisted Kimwipe, and immediately apply a small amount of the mixed silicone elastomer onto the cuticular opening to fill the gap between the head and the edge of the wrap film at the window. The silicone elastomer not only seals the cuticular opening but also helps restrict the movement of the brain during the recording of neural activities. Allow the fly to rest for ~5 min, letting it recover from the surgery.
9. Prepare a "stimulant fly." Cut the head and thorax from a cold-anesthetized fly with fine scissors. Insert a pulled glass capillary into the abdomen from the cut end, and then seal it with UV glue.

3.3 Recording of Neural Activities

1. Search for raw YFP fluorescence originating from the lateral protocerebrum in the brain of the fly preparation mounted on the microscope stage. To minimize photo-bleaching of YFP in

this process, the strength of excitation light should be as weak as possible.

2. Set the position of the treadmill as guided by lateral images of the fly obtained by an infrared camera. Make sure that the fly lands gently on the Styrofoam ball and that the legs of both sides evenly contribute to the walking motion on the ball (*see* **Note 9**).
3. Mount the stimulus fly on a holder. The stimulus fly will be brought into close proximity to a foreleg of the test fly using a micromanipulator under dim red light illumination.
4. Choose the recording site (the region of interest: ROI) from which real-time monitoring of the fluorescent signals will be done. Set the aperture of the excitation light in an appropriate range, with which the cuticular region around the head opening will not be illuminated (*see* **Note 10**). This helps improve the contrast by reducing the scattered light. Choose an appropriate ND filter and binning and gain values for the CCD camera such that exposure of the fly preparation for ~200 ms will provide sufficient fluorescent signals in the ROI (i.e., over 1,000 signal values using a 12-bit camera) for both the YFP and CFP channels.
5. As a tester male, choose one that is not hyperactive. In the case that the tester male candidates persist in running or grooming, wait ~5 min to allow them to calm down (*see* **Note 11**).
6. Start the recordings. Each recording session takes ~20 s when the image acquisition rate is 5–10 Hz. Briefly touch one of the tester fly forelegs with the stimulant fly by advancing the holder mounting the latter and then quickly withdrawing it by manual operation of the micromanipulator while monitoring the preparation in real time with the infrared camera (*see* **Note 12**).
7. A transient increase in the YFP signal and decrease in the CFP signal will be recorded upon touching the foreleg of the tester fly with a stimulant fly in successful trials. A good preparation will tolerate up to ten stimulation trials that together require ~1 h to complete.

3.4 Data Processing

1. Define the time of stimulus onset by detecting the moment of the first contact of the stimulant fly with a foreleg of the tester fly on the video record captured by the infrared camera. The timestamp on the video frame in which the first contact is recorded defines the time of stimulus onset (*see* **Note 13**).
2. Register the stack of images, using ImageJ, to cancel out any undesirable drifts in the position of the recording site due to inevitable brain movements.
3. Filter the images to reduce noise with a Gaussian filter (radius: 1–2).

4. Specify the ROI with YC2.1 expression as well as the background brain area without YC2.1 expression. Both are defined by circles of 10 pixels in diameter.
5. Subtract the fluorescent value at the background area from the value at the ROI for YFP or CFP in the respective channel to yield the crude signal value.
6. Calculate the ratio YFP/CFP frame by frame.
7. Calculate the average of the signal values for 5 frames prior to the stimulus onset for YFP, CFP, and YFP/CFP to determine the baseline values (F0 and R0 for the fluorescent outputs and their ratio, respectively). Determine the signal changes (dF or dR) by subtracting F0 or R0 from the raw signal values and calculate dF/F0 or dR/R0 (%) for use in data quantification.
8. In the data quantification, determine the peak response value (dR/R0) by averaging the maximal value and the values at its two flanking frames. The peak value corresponds to the largest dR/d0 within 3 min following the stimulus onset.

4 Notes

1. Increasing the copy number of transgenes encoding yellow cameleon or GAL4 would improve the signal-to-noise ratio in Ca^{2+}-imaging of neural activities, as it reduces the baseline fluorescence. Aged flies may have accumulated a larger amount of Ca^{2+} indicator proteins, which could enhance the signals.
2. Motorizing the diachronic filter exchange and the shutter control for excitation light will secure the preparation by reducing the need for manual operation under dark conditions. A customized microscopic stage with a larger working space under the objective lens facilitates functional arrangements of the fly preparation and instruments. A "vibration-free" table will be of benefit for reducing artifacts due to movements.
3. A CCD camera having higher sensitivity and more pixels with binning functionality is preferable. The dynamic range should exceed 12 bits.
4. Image processing requires a function to collect signals from the ROI in multichannel image stacks. Spatial filtrations to reduce noises or movement artifacts as well as image registration further assist in the analysis. ImageJ, for example, offers a free and expandable environment for batch image processing.
5. The success of the experiments will depend heavily on the state of the flies. Discard flies whose legs are insufficiently clean, because such flies tend to stick to grooming and show poor responses to sexual stimuli. Tester males need to be reared singly, but the food substrate of a vial containing a single male

easily becomes sticky due to the breeding of bacteria. This can be a source of deposits on fly legs. Frequent transfer of flies to new vials (e.g., every 1–2 days) makes them happier. Addition of antibiotics to food is a possible solution when the sticky leg problem is serious. Another factor that increases grooming is the loss of leg contact with the substrate—the Styrofoam ball in this case—which results in increased defecation and excrement deposits on the legs. It is therefore important to make the fly hold to the ball throughout the experiment.

6. Minimize the time used to place the fly under cold anesthesia (~2 min).
7. In making the tethered male preparation, the fly needs to be immobilized with UV glue. Reduce the amount of glue as much as possible. A thin tungsten wire is a convenient tool for handling a small amount of glue. The UV flash should be short (~5 s). After glue application, adjust the fly position quickly and cure the glue. Keep the polyethylene film tense while gluing it to the coverslip hole so that an opening in the film can be made easily.
8. In exposing the brain, special caution must be paid not to damage the air sacs. Impaired air supply can be vital.
9. Make sure that the flies in the store chamber are keeping their legs on the Styrofoam ball. If any fly has dropped the ball, pick it up and give it back to him with forceps.
10. To avoid photo-bleaching of the Ca^{2+} sensor, do not illuminate the brain-exposed preparation except when necessary.
11. After handling each fly, clean the scissors and forceps with a Kimwipe moistened with ethanol, which helps avoid any possible contamination of cuticular hydrocarbons in subsequent experiments.
12. In touching a leg of the tester fly for stimulation, we prefer to move the stimulant fly on the manipulator by manual rather than motorized operation in order to attain delicate control of the movement.
13. In analyzing the data, choose appropriate parameters of response (e.g., the peak amplitude, time course, and/or integrals) depending on the purpose of the experiment.

Acknowledgements

This work was supported by MEXT grants (23220007, 1802012, and 24113502 to D.Y.; 21770074 and 24570082 to M.K.; and 24770065 to S.K.), grant from the Strategic Japanese-French

Cooperative Program by JST to D.Y., a grant from the Tohoku Neuroscience GCOE program to D.Y., and a Life Science Grant from the Takeda Science Foundation to D.Y. We thank Hiromi Sato for secretarial assistance.

References

1. Kitamoto T (2001) Conditional modification of behavior in *Drosophila* by targeted expression of a temperature-sensitive *shibire* allele in defined neurons. J Neurobiol 47:81–92
2. Lima SQ, Miesenböck G (2005) Remote control of behavior through genetically targeted photostimulation of neurons. Cell 121:141–152
3. Luo L, Callaway EM, Svoboda K (2008) Genetic dissection of neural circuits. Neuron 57:634–660
4. Venken KJT, Simpson JH, Bellen HJ (2011) Genetic manipulation of genes and cells in the Nervous System of the Fruit fly. Neuron 72:202–230
5. Fotowat H, Fayyazuddin A, Bellen HJ, Gabbiani F (2009) A novel neuronal pathway for visually guided escape in *Drosophila melanogaster*. J Neurophysiol 102:875–885
6. Gong Z, Liu J, Guo C, Zhou Y, Teng Y, Liu L (2010) Two pairs of neurons in the central brain control *Drosophila* innate light preference. Science 330:499–502
7. Chiappe ME, Seelig JD, Reiser MB, Jayaraman V (2010) Walking modulates speed sensitivity in *Drosophila* motion vision. Curr Biol 20:1470–1475
8. Reiff DF, Plett J, Mank M, Griesbeck O, Borst A (2010) Visualizing retinotopic half-wave rectified input to the motion detection circuitry of *Drosophila*. Nat Neurosci 13:973–978
9. Cheng LE, Song W, Looger LL, Jan LY, Jan YN (2010) The role of the TRP channel NompC in *Drosophila* larval and adult locomotion. Neuron 67:373–380
10. Maimon G, Straw AD, Dickinson MH (2010) Active flight increases the gain of visual motion processing in *Drosophila*. Nat Neurosci 13:393–399
11. Claridge-Chang A, Roorda RD, Vrontou E, Sjulson L, Li H, Hirsh J, Miesenböck G (2009) Writing memories with light-addressable reinforcement circuitry. Cell 139:405–415
12. Aso Y, Siwanowicz I, Bräcker L, Ito K, Kitamoto T, Tanimoto H (2010) Specific dopaminergic neurons for the formation of labile aversive memory. Curr Biol 20: 1445–1451
13. Waddell S, Armstrong JD, Kitamoto T, Kaiser K, Quinn WG (2000) The *amnesiac* gene product is expressed in two neurons in the *Drosophila* brain that are critical for memory. Cell 103:805–813
14. Ruta V, Datta SR, Vasconcelos ML, Freeland J, Looger LL, Axel R (2010) A dimorphic pheromone circuit in *Drosophila* from sensory input to descending output. Nature 468:686–690
15. Lee T, Luo L (1999) Mosaic analysis with a repressible cell marker for studies of gene function in neuronal morphogenesis. Neuron 22:451–461
16. Kimura K, Hachiya T, Koganezawa M, Tazawa T, Yamamoto D (2008) Fruitless and Doublesex coordinate to generate male-specific neurons that can initiate courtship. Neuron 59:759–769
17. Kohatsu S, Koganezawa M, Yamamoto D (2011) Female contact activates male-specific interneurons that trigger stereotypic courtship behavior in *Drosophila*. Neuron 69:498–508
18. Rosenzweig M, Brennan KM, Tayler TD, Phelps PO, Patapoutian A, Garrity PA (2005) The *Drosophila* ortholog of vertebrate TRPA1 regulates thermotaxis. Genes Dev 19:419–424
19. Boyden ES, Zhang F, Bamberg E, Nagel G, Deisseroth K (2005) Millisecond-timescale, genetically targeted optical control of neural activity. Nat Neurosci 8:1263–1268
20. Chow BY, Han X, Dobry AS, Qian X, Chuong AS, Li M, Henninger MA, Belfort GM, Lin Y, Monahan PE, Boyden ES (2010) High-performance genetically targetable optical neural silencing by light-driven proton pumps. Nature 463:98–102
21. Sweeney ST, Broadie K, Keane J, Niemann H, O'Kane CJ (1995) Targeted expression of tetanus toxin light chain in *Drosophila* specifically eliminates synaptic transmission and causes behavioral defects. Neuron 14:341–351
22. Fan GY, Fujisaki H, Miyawaki A, Tsay RK, Tsien RY, Ellisman MH (1999) Video-rate scanning two-photon excitation fluorescence microscopy and ratio imaging with cameleons. Biophys J 76:2412–2420
23. Kimura K, Ote M, Tazawa T, Yamamoto D (2005) Fruitless specifies sexually dimorphic

neural circuitry in the *Drosophila* brain. Nature 438:229–233
24. Fiala A, Spall T (2003) In vivo calcium imaging of brain activity in *Drosophila* by transgenic cameleon expression. Sci STKE 2003:PL6
25. Nagai T, Yamada S, Tominaga T, Ichikawa M, Miyawaki A (2004) Expanded dynamic range of fluorescent indicators for Ca^{2+} by circularly permuted yellow fluorescent proteins. Proc Natl Acad Sci U S A 101: 10554–10559
26. Abramoff MD, Magelhaes PJ, Ram SJ (2004) Image processing with ImageJ. Biophoton Int 11:36–42

Chapter 20

Quantitative Assessment of Pheromone-Induced Dauer Formation in *Caenorhabditis elegans*

Scott J. Neal, Kyuhyung Kim, and Piali Sengupta

Abstract

Environmental conditions experienced during early larval stages dictate the developmental trajectory of the nematode *C. elegans*. Favorable conditions such as low population density, abundant food, and lower temperatures allow reproductive growth, while stressful conditions promote entry of second-stage (L2) larvae into the alternate dauer developmental stage. Population density is signaled by the concentration and composition of a complex mixture of small molecules that is produced by all stages of animals, and is collectively referred to as dauer pheromone; pheromone concentration is a major trigger for dauer formation. Here, we describe a quantitative dauer formation assay that provides a measure of the potency of single or mixtures of pheromone components in regulating this critical developmental decision.

Key words Dauer larva, Ascaroside, *C. elegans*, Pheromone

1 Introduction

C. elegans can develop via one of the two alternative developmental pathways [1]. This developmental decision is regulated by environmental cues experienced during early larval stages. Exposure of L1 larvae to overcrowding, high temperatures, and low food abundance causes them to develop as pre-dauer L2d larvae; continued stressful conditions result in irreversible commitment to the long-lived and stress-resistant alternative third larval stage called the dauer stage [2–5]. When conditions improve, dauer larvae molt and resume reproductive growth [6]. Conversely, favorable conditions promote continuous reproductive growth. Genetic analyses of mutants that fail to form dauers under stressful conditions, or enter into the dauer stage regardless of environmental cues, have identified the signaling pathways underlying the dauer developmental decision [7–13]. In brief, environmental cues sensed by sensory neurons regulate TGF-β and insulin signaling. These signals in turn regulate the production of steroid hormones and signaling via the DAF-12 nuclear hormone receptor. Hormone-bound

Kazushige Touhara (ed.), *Pheromone Signaling: Methods and Protocols*, Methods in Molecular Biology, vol. 1068, DOI 10.1007/978-1-62703-619-1_20,

DAF-12 promotes reproductive growth, whereas unliganded DAF-12 promotes dauer arrest (reviewed in [11, 13]).

A major trigger for dauer formation is a complex mixture of small molecules called ascarosides that are produced by all stages of animals, and whose concentration serves as a measure of local population density [2, 4, 14–20]. The role of secreted/excreted molecules in regulating dauer formation was demonstrated by showing that conditioned medium from wild-type worms was sufficient to induce dauer formation in well-fed larvae [2]. This chapter describes an assay in which the ability of partially purified conditioned medium, called crude pheromone (*see* Chapter 6), or chemically synthesized pure ascarosides, to induce dauer formation is assessed. Since food levels and ambient temperature significantly modulate the efficacy of pheromone-induced dauer formation [3, 4, 21], this assay is designed to carefully control for environmental variables so as to provide a reproducible quantification of the efficiency of single or mixtures of ascarosides in inducing dauer entry. A variation of this assay has been described [22].

2 Materials

Prepare all solutions using Milli-Q or equivalent (≥18 MΩ) ultrapure water. Use reagent-grade ethanol (≥99.5 %) for reconstituting and diluting synthetic pheromones. All other materials should be of molecular biology grade or better. This protocol assumes basic knowledge of standard *C. elegans* husbandry.

1. Dissecting microscope, heat block, 20 °C incubator, 25 °C incubator, water bath.
2. *E. coli* strain OP50, grown on Luria-Bertani (LB) medium.
3. 1 M potassium phosphate buffer pH 6.0: ~1:3 1 M K_2HPO_4:1 M KH_2PO_4, titrate with 1 M K_2HPO_4 to desired pH.
4. S-basal buffer: 0.1 M NaCl, 50 mM potassium phosphate buffer pH 6.0.
5. 5 mg/mL cholesterol in 95 % ethanol.
6. Dauer formation assay agar: 50 mM NaCl, 1.7 % Noble agar (*see* **Note 1**), 1 mM $CaCl_2$, 1 mM $MgSO_4$, 25 mM potassium phosphate buffer pH 6.0, 5 μg/mL cholesterol.
7. Crude and/or synthetic pheromone (*see* Chapter 6).

3 Methods

3.1 Overview of Assay Timeline

This protocol involves multiple steps that are carried out on several independent days. In this protocol, it is important to be considerate of sources of biological variation which might compromise assay results. For instance, there is increasing evidence of transgenerational inheritance of epigenetic information which may influence the results of behavioral or developmental assays [23–26]. It is thus important to control the developmental history of the parent animals that will contribute the embryos for the assay. Typically, worms should be maintained under well-fed conditions at 20 °C (*see* **Note 2**) for at least three generations prior to the assay. An overall summary of the protocol is shown in Fig. 1.

1. On day 1, pick five mid-L4-stage larvae to a fresh OP50-seeded nematode growth medium (NGM) plate around midday. Culture the worms at 20 °C. This parental plate should have at least 100 fairly synchronous well-fed L4 larvae on day 4.
2. On day 4, pick the required number of mid- to late-L4 larvae to fresh 60 mm OP50-seeded NGM plates; on day 5, five young adults will be required for each assay plate. Care must be taken not to induce crowding; typically no more than 70 L4 larvae should be subcultured to the same plate in order to minimize their exposure to endogenous pheromone. Culture these animals overnight at 20 °C in preparation for the assay.
3. Also on day 4, prepare the dauer agar assay plates and inoculate the OP50 culture for the heat-killed food source (Subheadings 3.3 and 3.4, respectively).
4. On day 5, apply the heat-killed OP50 cultures to the assay plates. Transfer five young adult animals to each assay plate and collect eggs (Subheading 3.5).
5. Incubate the assay plates at 25 °C and quantify dauer formation on day 8 or 9 (Subheading 3.6).

3.2 Preparation of Crude or Synthetic Pheromone

The methodology for preparation of crude pheromone is addressed elsewhere in this volume (*see* Chapter 6). Purification and synthesis of individual ascarosides have been described [14–20]. Crude or synthetic pheromones can be stably stored dry at −20 °C with some exceptions. For instance, PABA-C7 (ascr#8) [18] is less stable and is, therefore, stored at −80 °C. For use, dilutions of crude pheromone are made in water or 50:50 water:ethanol and may either be mixed with the molten assay agar or spread on the surface of the assay plate. Synthetic pheromones may be reconstituted at low millimolar concentrations in either dimethyl sulfoxide or ethanol. These solutions are stored at −20 °C and should be diluted in the original solvent when testing multiple concentrations in a dauer formation assay so as to keep the concentration of solvent constant in all assay plates. Our standard practice is to use ethanol as the solvent though in limited testing we have not observed any differences between the two solvents in this assay.

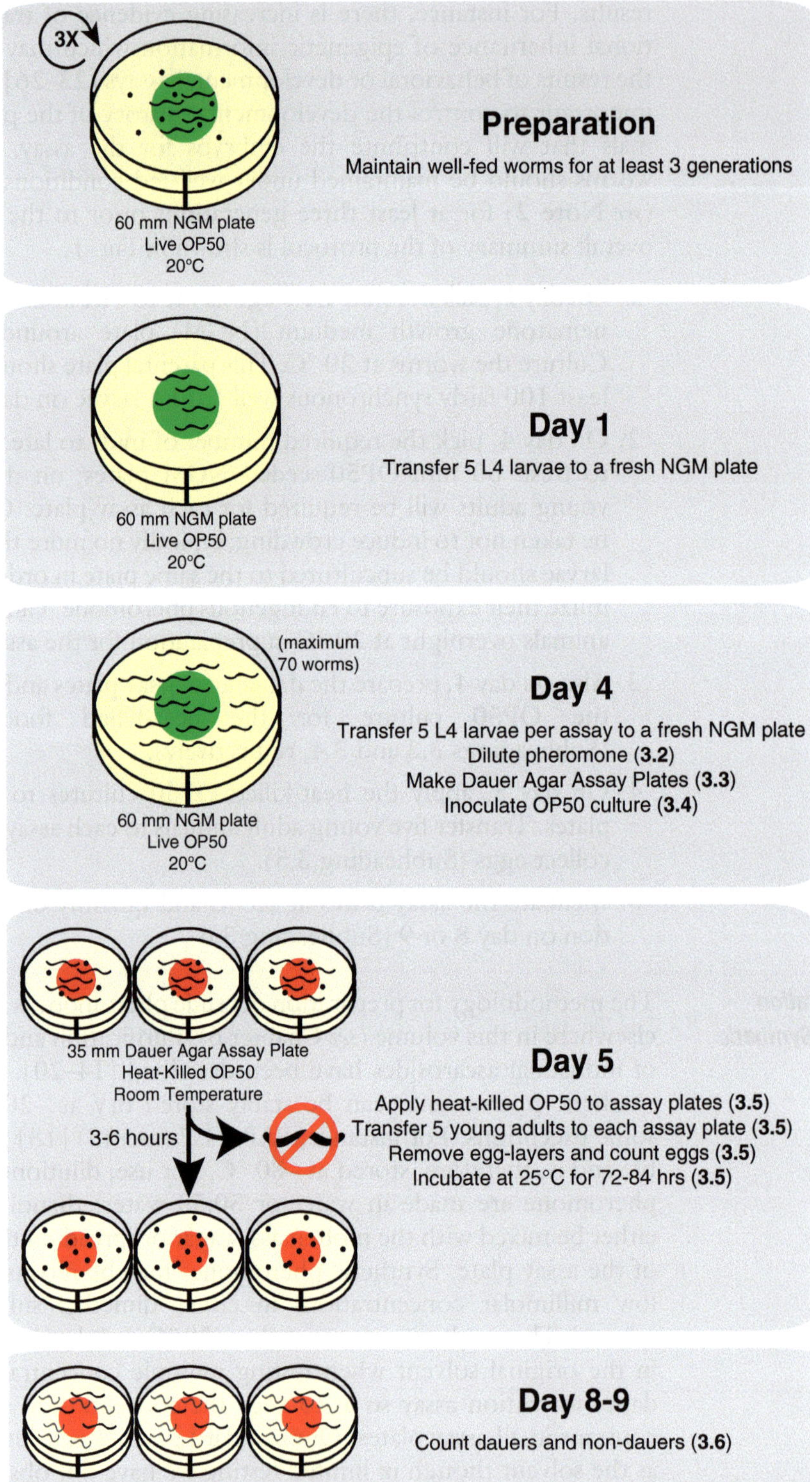

Fig. 1 Overview of the timeline and individual steps in the described protocol. *Numbers* in parentheses refer to the respective sections in Subheading 3 as described in the text. *Green* and *red* bacterial spots indicate live and heat-killed OP50 bacteria, respectively

3.3 Preparation of Dauer Agar Assay Plates

Use sterile technique throughout this procedure. Be sure to prepare sufficient agar for all assay plates while allowing for some loss due to evaporation and pipetting error. 3 mL of dauer assay agar is required per 35 mm assay plate (*see* **Note 3**).

1. In a clean, autoclave-safe bottle add (per 100 mL) 0.3 g NaCl, 1.7 g Noble agar, and an autoclave-safe stir bar. Add 100 mL of MilliQ H_2O and autoclave for 20 min on a liquid cycle. Tighten the cap and cool the sterilized solution to 60 °C in a water bath.
2. Place the cooled media on a stir plate and add (per 100 mL) 100 μL 1 M $CaCl_2$, 100 μL 1 M $MgSO_4$, 2.5 mL 1 M potassium phosphate buffer pH 6.0, and 100 μL 5 mg/mL cholesterol in 95 % ethanol. Stirring should be sufficient to mix in the cholesterol while avoiding bubbles or frothing.
3. Pheromone solutions should be prepared immediately before plate pouring, and for consistency should be prepared as a master mix. For each plate combine 6 μL of appropriately diluted pheromone in solvent (or solvent alone) with 94 μL of sterile MilliQ water. Array the empty 35 mm assay plates on a flat surface (*see* **Note 4**) and transfer 100 μL of the above pheromone master mix to the center of the empty plate.
4. Dispense 3 mL of assay agar into the pheromone-containing plates (*see* **Note 5**), and allow them to set overnight on the bench at room temperature (*see* **Note 6**). Ensure that there are no bubbles in the agar.

3.4 Preparation of Heat-Killed OP50

Although pheromone-induced dauer formation can be significantly suppressed by the presence of bacterial food, food is necessary to bypass the starvation-induced arrest at the L1 larval stage [27]. The typical food source used in most *C. elegans* laboratories is the *E. coli* strain OP50. Fresh cultures of OP50 may continue to grow on standard NGM plates, and this self-renewing food source suppresses dauer formation; therefore, either heat-killed or streptomycin-killed bacteria are used. For best results, OP50 should be freshly streaked onto an LB plate from a glycerol stock of the strain.

1. Inoculate a single OP50 colony into 35 mL of LB broth in a 50 mL conical tube. With the lid slightly loosened, grow the culture to stationary phase overnight at 37 °C with 225 rpm shaking.
2. Pre-weigh an appropriate number (generally 2–4) of 1.5 mL tubes to establish their tare weights. Transfer 1.45 mL of overnight culture into each tube and spin for 1 min at 8,000 × *g*. Discard the supernatant and transfer an additional 1.45 mL of overnight culture to the tube. Spin for 1 min at 8,000 × *g* and

aspirate the supernatant carefully. Spin the pellet for 1 min at maximum speed (≥13,000 × *g*) and carefully aspirate any supernatant with a pipette. Spin the pellet for 2 min at maximum speed and remove any final traces of supernatant with a P10 or a P20 pipetman.

3. Weight the tube and determine the pellet weight; the expected mass is between 8 and 10 mg. Fully resuspend the pellet to a concentration of 8 mg/mL in S-basal + 5 μg/mL cholesterol.
4. Heat-kill the bacteria by incubating in a heat block at 95 °C for 30 min with vortexing at 10-min intervals. Cool by placing at 4 °C for at least 30 min before use.
5. This reagent is best prepared fresh on the day of or the day before the assay (store overnight at 4 °C), but may also be kept for at most 1 week at 4 °C. Always vortex the bacterial suspension before use.

3.5 Setting Up the Assay

1. On the day of the assay, pipette 20 μL of fresh 8 mg/mL heat-killed OP50 (Subheading 3.4) onto the center of the assay agar. Care should be taken at all stages of this protocol not to pierce the surface of the assay agar to prevent animals from burrowing (*see* **Note 7**).
2. Briefly dry the food spot by incubating the plates in the fume hood; do not offset the lids, even if they are not vented (*see* **Note 8**). The plates are ready once the spot has completely dried and when there is no condensation on the lid. It generally takes 20–30 min for the food spot to dry fully. If a fume hood is not available the plates will dry on the bench but will require more time to do so.
3. Carefully transfer (*see* **Note 9**) five young adult hermaphrodites to each assay plate to begin the egg laying process (*see* **Note 10**). It is essential that only minimal live bacteria are transferred to the assay plates (*see* **Note 11**).
4. Allow the worms to lay approximately 75 eggs on each plate at room temperature. About 65–85 eggs are sufficient, although more consistent results are achieved with 75 eggs. Wild-type worms that have been correctly staged should lay sufficient eggs in about 3 h. It is useful to check the plates after the first hour and to return worms that have moved off the food back to the bacterial spot. When sufficient eggs are present on the plate, remove the adult worms and discard them. Be certain that all five adults are accounted for and removed, even those that may have begun to desiccate on the plate walls and those that may have burrowed (*see* **Note 7**).

5. Count the number of eggs on the plate and remove any eggs in excess of 85 (*see* **Note 12**); do not remove any of the heat-killed OP50 in this process.
6. Invert the plates in a loosely sealed plastic container and place them in a 25 °C incubator (*see* **Note 13**) for 72–84 h.

3.6 Scoring the Assay

The scoring of dauers can be subjective to untrained observers. Many physical hallmarks of true dauer larvae are not readily visible under a standard dissection microscope, and the use of a compound microscope is not compatible with large-scale experiments. Since true dauer larvae are resistant to 1 % sodium dodecyl sulfate (SDS) [1, 28], SDS selection can be used to identify dauers. However, it is difficult to accurately count all dauer larvae present on SDS-flooded plates, and we do not routinely perform this procedure. With some practice, the size, movement, body posture, and gut refraction phenotypes of dauer larvae [1, 10, 12, 29] can be readily used to discriminate them from non-dauer animals. In reporting the results from any assay it is important to state whether worms exhibiting pre-dauer arrest or characteristics of partial dauers are being excluded. Under the described conditions, it is atypical to observe either pre-dauer arrest or partial dauers in wild-type worms but the prevalence of such animals may be increased in some mutant backgrounds.

1. Remove the plates from the incubator and inspect them using a dissecting microscope. Provided with sufficient heat-killed food, wild-type worms will develop from embryo to late L4 stage within 72 h at the assay temperature of 25 °C. In a properly executed assay, 0–10 % of wild-type worms should arrest as dauers under negative control or "no-pheromone" conditions. When characterizing dauer formation in non-wild-type worms, it is important to allow sufficient time for their development, which might proceed at a slower rate than in wild-type worms. Therefore, the assay is routinely scored 84 h after egg laying. This additional time has no effect on the results for wild-type worms since dauers remain arrested and animals that have bypassed the dauer pathway remain easily distinguishable. If scoring at 84 h, ignore any next-generation L1 larvae that may have hatched.
2. Count the number of dauer larvae and the number of young adult worms on the plate. The counting process is simplified by drawing a grid on the base of the plate. Note that some worms may be burrowed into the agar and that dauers may be present on the sidewalls of the plate.
3. Compare the number of worms counted to the number of eggs recorded on day 5. In wild-type worms, at least 95 % of eggs should hatch and develop normally. Discard data from

plates with significant contamination and from those with too many (>85) or too few (<65) worms.

4. The number of dauers is presented as a proportion of the total number of worms (dauers and adults) on the plate, and is correlated with the log of the concentration of pheromone in the assay plate. A complete dataset should consist of replicate data from at least three independent experiments performed on different days.

4 Notes

1. Noble agar increases the rate of dauer formation by about 10 % and gives more consistent results. The concentration of agar may be increased up to 2 % to suppress burrowing behavior (*see* **Note 7**), although this may affect the baseline rate of dauer formation.
2. Culturing animals at 20 °C is optimal for growth and fecundity, especially for mutant strains with compromised health. We have not observed any quantitative differences in wild-type dauer formation rates when the parental worms are cultured at 20 °C versus 25 °C.
3. Depending on the actual dimensions of the Petri plates used, it might be necessary to increase the volume of assay agar in order to fully cover the bottom surface of the plate. To achieve consistent results the agar must evenly cover the bottom surface of the plate. However, if the agar is too thin it may dry excessively or crack over the course of the assay. Remember to adjust the pheromone volume to maintain the desired concentration.
4. It is useful to array the assay plates on a small flat tray to facilitate their movement from the bench to the fume hood when it is time to dry the food spots.
5. When dispensing the molten agar on top of the 100 μL of pheromone in the assay plate it is not necessary to stir or swirl the media. The mixing that occurs during dispensation is sufficient to distribute the pheromone throughout the medium. Swirling the cooling medium tends to leave a meniscus of dried agar around the plate, allowing worms to crawl up the sidewalls and eventually desiccate.
6. Plates should be prepared by mid-afternoon and be given 16–18 h to dry completely on the bench. There should be no condensation on the lid when the assay is started. If the assay plates are too humid the egg-laying worms tend to move off the food and crawl up the wall of the plate and/or burrow (*see* **Note 7**).

7. If an egg-laying worm burrows into the agar it does not have access to the food on the surface and will typically lay fewer eggs. Eggs laid in the agar are also difficult to see and count and will develop abnormally with respect to this assay. Extensive burrowing on a plate also typically coincides with reduced rates of dauer formation. Occasionally an egg-laying worm will burrow into the agar and must be removed prior to counting the eggs. Briefly vortexing or tapping the plate will sometimes draw the worm to the surface. The worm should only be extracted/dug from the plate as a last resort since the damage to the agar will encourage the larvae to burrow as well.
8. It is not advisable to offset the lids of the assay plates to expedite drying. This may lead to dust particles landing on the plate and increases in fungal contamination. Overdrying of the plates may lead to cracking of the agar and will cause the worms to burrow (*see* **Note 7**).
9. The shape of the tip of the platinum wire used to transfer worms is important in this assay—it must not scratch or gouge the assay agar surface during transfer and it must also carry over minimal live bacteria (*see* **Note 11**). A smoothed spatula-shaped wire tip is useful since this allows worms to be scooped from underneath without the use of bacterial "glue." This is also useful when the egg-laying worms must be removed from the assay plate without removing significant amounts of the heat-killed bacteria.
10. The use of five gravid hermaphrodites in this assay is intended to allow the correct number of embryos to be deposited on a plate in a reasonable time frame in order to have a growth-synchronized brood. Using varying number of egg-layers will affect the assay in two ways: it will affect the amount of food that is consumed prior to the hatch of the brood and it will alter the concentration of endogenous pheromone. Also note that extended egg-laying sessions may result in the earliest-laid embryos hatching before the eggs are counted.
11. Live OP50 is more attractive to egg-laying worms than the heat-killed OP50. Eggs will tend to be tightly clustered in patches of live food whereas they are normally more randomly distributed on plates featuring only heat-killed food. Live food is a potent suppressor of dauer formation in this assay and will negatively impact results.
12. Care should be taken to only count fertilized eggs when assessing the number of embryos on each assay plate. One-day-old adult wild-type worms will lay few if any unfertilized oocytes whereas some mutant strains will lay many. Since population density is an important factor in this assay as is the concentration of available food, it is important that the correct number

of L1 larvae hatch on each assay plate. Inspect the full surface of the plate, including around the edges of the agar, when counting fertilized eggs.

13. Temperature significantly modulates dauer formation and must be carefully controlled. A sensitive thermometer should be used to calibrate the 25 °C incubator and in addition, the placement of assay plates within the incubator should be consistent. Since humidity affects dauer formation to some extent, care should also be taken to house the plates in similar containers and to not overly stack the assay plates within the container. Natural convection incubators are used as opposed to forced-air incubators.

Acknowledgments

Funding for this work was provided by the National Science Foundation (IOS 0842452 and IOS 1256488 to P.S.), the Human Frontiers Science Program (RGY0042/2010 to P.S.), and the DGIST Convergence Science Center Program of the Ministry of Education, Science and Technology (11-BD-04 to K.K.). S.J.N. was supported by a postgraduate scholarship (PGS-D3) from the Natural Sciences and Engineering Research Council of Canada and by the Brandeis National Committee.

References

1. Cassada RC, Russell RL (1975) The dauer larva, a post-embryonic developmental variant of the nematode *Caenorhabditis elegans*. Dev Biol 46:326–342
2. Golden JW, Riddle DL (1982) A pheromone influences larval development in the nematode *Caenorhabditis elegans*. Science 218:578–580
3. Golden JW, Riddle DL (1984) The *Caenorhabditis elegans* dauer larva: developmental effects of pheromone, food, and temperature. Dev Biol 102:368–378
4. Golden JW, Riddle DL (1984) A *Caenorhabditis elegans* dauer-inducing pheromone and an antagonistic component of the food supply. J Chem Ecol 10:1265–1280
5. Schaedel ON, Gerisch B, Antebi A, Sternberg PW (2012) Hormonal signal amplification mediates environmental conditions during development and controls an irreversible commitment to adulthood. PLoS Biol 10(4):e1001306
6. Klass M, Hirsh D (1976) Non-ageing developmental variant of *Caenorhabditis elegans*. Nature 260:523–525
7. Riddle DL, Swanson MM, Albert PS (1981) Interacting genes in nematode dauer larva formation. Nature 290:668–671
8. Thomas JH, Birnby DA, Vowels JJ (1993) Evidence for parallel processing of sensory information controlling dauer formation in *Caenorhabditis elegans*. Genetics 134: 1105–1117
9. Vowels JJ, Thomas JH (1992) Genetic analysis of chemosensory control of dauer formation in *Caenorhabditis elegans*. Genetics 130: 105–123
10. Riddle DL, Albert PS (1997) Genetic and environmental regulation of dauer larva development. In: Riddle DS, Blumenthal T, Meyer BJ, Priess JR (eds) *C. elegans* II. Cold Spring Harbor Press, Cold Spring Harbor, Plainview, NY, pp 739–768
11. Hu PJ (2007) Dauer. WormBook:1–19. http://www.wormbook.org/chapters/www_dauer/dauer.html
12. Albert PS, Riddle DL (1988) Mutants of *Caenorhabditis elegans* that form dauer-like larvae. Dev Biol 126:270–293

13. Fielenbach N, Antebi A (2008) *C. elegans* dauer formation and the molecular basis of plasticity. Genes Dev 22:2149–2165
14. Butcher RA, Fujita M, Schroeder FC, Clardy J (2007) Small molecule signaling of dauer formation in *C. elegans*. Nat Chem Biol 3:420–422
15. Butcher RA, Ragains JR, Kim E, Clardy J (2008) A potent dauer pheromone component in *Caenorhabditis elegans* that acts synergistically with other components. Proc Natl Acad Sci U S A 105:14288–14292
16. Butcher RA, Ragains JR, Clardy J (2009) An indole-containing dauer pheromone component with unusual dauer inhibitory activity at higher concentrations. Org Lett 11:3100–3103
17. Srinivasan J, von Reuss SH, Bose N, Zaslaver A, Mahanti P, Ho MC, O'Doherty OG, Edison AS, Sternberg PW, Schroeder FC (2012) A modular library of small molecule signals regulates social behaviors in *Caenorhabditis elegans*. PLoS Biol 10:e1001237
18. Pungaliya C, Srinivasan J, Fox BW, Malik RU, Ludewig AH, Sternberg PW, Schroeder FC (2009) A shortcut to identifying small molecule signals that regulate behavior and development in *Caenorhabditis elegans*. Proc Natl Acad Sci U S A 106:7708–7713
19. Jeong PY, Jung M, Yim YH, Kim H, Park M, Hong E, Lee W, Kim YH, Kim K, Paik YK (2005) Chemical structure and biological activity of the *Caenorhabditis elegans* dauer-inducing pheromone. Nature 433:541–545
20. von Reuss SH, Bose N, Srinivasan J, Yim JJ, Judkins JC, Sternberg PW, Schroeder FC (2012) Comparative metabolomics reveals biogenesis of ascarosides, a modular library of small-molecule signals in *C. elegans*. J Am Chem Soc 134:1817–1824
21. Ailion M, Thomas JH (2000) Dauer formation induced by high temperatures in *Caenorhabditis elegans*. Genetics 156: 1047–1067
22. Lee J, Kim KY, Joo HJ, Kim H, Jeong PY, Paik YK (2011) Methods for evaluating the *Caenorhabditis elegans* dauer state: standard dauer-formation assay using synthetic daumones and proteomic analysis of O-GlcNAc modifications. Methods Cell Biol 106:445–460
23. Gu SG, Pak J, Guang S, Maniar JM, Kennedy S, Fire A (2012) Amplification of siRNA in *Caenorhabditis elegans* generates a transgenerational sequence-targeted histone H3 lysine 9 methylation footprint. Nat Genet 44:157–164
24. Rechavi O, Minevich G, Hobert O (2011) Transgenerational inheritance of an acquired small RNA-based antiviral response in *C. elegans*. Cell 147:1248–1256
25. Greer EL, Maures TJ, Ucar D, Hauswirth AG, Mancini E, Lim JP, Benayoun BA, Shi Y, Brunet A (2011) Transgenerational epigenetic inheritance of longevity in *Caenorhabditis elegans*. Nature 479:365–371
26. Katz DJ, Edwards TM, Reinke V, Kelly WG (2009) A *C. elegans* LSD1 demethylase contributes to germline immortality by reprogramming epigenetic memory. Cell 137:308–320
27. Johnson TE, Mitchell DH, Kline S, Kemal R, Foy J (1984) Arresting development arrests aging in the nematode *Caenorhabditis elegans*. Mech Ageing Dev 28:23–40
28. Swanson MM, Riddle DL (1981) Critical periods in the development of the *Caenorhabditis elegans* dauer larva. Dev Biol 84:27–40
29. Popham JD, Webster JM (1979) Aspects of the fine structure of the dauer larva of the nematode *Caenorhabditis elegans*. Can J Zool 57:794–800

Chapter 21

Acute Behavioral Responses to Pheromones in *C. elegans* (Adult Behaviors: Attraction, Repulsion)

Heeun Jang and Cornelia I. Bargmann

Abstract

The pheromone drop test is a simple and robust behavioral assay to quantify acute avoidance of pheromones in *C. elegans*, and the suppression of avoidance by attractive pheromones. In the pheromone drop test, water-soluble *C. elegans* pheromones are individually applied to animals that are freely moving on a large plate. Upon encountering a repellent, each *C. elegans* animal may or may not try to escape by making a long reversal. The fraction of animals that make a long reversal response indicates the repulsiveness of a given pheromone to a specific genotype/strain of *C. elegans*. Performing the drop test in the presence of bacterial food enhances the avoidance response to pheromones. Attraction to pheromones can be assayed by the suppression of reversals to repulsive pheromones or by the suppression of the basal reversal rate to buffer.

Key words *C. elegans*, Pheromone, Ascaroside, Long reversal, Repulsion, Suppression of reversal, Attraction

1 Introduction

The pheromone drop test is adapted from a more general drop test for acute avoidance of water-soluble repellents [1, 2]. *C. elegans* moves forward through sinusoidal locomotion and occasionally changes direction by making a transient reversal or by making a sharper acute turn. A long reversal, often followed by a sharp turn of the entire body into an omega shape, is a common escape strategy upon encountering repellents. In the pheromone drop test, a population of animals is transferred to a test plate on which they crawl freely. A volume of a few nanoliters of pheromone dissolved in buffer is delivered to the side of each animal while it moves forward. As the drop touches the animal, it surrounds the entire animal by capillary action and is sensed by the amphids, the *C. elegans* sensory organs in the anterior. If the worm senses a repulsive stimulus, it makes a long reversal. The presence or the absence of the long reversal response in each animal is recorded, and the fraction

Kazushige Touhara (ed.), *Pheromone Signaling: Methods and Protocols*, Methods in Molecular Biology, vol. 1068, DOI 10.1007/978-1-62703-619-1_21, © Springer Science+Business Media, LLC 2013

of animals with a long reversal response among the total number of animals tested is calculated.

The known *C. elegans* pheromones belong to a family of small molecules with an ascarylose sugar backbone, and are called ascarosides. Some of the identified ascarosides including ascr #3 (C9) are repulsive to wild-type hermaphrodites in the drop test [3]. For a negative control and basal level correction, buffer alone is also tested. For positive controls, a high-osmolarity glycerol solution can be used as a repellent, and isoamyl alcohol, an attractive odor, can be used as an attractant.

2 Materials

Prepare all solutions using ultrapure water that is distilled, filter purified with 0.2 μm membrane, and autoclaved.

1. M13 buffer: 30 mM Tris–HCl, pH 7.0, 100 mM NaCl, 10 mM KCl. To make a 10× M13 stock, weigh 3.634 g Tris–HCl, 5.844 g NaCl, and 745.5 mg KCl, and dissolve in water. Adjust pH to 7.0, and make up to 100 mL with water. Filter through a 0.2 μm membrane. Store at room temperature. Dilute 1:10 in water with pheromone immediately before use.
2. Pheromone ascaroside stock: *C. elegans* pheromones are not commercially available as of this writing, but methods for purification and synthesis of individual ascarosides have been developed (*see* chapter by Rebecca Butcher). Dissolve each synthesized ascaroside in ethanol to the final concentration of 10 mM. Tightly seal with Parafilm in small aliquots in glass vials and store at 4 °C.
3. Glycerol solution: Prepare 1 M glycerol. Weigh 9.209 g glycerol. Dissolve in water, and make up to 100 mL with water. Store at room temperature.
4. Nematode growth media (NGM) plates: Prepare agar, peptone, 5 mg/mL cholesterol in ethanol, 1 M $CaCl_2$, 1 M $MgSO_4$, 1 M KH_2PO_4, and 1 M K_2HPO_4. To make 5 mg/mL cholesterol, add 2.5 g of cholesterol to 500 mL of ethanol, followed by overnight incubation at 37 °C. Prepare 1 M KPO_4 buffer (pH 6.0) by mixing 13.2 mL of 1 M KH_2PO_4 with 86.8 mL of 1 M K_2HPO_4 in a separate sterile beaker, and use fresh within 1–2 days. To make 1 L of NGM, add 1 mL of 5 mg/mL cholesterol in ethanol, 3 g NaCl, 22 g agar, and 2.5 g peptone to a 2 L Erlenmeyer flask. Add 975 mL water and stir well. Cover flask with aluminum foil, and autoclave for 50 min. Cool flask to 52 °C with stirring. Add 1 mL of 1 M $CaCl_2$ and 1 mL of 1 M $MgSO_4$ and mix well. Add 25 mL of 1 M KPO_4 (pH 6.0) to the flask. Using sterile procedures and a peristaltic pump, dispense 24 mL of the NGM solution into

a 90 mm diameter Petri plate, and 10 mL into a 55 mm diameter Petri plate [4]. Make NGM plates in large quantities and store in the cold room at 4 °C.

5. OP50 culture: From a single colony, grow a culture of the *E. coli* strain OP50 in 100 mL LB media at room temperature for 3–4 days. Measure OD 600 and use at a density range of OD 600 = 0.4–0.7. Store the bottle at 4 °C for up to 2 weeks.
6. Platinum wire pick: Prepare platinum wire 0.255 mm in diameter (Tritech Research) and Borosilicate Glass Pasteur Pipette 5.75 in. in length (Fisherbrand) or equivalent. Cut 2–3 cm of platinum wire and insert one end into the narrow end of a glass Pasteur pipette, and flame the glass to melt it around the platinum wire. Flatten a few mm at the free end by pressing the wire between pliers, leaving a free flat wire tip.
7. Glass capillary: 10 μL Glass microcapillary pipettes (Kimble) or equivalent.
8. Filtered plastic tips: 1–200 μL Graduated filter tips (USA Scientific, INC) or equivalent.
9. Unfiltered plastic tips: 1–200 μL Natural beveled tips (USA Scientific, INC) or equivalent.
10. Tubing: Rubber tubing about 40–60 cm in length and 1.5 mm in inner diameter.

3 Methods

3.1 Pheromone Drop Test (Off Food)

1. Four days prior to the assay, pick three animals in the L4 larval stage onto a new NGM plate (40 mm diameter) that was previously seeded with 100–200 μL *E. coli* OP50 culture. Seed plates 1–3 days before adding worms. The progeny of these animals will be used for the drop test.
2. Grow the animals for 4 days at 20 °C (*see* **Note 1**).
3. On the day of the assay, take out unseeded NGM plates (90 mm diameter) from 4 °C. Dry at room temperature for 1–2 h with the lid open to remove excessive moisture. Assays will be performed on these plates (*see* **Note 2**).
4. Remove the animal growth plates from the 20 °C incubator. The majority of the animals should have reached the young adult stage.
5. Transfer 30 young adult animals to a new unseeded NGM plate (55 mm diameter) at room temperature with the platinum wire, transferring as little food as possible (*see* **Note 3**).
6. Let the animals crawl away from any food that was transferred with them.

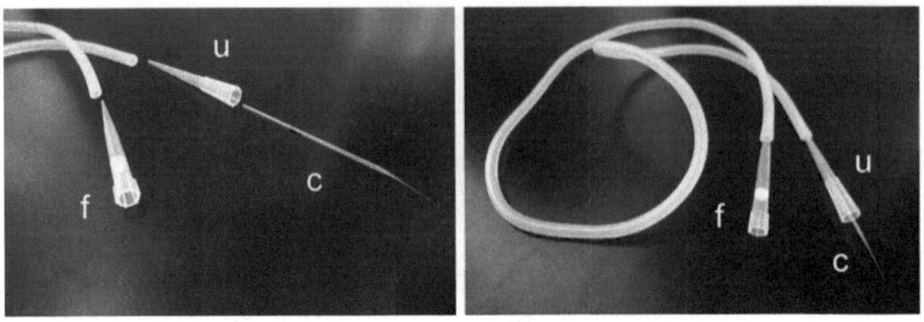

Fig. 1 The pheromone delivery system. The filtered tip (f) connects the *upper end* of the tubing to the mouth and prevents saliva from going into the tubing. The unfiltered tip (u) connects the *lower end* of the tubing to the drawn-out glass microcapillary that contains pheromones diluted in M13 buffer (c)

7. Transfer animals to a second unseeded NGM assay plate with the platinum wire, using no food. Animals should be food-free (*see* **Note 3**).
8. Wait for 30–60 min to precondition the worms.
9. During preconditioning, make a serial dilution of ascaroside in M13 buffer in Eppendorf tubes (*see* **Note 4**).
10. Pull the glass capillary by hand on a flame and break it to reduce the diameter of the tip.
11. Set up the pheromone delivery system using a pulled glass capillary, a rubber tube, an unfiltered tip, and a filtered tip (*see* Fig. 1). Push the broad end of the capillary into the unfiltered tip firmly to seal the joint.
12. Place the narrow end of the capillary into the ascaroside solution. The ascaroside solution will enter the glass capillary by capillary action.
13. Apply a droplet of ascaroside by mouth pressure through the filtered tip to a forward moving worm. Score the positive and negative responses. A long reversal is defined as a backward movement that equals or is larger than the half-length of the worm, initiated within 4 s after the stimulus touches the worm. Most long reversals begin within a second (*see* Fig. 2).
14. Repeat **step 13** for additional worms. Moving from one side of the plate to the other side and testing only 20 worms per plate help prevent testing the same animal twice.
15. The fraction reversing is (number of animals that make a long reversal)/(number of total animals tested). A fraction reversing of 1 represents complete avoidance and 0 represents no response (*see* **Note 5**).
16. The same assay is performed with a drop of M13 buffer alone on a similar number of animals. The fraction reversing is obtained as above. The fraction reversing of wild-type N2 hermaphrodite animals to buffer is typically about 0.1–0.2.

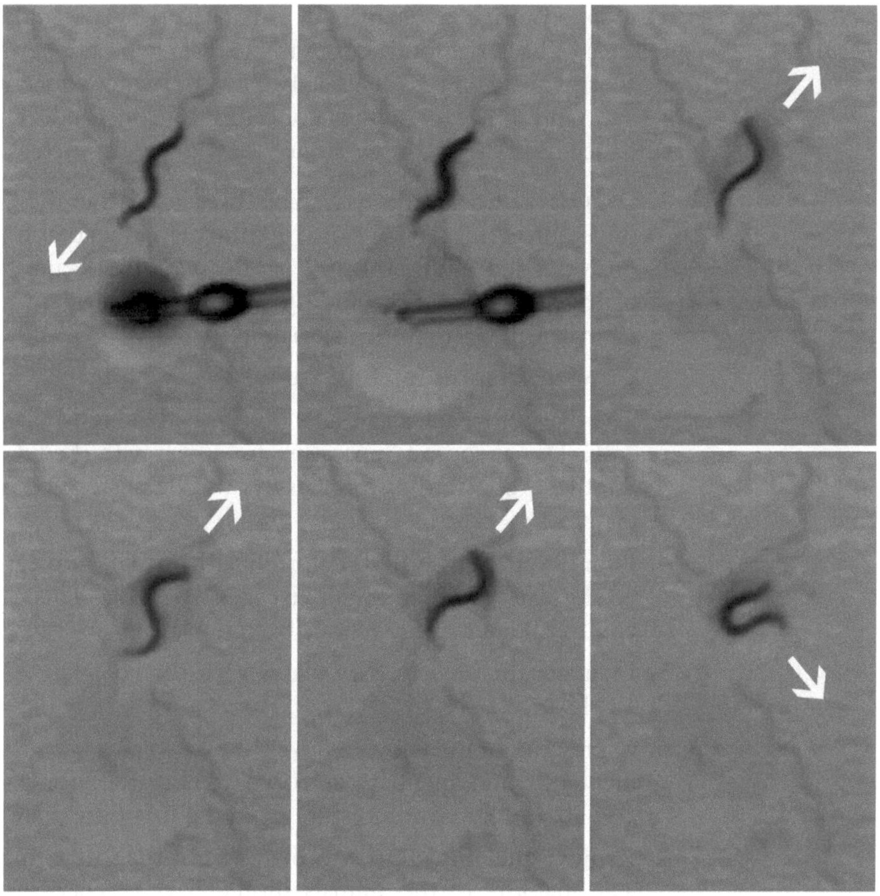

Fig. 2 The long reversal response of *C. elegans* in the pheromone drop test. Time runs from *left* to *right* in the *top row* and then the *bottom row*; total time elapsed is about 5 s. *Top left panel*: 10 nM of ascr #3 (C9) is applied to the side of a wild-type N2 hermaphrodite worm moving forward in the direction of the *arrow* (toward *lower left*). As the stimulus touches the worm (*second panel*), the animal initiates a long reversal toward the *upper right* (*arrows, third to fifth panels*). *Bottom right panel*: Animal ends the long reversal with a sharp turn, and changes direction of movement toward the *lower right* (*arrow*)

17. The avoidance index is the effect size of (fraction reversing to pheromone) − (fraction reversing to buffer alone). An avoidance index of 1 represents complete avoidance, but is rarely observed because of background response to buffer. Typical repulsion value ranges from 0.3 to 0.7. A neutral response is zero. An attraction response is represented as a negative number, but cannot easily be tested in the absence of food due to the low basal reversal rate to buffer.
18. As a positive control, you may perform the same assay with high-osmolarity glycerol solution in the capillary (1 M). The fraction reversing of wild-type N2 hermaphrodite animals is about 0.9. Therefore, the avoidance index of glycerol (with subtraction of the buffer-alone response) in wild-type N2 hermaphrodites is about 0.7–0.8.

3.2 Pheromone Drop Test (on Food)

Acute pheromone avoidance is enhanced 10–100-fold by the presence of bacterial food. For example, wild-type N2 hermaphrodite worms avoid ascr #3 (C9) at 100 nM or higher concentrations of food, and 1 nM or higher concentration on food. The basal avoidance response to buffer alone is also generally increased.

1. Four days prior to the assay, pick three animals in the L4 larval stage onto a new NGM plate (40 mm diameter) that was previously seeded with 100–200 μL *E. coli* OP50 culture. Seed plates 1–3 days before adding worms. The progeny of these animals will be used for the drop test.
2. Grow the animals for 4 days at 20 °C (*see* **Note 1**).
3. The day before the assay, take out unseeded NGM plates from 4 °C. Dry at room temperature for 1–2 h with the lid open to remove excessive moisture. Assays will be performed on these plates.
4. Pour about 6 mL OP50 culture in LB media onto each dried NGM plate, swirl, and remove excessive media. This will allow you to obtain a homogenous bacterial lawn.
5. Incubate the seeded assay plates with the lid on at 37 °C for 14–18 h.
6. On the day of the assay, remove the assay plates from the 37 °C incubator and incubate at room temperature for 1–2 h to equilibrate. Remove moisture from the lid of the plate.
7. Remove the animal growth plates from the 20 °C incubator. The majority of the animals should have reached the young adult stage.
8. Transfer 30 young adult animals to the assay plate with the platinum wire, using minimal food.
9. Wait for 30–60 min to precondition the worms.
10. During preconditioning, make a serial dilution of ascaroside in M13 buffer in Eppendorf tubes (*see* **Note 4**).
11. Pull the glass capillary by hand on a flame and break it to reduce the diameter of the tip.
12. Set up the pheromone delivery system using a pulled glass capillary, a rubber tube, an unfiltered tip, and a filtered tip (*see* Fig. 1). Push the broad end of the capillary into the unfiltered tip firmly to seal the joint.
13. Place the narrow end of the capillary into the ascaroside solution. The ascaroside solution will enter the glass capillary by capillary action.
14. Apply a droplet of ascaroside by mouth pressure through the filtered tip to a forward moving worm. Score the positive and negative responses. A long reversal is defined as a backward movement that equals or is larger than the half-length of the

worm, initiated within 4 s after the stimulus touches the worm. Most long reversals begin within a second (*see* Fig. 2).

15. Repeat **step 14** for additional worms. Moving from one side of the plate to the other side and testing only 20 worms help prevent testing the same animal twice.
16. The fraction reversing is (number of animals that make a long reversal)/(number of total animals tested). A fraction reversing of 1 represents complete avoidance and 0 represents no response (*see* **Note 5**).
17. The same assay is performed with a drop of M13 buffer alone on a similar number of animals. The fraction reversing is obtained. The fraction reversing to buffer of wild-type N2 hermaphrodite animals on food is typically about 0.3.
18. The avoidance index is the effect size of (fraction reversing to pheromone) − (fraction reversing to buffer only). An avoidance index of 1 represents complete avoidance, but is rarely observed because of background response to buffer. Typical repulsion value ranges from 0.2 to 0.6. A neutral response is zero. An attraction response is represented as a negative number (*see* Subheading 3.3).
19. As a positive control, you may perform the same assay with high-osmolarity glycerol solution (1 M). The fraction reversing of wild-type N2 hermaphrodite animals is about 0.7. Therefore, the avoidance index of glycerol (with subtraction of buffer-alone response) in wild-type N2 hermaphrodite animals is about 0.4.

3.3 Pheromone Drop Test on Food for Measuring Attraction

According to the biased random walk model for *C. elegans* chemotaxis, the suppression of reversals may indicate attraction [5, 6]. Attraction to pheromones can be measured as a reduction in spontaneous reversal responses in the drop test on food, taking advantage of the increased basal reversal rate to buffer in the presence of food. When given attractive ascaroside(s) in a drop test, the worm will make fewer reversals compared to the buffer alone. This will result in a negative value of the avoidance index. For example, for wild-type male animals in the presence of food, the fraction reversing to 100 nM ascr #3 (C9) in buffer is 0.3, and the fraction reversing to buffer alone is 0.5. The resulting avoidance index is about −0.2, indicating attraction of males to 100 nM ascr #3 (C9) [3]. The maximal attraction response is defined by the response to buffer alone or another control; typical attraction values on food range from −0.1 to −0.4.

As a positive control for reversal suppression, the attractant isoamyl alcohol can be used. A 1:10,000 dilution of isoamyl alcohol suppresses the basal reversal frequency to M13 buffer on food.

Alternative assays for pheromone attraction or repulsion have been developed based on the accumulation of animals on agar impregnated with pheromones, compared to accumulation on agar alone [7, 8].

4 Notes

1. Pheromone drop test results can be sensitive to the growth conditions and the density of the worms. When strains with smaller brood size or slower growth rate are tested along with other strains, try to achieve similar population density on the day of the assay by picking more parental animals or picking a few days earlier.
2. Dryness of the plate is best if the moving worms can leave a visible track on the agar that will disappear in few minutes.
3. Using a platinum wire pick with an extensively flattened, wide end helps to scoop up worms without food. Briefly flame-sterilize the platinum wire between animals.
4. Prepare a fresh pheromone solution each day of the assay.
5. Animals can also exhibit other behavioral responses in the drop test, including short reversals, acceleration, or dwelling (short back-and-forth movements). These responses can be scored additionally and analyzed as needed.

References

1. Hilliard MA, Bargmann CI, Bazzicalupo P (2002) *C. elegans* responds to chemical repellents by integrating sensory inputs from the head and the tail. Curr Biol 12:730–734
2. Hart AC (July 3, 2006) Behavior. In: WormBook (ed) The *C. elegans* research community, WormBook. doi: 10.1895/wormbook.1.87.1, http://www.wormbook.org
3. Jang H et al (2012) Neuromodulatory state and sex specify alternative behaviors through antagonistic synaptic pathways in *C. elegans*. Neuron 75(4):585–592
4. Stiernagle T (February 11, 2006) Maintenance of C. elegans. In: WormBook (ed) The *C. elegans* research community, WormBook. doi/10.1895/wormbook.1.101.1, http://www.wormbook.org
5. Pierce-Shimomura JT, Morse TM, Lockery SR (1999) The fundamental role of pirouettes in *Caenorhabditis elegans* chemotaxis. J Neurosci 19:9557–9569
6. Luo L et al (2008) Olfactory behavior of swimming *C. elegans* analyzed by measuring motile responses to temporal variations of odorants. J Neurophysiol 99:2617–2625
7. Srinivasan J et al (2008) A blend of small molecules regulates both mating and development in *Caenorhabditis elegans*. Nature 454: 1115–1118
8. Macosko EZ et al (2009) A hub-and-spoke circuit drives pheromone attraction and social behaviour in *C. elegans*. Nature 458: 1171–1175

Chapter 22

Behavioral Analysis of Pheromones in Fish

Peter W. Sorensen

Abstract

Pheromones are chemicals that pass between members of the same species which have inherent meaning. Because most fish pheromones are mixtures, and their actions can be complex, behavioral assays are required to identify them. This chapter describes a few strategies and two specific methods (one for measuring attraction and another for sexual arousal) that can serve this purpose in fishes that live in nonflowing water such as the carps.

Key words Fish pheromone, Bioassay, Carp, Attraction, Arousal, Priming, Releasing

1 Introduction

Pheromones can be defined as chemicals that serve as evolved cues which pass between members of the same species and elicit adaptive behavioral and/or endocrinological responses [1, 2]. Pheromones are thus defined and identified by their biological activity and bioassays are required to first prove their existence and their quantify them [3]. This is especially true for pheromones in fishes as they commonly use mixtures of relatively unspecialized bodily metabolites as pheromones, making simple biochemical approaches ineffective on their own. Pheromones have traditionally been placed into two categories based on their activity: "primers" that evoke critical developmental and/or endocrinological responses, and "releasers" that evoke dramatic and rapid and adaptive behavioral responses [1, 2]. Recent studies suggest that some pheromones may have both types of actions although generally just one of these actions may dominate [4]. Although priming pheromones which modulate growth rate, body shape, and rate of sexual maturation have been described in a wide variety of fishes [4], only one priming pheromone, 17α,20β-dihydroxy-4-pregnen-3-one (17,20βP), has been unequivocally identified [3]. Accordingly, this chapter focuses on behaviorally active releaser pheromones for which about half a dozen have been identified and which are also generally of greater interest [4].

Kazushige Touhara (ed.), *Pheromone Signaling: Methods and Protocols*, Methods in Molecular Biology, vol. 1068,
DOI 10.1007/978-1-62703-619-1_22,

Behavioral actions of fish pheromones are generally both complex as well as species and life stage specific [2, 3]. Releasing effects of pheromones in fishes can include (1) species recognition and aggregation [6]; (2) alarm [7]; (3) sexual attraction [2–4]; and (4) sexual arousal [2–4]. Depending on the species and situation, these activities may not be mutually exclusive (i.e., a sexual signal could also be species specific). Pheromones may also complement the activities of other types of waterborne chemicals that fish detect and learn to recognize. To effectively measure a pheromone, a bioassay must identify and focus on the unique (distinctive) activities it stimulates in conspecifics; assays are thus species and situation specific. This chapter briefly reviews variables associated with strategies for bioassay design and then provides details on sex pheromones (and assays) used for carps, large riverine fishes in the minnow family (*Cyprinidae*) which include the goldfish (*Carassius auratus*) and common carp (*Cyprinus carpio*). These species are useful models both because their behaviors and physiologies typify many fishes and their pheromones are well known [2–4].

1.1 Fish Behavior and Ecological Context, and Their Roles in Pheromone Function as Well as Assay Design

An assay must test relevant aspects of a species' behavior in realistic but simple (i.e., quantifiable) manners. Developing a new assay can be challenging because there are close to 30,000 species of fish, and all have their own distinctive ecologies, and physiologies. Because a good assay should only be measuring one behavioral cue (i.e., the pheromone), all other variables (e.g., ecological requirements of the test species) must be understood and controlled. Depending on the species, fish pheromone function may be influenced by several ecological factors including (1) photoperiod/season (reproduction may be highly seasonal); (2) time of day (fish often reproduce only at certain times of day); (3) light and sound levels and spectra; (4) water chemistry and salinity; (5) water temperature; (6) depth; (7) water flow; and (8) substrate (for hiding or constructing nests). Not all variables are important to all species and it can be insightful to study wild fishes where they evolved to before examining and assaying pheromones in the laboratory. Much work has been performed in the carps which need 18 °C water temperature to be reproductively active (more detail below) and still-to slowly flowing water with spawning substrate. Another example is the sea lamprey, *Petromyzon marinus*, which only responds to its migratory pheromone in natural river waters with appropriate chemistries and temperatures, and then at night when still maturing [8]. In contrast, sexually mature, sea lamprey respond only to a gill-derived sex pheromone but at any time of day [9].

Not only is it important to understand the ecological requirements for fish when testing their behavior, but it is also essential to understand physiological underpinnings of sexual drive: fish do not respond to sexual cues unless sexually receptive. In some species sexual receptivity is short-lived, and it nearly always is species and situation specific [4]. Thus, in most male fishes, sexual receptivity correlates with spermiation and surges in hormones; for example, sexual activity of male common carp is greatly enhanced by exposure to 17,20βP [4]. In most externally fertilizing female fishes including the carps, female sexual receptivity generally coincides with ovulation and elevated levels of circulating prostaglandin $F_2\alpha$ ($PGF_2\alpha$).

A third set of factors that determines both pheromonal function and assay design is the suite of social behaviors exhibited by a species. For example, whether a fish spawns in large groups, or in pairs, or builds and defends its own nests, correlates with whether its pheromone might be expected to stimulate aggregation; pheromone assays should examine relevant attributes. In the case of the carps, small groups of males aggressively compete for access to receptive females while releasing pheromones in their urine; bioassays for this species thus test carps in groups (see below). Context, or the spectrum of other sensory cues presented to animals, is also highly relevant. For example, male common carp only respond strongly to sex pheromones when presented in the context of body odor (in which it is normally found) and it must be presented in sex pheromone assay [10]. Finally, exposure regime and how pheromones are released are also very important. For example, most carp pheromones are released in urine (vs. gills or feces) whose release is pulsed and are thus encountered a discrete pulse with high-concentration gradients that contain temporal information. Bioassays for carp sex pheromones add pheromone as a concentrated stream (see below).

1.2 A Note on the Reproductive Biology and Chemical Ecology of Goldfish and Common Carp

Like many fishes, both the goldfish and common carp (herein considered "carps") are seasonal spawners which typically spawn once a year [4, 11]. Final maturation in female carps is triggered by rising water temperatures, spawning substrate (floating vegetation), and hormonally derived preovulatory priming pheromones (e.g., 17,20βP), so females tend to ovulate in a reasonably synchronized fashion at daybreak (very low light). Upon ovulation, females experience a surge in $PGF_2\alpha$ which stimulates their sexual receptivity and is then released in pulses as a urinary sex pheromone (assays for which we describe herein). Males appear to have evolved to find (be attracted to) this pheromone and then respond strongly to it when very close (in the presence of strong-concentration gradients) by exhibiting strong courtship and competitive behaviors. Interestingly, responses to this cue are also short-lived, perhaps because males that respond for an overly long length of time attract the attention of predators. The hormones and pheromones used by the carps seem to typify many fishes, including all members of the family Cyprinidiae, and their reproductive behaviors are relatively typical of this group; variants of these assays should work well for many other fishes including the zebrafish, *Danio rerio*.

1.3 Introduction to the Nature, and Measurement of Sexual Attraction

Not surprisingly, many species of fish including carps are attracted to conspecifics in particular stages of reproductive maturity, most notably ovulation. Aggregation and sex pheromones often (but not always) play a key role in attraction which is commonly measured. Attraction is a consequence of changes in movement patterns which result in a change in a fish's distribution. These changes are generally species specific and may also include increased activity, or arousal (discussed below) as well as movement direction. A key factor to consider in attraction assays is water flow; the carp have evolved in relatively still waters and evolved the ability to locate the diffuse pheromone plumes found within them. In contrast, other species have evolved in flowing waters (e.g., sea lamprey, trout) and to move upstream (i.e., exhibit positive rheotaxis) in response to the relatively well-defined filamentous pheromone plumes found within them. Herein, we describe a still-water assay.

1.4 Introduction to the Nature and Measurement of Sexual Arousal

Upon encountering pheromones released by sexually mature and active female conspecifics, many mature fish become active (aroused), and start courting females (exhibit species-specific and stereotypical sets of behaviors designed to stimulate them). Although species specific, these behaviors typically include increases in nudging (physical contact) females' vents (or sometimes gills) where urine laden pheromones are generally released, and chasing and pushing females into areas where they might spawn (for carp this is floating vegetation). Time of day, water temperature, light, and the presence of spawning substrate are important contextual cues. We have developed an assay for arousal that works well for still-water fishes that spawn in open water. It models the sudden appearance of a pheromone-laden plume in regions of the tanks with floating vegetation and is relatively facile and quick to implement Variations of this assay should be relevant for many related species that spawn and use pheromones in similar manners (see below).

2 Materials

2.1 An Assay to Test for Sexual Attraction in Carps

1. Quiet, dark area with a controlled photoperiod, dimmable overhead lights (*see* **Note 1**).
2. 1–2 m circular tank (test arena) with a gravel-covered bottom, central drain, predetermined test areas, and water inputs at opposite ends (*see* **Notes 2** and **3**).
3. Overhead infrared lights and an overhead low-light camera (*see* **Note 4**).
4. Video monitor, DVD recorder, and optional data acquisition system (*see* **Note 5**).
5. At least 40 sexually mature, receptive male fish (*see* **Note 6**).

6. Two peristaltic pumps, odorless Tygon tubing, beakers to hold odors, and airstones (*see* **Note 7**).
7. Control, pheromone, and food odors (*see* **Note 8**).
8. A supply of dechlorinated or natural well water at 18–20 °C.

2.2 An Assay to Test for Sexual Arousal in Carps and Other Still-Water Fishes

1. Quiet, dark area with a controlled photoperiod and overhead lights (*see* **Note 9**).
2. Glass aquaria (typically 70–100 L) with gravel on bottoms (*see* **Notes 10** and **11**).
3. At least 50 sexually mature, receptive male fish (*see* **Note 12**).
4. Peristaltic pumps, odorless Tygon tubing, beakers, and airstones or filters (*see* **Note 13**).
5. Control, pheromone, and food odors (*see* **Note 14**).
6. A supply of dechlorinated or well water at 18–20 °C.

3 Methods

3.1 An Assay to Test for Sexual Attraction

Carps and their relatives are extremely sensitive to low-frequency sounds and lights, so a relatively isolated area must be identified for this assay. It is desirable that behavioral scoring be conducted "blind" (i.e., by individuals unaware of treatment identity, to prevent bias). This can be accomplished by having someone other than the observer make up test odors, or by recording all trials. Four to six trials can often be run a day in the following manner (11 trials is usually a good sample size for an experiment):

1. Drain test assay arena and refill with high-quality water at the correct temperature.
2. Gently move and add test subjected to the assay tank, allow them to acclimate for a predetermined time, and turn off water (*see* **Note 15**).
3. Turn peristaltic pumps on and start observing fish distribution across test areas in the arena(s) for the "Pretest period." (*see* **Note 16**).
4. After the pretest period, determine which test area the fish had spent most of their time in and then change odor inputs so that the test odor (pheromone) is added to the area where they had spent the least amount of time. An appropriate control (or another odor for head-to-head tests to measure relative strength) should simultaneously be added to the other area (*see* **Note 17**).
5. Observe and record the distribution of all fish for the "Test Period" following the same procedures used for the pretest period.

6. Add food odor to the least used side and watch for a final positive control period (*see* **Note 18**).
7. Repeat at least ten more times using different groups of fish.
8. Place fish in a recovery tank for future experiments.
9. Analyze data (*see* **Note 19**).

3.2 An Assay to Test for Sexual Arousal

Carps and their relatives are scramble-spawners, so when they become aroused, inspect and chase each other actively. Specific behavioral attributes that you may want to score should be determined in advance by watching naturally spawning fishes. It is desirable that behavioral scoring be conducted “blind” (i.e., by individuals unaware of treatment identity, to prevent possible bias). Trials can be recorded but it is usually much easier to watch. Up to 11 trials (a convenient number which also has good statistical power) can be run a day following these protocols:

1. Drain test tanks and refill with high-quality chlorine-free water at the correct temperature; water inflow may be allowed to continue throughout the experiment.
2. Gently move and add groups of test subjects to tanks and allow them to acclimate for a predetermined time (*see* **Note 20**).
3. Briefly monitor behavioral activity of the test fish (*see* **Notes 21** and **22**).
4. If appropriate, start the “Pretest”: turn peristaltic pump on to add control blank odor and then start observing fish behavior and activity. Watch fish for a predetermined test period while noting and recording behavioral attributes of interest (*see* **Notes 23** and **24**).
5. At the conclusion of pretest period, change odor input so that pheromone is added instead of the control (*see* **Note 25**).
6. Start the “Test Period” and watch the distribution of all following the same procedures used for the pretest period.
7. Add food odor to the least used side and watch for a final period (*see* **Note 26**).
8. Repeat at least ten more times using different groups of fish.
9. Analyze data (*see* **Note 27**).

4 Notes

1. It is important to test fish in areas that are quiet and vibration free as many fishes are very sensitive to these. Light levels should be less than 1 lx and follow natural photoperiods during the spawning season (16 L:8D for carps).

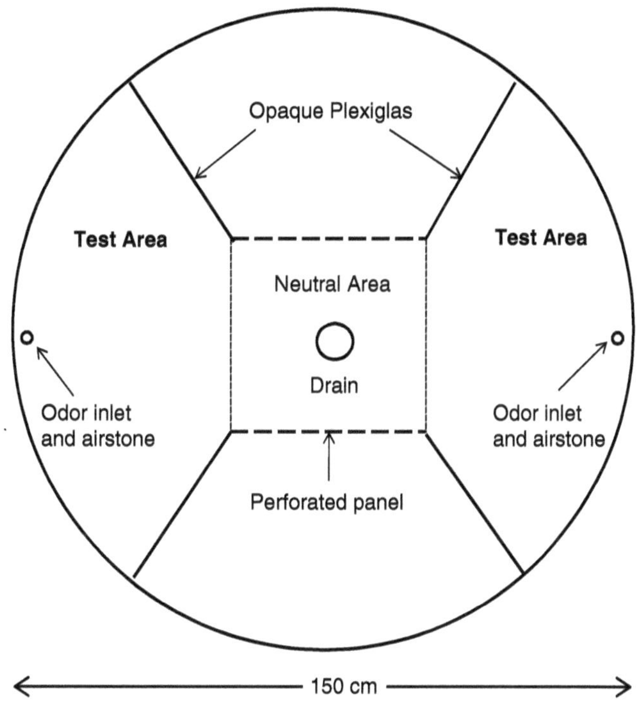

Fig. 1 Schematic diagram of the pheromone attraction assay for carp viewed from above

2. An assay tank (arena) design for common carp is shown in Fig. 1. It can be modified to simultaneously optimize both odor dispersal (this can be studied by injecting fluorescent dyes) and the swimming patterns of fish (it is desirable that test fish swim throughout the entire maze during the pretest [this might be evaluated in pilot studies, see below]). Thus, while we sometimes use partitions in test arenas to reduce odor dispersal from one side to other, for fishes that do not swim spontaneously (Asian carps [*Hypophthalmicthys* sp.]), we remove the partitions and shorten test periods. The test period also need not be longer than the period of time that fish might be expected to respond to the presence of a sex pheromone in the absence of a pheromone donor (in goldfish this is 15 min [this may be determined empirically]). Gravel can also be added to the bottom of assay tanks used for bottom-feeding fish as it makes them more comfortable and prone to explore all regions of the tank. Generally we set water depth at 10 cm because fish are often comfortable with this depth and the cameras can see into it. Dark cloth curtains are placed around arenas. Water inlets should be designed so that they can easily and quietly be turned off; otherwise supply incoming dechlorinated or tap water at rates between 0.1 and 1.0 L/min. More details of the assay tank design are found in refs. 10, 12.

3. We recommend setting up at least four mazes so that experiments can proceed quickly and multiple trials can be run in a day.
4. We use pairs of high intensity deep-infrared 12VDC LED lights (most fish see minimally into the infrared) as light sources. These are placed on an overhead frame and an oblique crossing pattern to minimize glare. Matching low-light security cameras with good-quality polarized optics with high resolution are used to watch fish. A fish eye lens may be useful. If you are interested in fish that do not require dim lighting, full-spectrum and brighter lights might also allow you to track individual fishes using their pigmentation or even colored tags become possible.
5. A simple black-and-white monitor is generally adequate for low-light infrared observation. It should be placed at least a meter away from the assay tanks. We often mark regions of tank on the screen so that it can be scored live but always record to a DVD to allow for reanalysis. Automated data acquisition systems may be helpful, especially if multiple arenas and color are used because these systems can track individuals. However, we have had some difficulties with these (they may have trouble tracking individual fish under dim light when they cross), so it is good to test such systems.
6. It is essential to use healthy, sexually mature, and receptive fish as subjects. Ideally fish will be obtained directly from the wild or suppliers as pet store fish are often treated with antifungal products that destroy their olfactory systems (electro-olfactogram [EOG] recording can confirm this [13]). We use fully spermiated males that we know will spawn with females if offered the opportunity. All-male groups are used as it reduces variance (sometimes a single female is distracting). In some cases, we "prime" males by pre-exposing the priming pheromones (0.1 nM 17,20βP) the night before because this greatly enhances their tendency to respond to sex pheromones and females [10]. If necessary, fish may be reused but we wait at least 3 weeks before doing so [10] and we do while tracking them and alternating treatments, so experience is unlikely to be a factor. Typically subjects are kept in conditions identical to those of the assay arena. All experiments are run at standard and relevant times of day and spawning temperatures (e.g., 18–20 °C for goldfish and common carp). We starve fish for 24 h to assure a constant level of motivation.
7. Odors are pumped into each end of the assay tanks using high-quality peristaltic pumps (typically 10 ml/min). Test and control odors are typically drawn from pre-cleaned (with clean tank water) 4-L beakers via plastic tubing and then injected into the tanks via PE tubing associated with airstones. Airflow through the latter is adjusted to provide a minimal but constant level of controlled mixing. Generally pumps run throughout the entire

experiment; we simply alternate between blanks, controls, and test odor inputs.

8. Food odor (paper-filtered odor of fish food) is often added at the experiment's conclusion to confirm the health and responsiveness of the fish [10, 12]. Odors are always freshly made up (that day) in the same water that is used to supply the test assays. If solvent carriers (e.g., ethanol) are needed to dissolve pheromones, the carrier is kept to a concentration of less than 0.1 ml/L to reduce the chance of it being detected and then it is as a control treatment. Beakers that house the odors must be cleaned and then rinsed extensively with water use with the assay to remove residual soap and amino acids (EOG recording can confirm cleanliness).
9. Conditions are similar to those required for attraction (*see* **Note 1**) except that lighting levels are usually greater to allow fish to be observed from the side and the holding facility may be more modest.
10. Standard 20 gal (80 L) glass aquaria work well for fishes less than 10 cm in length. We cover all sides except the front which we mark with lines using electric tape (to note movement). A drain is often drilled in the top edge of one side and water flow comes through the top (which may be screened; egg crate from lighting fixtures works well) (Fig. 2). We typically set tanks at a height of 1.5 m so that they can be easily viewed by observers on stools and install black fabric viewing curtains in front of them with holes for viewing. More detail is found in refs. 5, 13.
11. We recommend setting up 6–10 aquaria as testing is relatively rapid and multiple tests can be performed in a day.
12. We preselect males for these relatively simple experiments by determining that they will spawn with $PGF_2\alpha$-injected female first [13]. They could also be pretreated with priming pheromone (*see* **Note 6**). Generally we starve fish for 24 h before testing to ensure constant levels of motivation. By restricting tests to male-only groups, variance in responsiveness is reduced.
13. Odor addition is similar to the attraction assay (*see* **Note 7**) except that there is only one odor injection location per tank and we generally place it in floating artificial vegetation (green yarn tied to floats for carps).
14. Odors are made up exactly as for the attraction assay (*see* **Note 8**).
15. Pilot tests that do not add any odors should be conducted to determine acclimation time and group size: the objective is to find a combination of both so that all fish are exploring all areas of the test arena at least once every 5 min before trials begin. Usually movement patterns follow a bell curve with acclimation

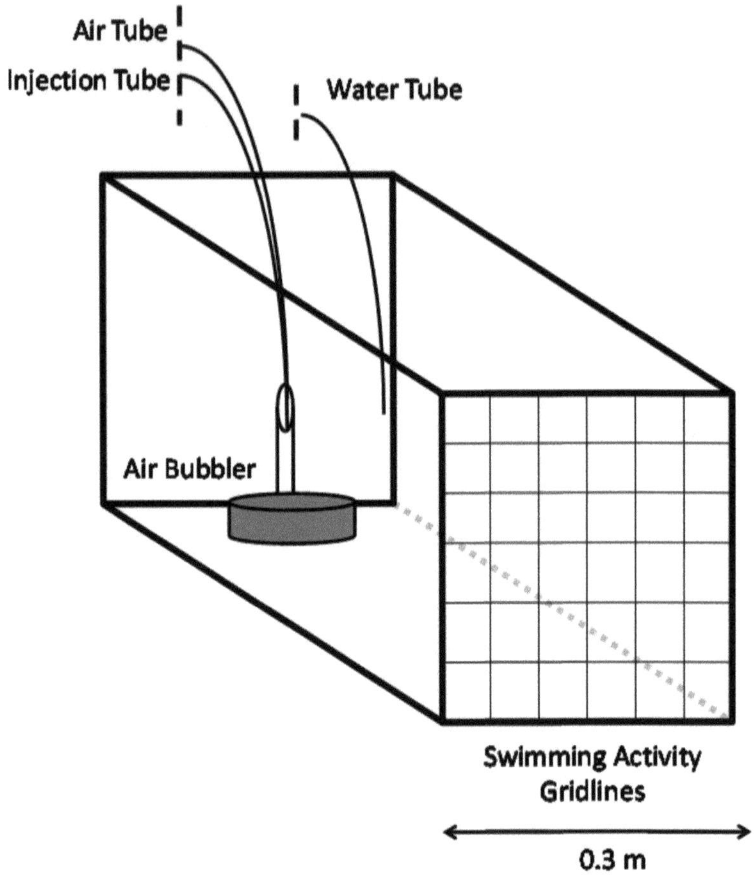

Fig. 2 Schematic diagram of tank assay to measure sexual arousal. It is viewed from the side. For fish that routinely move the length of the aquarium (carp), we draw lines on the long side and watch from that side

time—we find that 1 h works well for groups of five common carp and 20 min for groups of three silver carp. Pilot tests should also be conducted with dyes to confirm that odors stay in one-half of the arena(s) for at least 10 min (pump rates and aeration can be adjusted) and that turning peristaltic pumps on/off does not startle fish.

16. Pretest and test period duration (which should be the same) and group size are determined by pilot tests as the period of time needed to ensure that most fish in most trails actively visit all areas of the arena and spend at least a third of their time in a region where the odor could be added (we typically use five fish and 10–15 min for carp). During the pretest the location of all fish throughout the test arena is noted at preset intervals that allow one to gauge overall distribution (we use 15-s intervals). If desired, information on swimming activity (and perhaps some readily discernible behavioral attributes) can also be garnered

by replaying DVDs and noting fish crossing into specific regions as designated by lines drawn in the assay arena or on monitor [12]. It is desirable to record data in short intervals (5 min) so that data can be reanalyzed in different ways if responses prove to be short-lived.

17. Odor is added to the least used side to create a consistent conservative bias. Odors are changed by changing inlet tubes in beakers used to feed experiment. A control is always tested as a treatment group. If fish show a strong bias in the maze (75 % time during control) or do not move, the trial is terminate and not scored.
18. This step is an optional positive control to confirm fish responsiveness (they should always feed).
19. Each trial is independently analyzed live or ideally by reviewing DVDs [6, 10]. To accomplish this, the number of carp observed in each test area of each maze at predetermined intervals during both the pretest period and then the test period is noted. Fish position is based on the location of its eye. The number of fish is then summed for each period, and divided by the total possible score to yield percent time that all fish spent in the experimental area (the area or the side to which the test odor was added) during each period. Pretest values are then subtracted from test values for each trial to yield "relative attraction," a measure of change in time spent in the test area during odor addition. The use of relative attraction normalizes the data and removes any possible side bias. (Note: If the assay design includes a neural area, then expected time in each area is not 50 %.) An average value for relative attraction is next calculated for all trials of each experiment and differences between treatments compared using one-way ANOVAs. If significance is indicated ($P < 0.05$), individual values are compared to each other and the blank control using a Holm–Sidak adjustment. In addition, values for blank control are also tested against zero using t-tests to confirm a lack of maze bias. Head-to-head experiments (i.e., one odor versus another) are evaluated by comparing relative attraction values with t-tests after confirming normality with a Kolmogorov–Smirnov test.
20. Pilot tests that do not add any odors should be conducted to determine acclimation time and group size: the objective is to find a combination of both so that all fish are exploring all areas of the tanks and only interacting weekly when trials begin. Usually movement patterns follow a bell curve with acclimation time—we find that adding groups of five male fish the night before experiments works well. Pilot tests should also be conducted with dyes to confirm that odors stay in a well-defined area of the tank. Modifying aeration rate and method can achieve this purpose. Adding artificial vegetation can also

help as does adding more fish (fish seem to learn individual identities and their reproductive condition if small groups are used, and then may cease responding to new cues).

21. Experiments should use a set temperature and be conducted at a set time of day to mimic natural spawning conditions (an hour after lights on for carps and 18–20 °C).
22. If test fish are extremely active at the time the experiment is due to start (i.e., fish are chasing each other more than 25 % of the time), the trial is aborted. In very mature goldfish this is relatively common and longer acclimation times may reduce the frequency with which it occurs.
23. Test period duration is determined by pilot tests as the period of time needed to ensure that most fish in most trails actively visit all areas of the tank and inspect each other a few times (we typically use 10 min for carp). The duration of the test period should match that of the pretest but might be the duration of the response they exhibit to pheromones in the absence of a female in pilot experiments.
24. Pilot tests should determine behavioral attributes of interest. These ideally will be determined by watching naturally ovulated and spermiated fishes spawn in advance, ideally in the wild. For carps, it is relevant to record "locomotor activity" (the number of times fish cross lines drawn across face of test aquaria using their pupils as reference point), "nudges" (instances when fish come into physical contact with each other [inspecting their vents or gills]), "pushes" (when fish displace each other), "chasing" (when fish swim quickly after another [either chasing incidents or time spent chasing can be monitored]), and "feeding activity" (when carp pick up stones and other bottom; a positive control as aroused fish do not feed as much). More detail on these attributes is found in refs. 12–14. Activity may be recorded using a manual recorder or a computer data recorder. It is desirable to record in short segments (5 min) so that data can be reanalyzed in different ways if responses are short-lived.
25. Odors are changed by changing inlet tubes in the supply beakers.
26. This step is an optional positive control which is not analyzed and may be omitted if one is sure that the fish are healthy and responsive.
27. Each trial and set of behaviors is independently scored and then analyzed. One way is to use nonparametric procedures [13] if experimental variance is very high. Another way is to normalize the data and examine changes in behaviors. Here, one measures the number of times each behavioral action is performed by all fish in each group (i.e., fish identity is not noted) for each trail,

first during the pretest period and then the test period. Percent change is then calculated for each behavior and treatment, and then comparing all of these values (including a blank control) by one-way ANOVA if data are first found to be normally distributed and to meet assumptions of homogeneity of variance. Blank controls must be included as a treatment group and post hoc follow-up tests performed as for the attraction tests (*see* **Note 19**). If desired, possible pump effects and maze bias effects can be evaluated using raw data in Wilcoxon rank sum tests.

References

1. Wyatt T (2003) Pheromones and animal behaviour. Cambridge University Press, Cambridge
2. Sorensen PW (in press) Introduction to pheromones and related cues in fish In: Sorensen PW, Wisenden, BW (eds) Fish pheromones and related cues. Wiley Blackwell, Iowa
3. Sorensen PW, Hoye TR (2010) Pheromones in vertebrates. In: Lew M, Hung-Wen L (eds) Comprehensive natural products II. Elsevier, Oxford, pp 225–262
4. Stacey NE, Sorensen PW (2011) Hormonal pheromones. In: Encyclopedia of fish physiology: from genome to environment, vol 2. Academic, San Diego, pp 1513–1562
5. Sorensen PW, Stacey NE, Chamberlain KJ (1989) Differing behavioral and endocrinological effects of two female sex pheromones on male goldfish. Horm Behav 23:317–332
6. Levesque H, Scaffidi D, Polkinghorne CA, Sorensen PW (2011) A multi-component species identifying pheromone in the goldfish. J Chem Ecol 37:219–227
7. Mathuru AS, Kibat C, Cheong WF, Shui S, Wenk MR, Friedrich RW, Jusuthasan S (2012) Chondroitin fragments are odorants that trigger fear behavior in fish. Curr Biol 22:1–7
8. Sorensen PW, Fine JM, Dvornikovs V, Jeffrey CS, Shao F, Wang JZ, Vrieze LA, Anderson KR, Hoye TR (2005) Mixture of new sulfated steroids functions as a migratory pheromone in the sea lamprey. Nat Chem Biol 1:324–328
9. Li W, Scott AP, Siefkas MJ, Yan H, Liu Q, Yun S-S, Gage DA (2002) Bile acid secreted by male sea lamprey that acts as sex pheromone. Science 296:138–141
10. Lim H, Sorensen P (2011) Polar metabolites synergize the activity of prostaglandin F2α in a species-specific hormonal sex pheromone released by ovulated common carp. J Chem Ecol 37:695–704
11. Kobayashi M, Sorensen PW, Stacey NE (2002) Hormonal and pheromonal control of spawning in goldfish. Fish Physiol Biochem 26:71–84, (22):1–7
12. Maniak PJ, Lossing R, Sorensen PW (2000) Injured Eurasian ruffe, *Gymnocephalus cernuss*, release an alarm pheromone which may prove useful in their control. J Great Lakes Res 26: 183–195
13. Sorensen PW, Hara TJ, Stacey NE, Goetz FW (1988) F prostaglandins function as potent olfactory stimulants that comprise the postovulatory female sex pheromone in goldfish. Biol Reprod 39.1039–1050
14. DeFraipont M, Sorensen PW (1993) Exposure to the pheromone 17α,20β-dihydroxy-4-pregnen-3-one enhances the behavioural spawning success, sperm production and sperm motility of male goldfish. Anim Behav 46:245–256

Chapter 23

Analysis of Male Aggressive and Sexual Behavior in Mice

Takefumi Kikusui

Abstract

Pheromone and odor signals play a pivotal role in male mouse reproductive behaviors, such as sexual and aggressive behavior. There are several methods used to assess male behaviors, each of which examines a unique aspect of the biological function of mice. There are two major ways of assessing male aggressive behavior in mice, one is using isolation-induced aggression, and the other is territorial aggression in pair-housed males. To analyze male sexual behavior, a female mouse that is hormone-primed with estradiol and progesterone is usually introduced into a male home range, and mounting, intromission, and ejaculation behaviors are observed for 1 h. Here, we summarize the detailed protocols for assessing male behaviors.

Key words Male aggression, Sexual behavior, Mounting, Attacking, Ejaculation, Territorial behavior

1 Introduction

Male and female reproductive behavior is mainly regulated by olfactory information. Chemical signals, including pheromones, are the most important factors that stimulate aggression and sexual behavior in male mice [1]. Aggressive behavior in mice is defined as physical attacking, and including the potential to bite other animals. Predation, infanticide, defense, and offense are types of aggressive behaviors in mice [2]. However, this section is concerned with offensive attacking and territorial marking as forms of territorial defense.

Whereas male chemosignals induce aggressive behavior in male mice, female chemosignals enhance male sexual behavior. Male sexual behavior includes precopulatory behavior (sniffing the body and anogenital area and ultrasonic vocalization) and copulatory behavior (mounting, intromission, and ejaculation), both of which are highly dependent on female chemosignals.

Male mice emit ultrasonic vocalizations in the presence of females or when they are stimulated by a female's urinary pheromones [3]. Recent studies have demonstrated that ultrasonic song vocalizations of male mice have behavioral features similar to the songs of songbird, including discrete syllables with temporal

Kazushige Touhara (ed.), *Pheromone Signaling: Methods and Protocols*, Methods in Molecular Biology, vol. 1068,
DOI 10.1007/978-1-62703-619-1_23, © Springer Science+Business Media, LLC 2013

sequencing, repeated phrases, and variability among individuals [4]. Sexual activity and motivation are related to ultrasonic vocalization calls, and these calls can stimulate female sexual behavior. Therefore, we also describe methods of analysis of ultrasonic song vocalizations.

2 Materials

2.1 Male Aggressive Behavior

2.1.1 Subject Males in Aggression Analysis

Strain differences that affect the level of offensive attack in mice were initially reported 50 years ago [5]. Extensive information on strain distributions and male offense are available [6–8]. For example, it has been reported that DBA/1 and DBA/2 males are more aggressive than C57BL/6 or C57BL/10 males [9]. It is notable that these strain differences in aggression depend on life history, test situation, and opponent type [10] (*see* **Note 1**). Molecular biologists and behavioral neuroscientists are encouraged to consult the referenced literature, at the beginning stages of their collaboration, to choose the optimal strain, with background genes that are most appropriate for their specific research goals. This is particularly important in genetically engineered mouse studies, for which there is no ideal strain recommended across all behavioral paradigms for all null mutations. Normally, C57BL/6 or 129SV strains are used to generate transgenic lines, but 129SV males show higher aggression than C57BL/6 males [11]; additionally, there have been reports indicating that gene(s) on the Y-chromosome can contribute to strain differences in inter-male aggression [12]. Thus, it is important to consider purifying the genetic background of transgenic mice in order to assess the aggression levels in transgenic mice (*see* **Note 2**).

2.1.2 Opponent Mice in Aggression Analysis

One of the most important issues in aggression analysis is selecting the appropriate opponent mouse. The size, age, strain, and behavior of an opponent mouse can lead to a completely different type of aggression behavior in the subject mouse [13]. If the opponent mouse attacks back, the subject resident mouse thereafter tends to show reduced aggression. The opponent should have the following characteristics: (1) very low aggressiveness, (2) younger or lighter than the subject mouse, (3) sexually mature intact male (older than 6–7 weeks), (4) group housed with male siblings, and (5) not sexually and aggressively experienced.

Male chemosignals are the most important factors for inducing aggression, and these chemicals are dependent on gonadal testosterone secretion [14]. Therefore, intact male mice are preferable as the opponents. However, testosterone can enhance the opponent's aggressiveness. To solve this issue, the most commonly used opponents are anosmic mice, juvenile and subordinate mice, and A/J strain males (*see* **Note 3**). Anosmic mice have undergone surgical removal of their olfactory bulbs, whereas juvenile and submissive intruders are those that have experienced being bitten; neither of these groups attack the residents and are used as opponents. Olfactory information is necessary for the induction of aggression in mice, as supported by the fact that olfactory balbectomy completely abolishes aggression [15, 16]. Therefore, anosmic mice are not aggressive, even though they are gonadally intact and can emit male chemosignals. A/J mice are extremely unaggressive [17]; therefore, using an A/J mouse as the opponent is also preferable. However, even though they are not aggressive, each type of opponent shows different attack responses and bite avoidance behavior. These differences in opponent behavior lead to differences in the aggression levels of the resident mice [13].

The housing conditions of an opponent also influence its aggression. Isolation for more than 2–4 weeks causes an increase in male aggressiveness; therefore, the opponent should be housed with siblings.

2.1.3 Housing and Hormonal Conditions in Aggression Analysis

There are many factors that influence male aggression, including hormone levels, prior experience of aggression, isolation and grouping, dominance and subordination, a high or a low incidence of the stress experience, and food deprivation. There are two major sets of conditions for housing the subject mice used for aggression analysis, namely, social isolation-induced aggression and pair-housed territorial aggression. It has been found that fighting and threatening behavior increases with progressive isolation, up to an asymptote at 56–58 days. In pair-housed territorial aggression, the residents usually show stable aggression 4–8 weeks after pairing [18]. The earlier the age at which differential housing is imposed, the greater are the behavioral differences between animals under each housing condition [19].

Isolation confers certain behavioral characteristics, similar to those of pair-housed resident mice, such as aggressiveness and the tendency to attack other males. However, isolated mice are relatively less aggressive than pair-housed resident mice. There is a hypothesis that aggressive behavior induced in isolation is related to pathological changes [20]. Therefore, one should be cautious in using mice with isolation-induced aggression because this type of aggression may be dependent on isolation-related stress responses, as well as on chemosignal information (*see* **Note 4**).

Either isolation-induced aggression or pair-housed territorial aggression in laboratory mice can be dramatically reduced by castration. Decline by castration may, however, be restored or even accentuated by androgen replacement [9]. A recent transgenic mouse study demonstrated that both estrogen receptors and androgen receptors are involved in the maintenance of aggression [21]. Both adult androgen levels and neonatal testosterone levels influence aggression, and are referred to as organizational effects. Other hormones, such as glucocorticoid, can also modulate the aggression levels. Mice that have been isolated for longer periods, and have fought in aggression tests, have higher corticosterone titers than comparable group-housed animals. Ideally, these hormone levels should be equal in each mouse to assess the effects of chemosignals on aggression. At the very least, researchers need to consider the individual differences in aggressive behavior in their experiments.

2.2 Male Sexual Behavior

2.2.1 The Subject Male in Sexual Behavior Analysis

When assessing male sexual behavior, sexually naïve males are normally used as the subject because the experience of mating changes male sexual motivation and chemosensory sensitivity. For example, if almond odor is paired with the reward state induced by ejaculation in males, they will ejaculate more frequently with almond-scented females when given the choice of copulating with scented and unscented females [22].

There are large strain differences in sexual behavior in mice [23]. These differences in inbred mouse strains represent a wonderful opportunity for genetic analyses. However, not all strain differences are solely genetic in origin, and the early developmental environment can shape male sexual behavior. Thomas E. McGill initially compared 16 characteristics of male sexual behaviors among 3 inbred strains (C57BL/6, BALB/c, and DBA/2J). C57BL/6J males had shorter latencies between mounting and mounting with thrusting and shorter intervals between intromissions. Furthermore, these males displayed more intromissions and thrusts than DBA/2J males. Interestingly, one existing hypothesis suggests that genes on the Y chromosome regulate sexual behavior independently of gonadal hormones [24]. One particular F_1 hybrid, C57BL/6J dams by DBA/2J sires, resulted in a male hybrid (B6D2F1) capable of sustaining copulatory behavior for up to a year after castration. Males produced by the reciprocal cross (DBA/2J dams and C57BL/6J sires) did not have this characteristic, nor did F_1 males from BALB/c by DBA/2J matings [25], suggesting that the interaction of genes on the Y chromosome and other genes on autosomes modulate male sexual behavior.

As described above in the aggression section, C57BL/6 or 129SV strains are usually used to generate transgenic lines, and thus it is appropriate to consider that the degree of genetic background of the transgenic mice is critically related to male sexual behavior. For example, estrogen receptor-alpha knockout mice, in a mixed background of 129SV and C57BL/6, showed copulation with females and deposited sperm plugs; however, backcrossing these to a C57BL6 line diminished this behavior [23]. Thus, it is important to consider purifying the genetic background of transgenic mice when assessing aggression levels in mice (*see* **Note 2**).

2.2.2 The Opponent Female in Sexual Behavior Analysis

The most common method of sexual behavior analysis is to use a hormone-primed female as an opponent. Subject males are initially tested with ovariectomized females, which are treated with estradiol and progesterone beforehand, of the same strain, or with BALB/c female mice, which tend to show greater sexual receptivity. The dosage and timing of steroid treatment is as follows: estradiol benzoate (EB) dissolved in corn oil, ranging from 0.5 to 10 μg, injected twice at 48 and 24 h before testing, followed by progesterone 200–500 μg dissolved in corn oil, injected once between 12 and 4 h before testing. These treatments fully recover female receptivity for male copulatory behavior. Using sexually experienced females is preferable, because sexually naïve females show relatively lower receptivity and higher avoidance behavior with the male mice (*see* **Note 5**). Estrogen can be administered to ovariectomized females by subcutaneous implantation of an estrogen-containing tube, which is usually a silicon tube, 1 cm in length and 1 mm in inner diameter, filled with 17β-estradiol of 200–500 μg/mL mixed in silicon adhesive or oil. However, progesterone should be administered by injection (*see* **Note 6**).

Alternatively, an intact female can be an opponent of male mice. In this case, researchers need to monitor vaginal smears to confirm the estrous cycle. Female mice normally tend to display signs of estrus, including mating behavior, every 4 or 5 days, and the estrous cycle has been divided into as few as four phases: diestrus, proestrus, estrus, and metestrus. Cellular characteristics of vaginal smears reflect changes in the structure of the vaginal epithelium during the cycle. In the stage of metestrus, a large number of cornified cells and leukocytes are observed. In the diestrus stage, the vaginal contents consistently lack cornified cells, whereas leukocytes are very plentiful. In proestrus, the vaginal smear is devoid of leukocytes and characterized by nucleated epithelial cells. The stage of estrus is characterized by marked cornification of the cells and the disappearance of leukocytes. At the end of estrus, the cornified layer is sloughed off, and invasion by leukocytes occurs [26]. The advantage of using an intact female is that the displayed behavior reflects a natural behavior. Conversely, there is a disadvantage that not all females show receptive behaviors towards male mice, and this probably causes variation in male sexual behavior (*see* **Note 5**).

2.2.3 Equipment for Analyzing Male Ultrasonic Song Vocalization

Ultrasound recording: All experiments are performed in a soundproof chamber under a dim red light. Ultrasonic sounds are detected using a condenser microphone, such as UltraSoundGate CM16/CMPA (Avisoft Bioacoustics, Berlin, Germany), designed for recordings between 10 and 200 kHz. The microphone was connected to an A/D converter (for example, UltraSoundGate 116; Avisoft Bioacoustics, Berlin, Germany) with a sampling rate of 300 kHz, and acoustic signals were transmitted to a sound analysis system (for example, SASLab Pro; Avisoft Bioacoustics, Berlin, Germany).

During the recording, a subject male mouse was housed individually in a testing cage for 0.5–2 h. The test cage was placed in the soundproof chamber, and a female mouse, devocalized by unilateral sectioning of the inferior laryngeal nerve [27], was introduced into the test cage. The ultrasound was recorded and the data were later analyzed using audio software (for example, SASLab Pro; Avisoft Bioacoustics, Berlin, Germany and Adobe Audition) (*see* **Note 7**).

2.2.4 Housing and Hormonal Conditions in Male Sexual Behavior Analysis

The subject mice are singly or group housed before the test. As mentioned above, a long period of isolation, such as more than 5–8 weeks, is related to pathological changes in behavior [20]. Therefore, individual housing for less than 5 weeks is preferable. Group housing of male siblings cannot result in certain problems associated with social dominance and subordination within the group; however, unfamiliar males housed in groups are not suitable for behavioral analysis because the social hierarchy changes their sexual behavior considerably, such that it is higher in dominants and lower in subordinates [28].

The degree of sexual behavior is considerably dependent on androgen levels, both in neonates and adults. Male precopulatory and copulatory behavior is critically reduced by castration, and almost fully recovered by testosterone treatment [23]. Therefore, comparison of male sexual behavior needs an equal level of circulating testosterone both in neonates and adults, particularly in transgenic mice. Ideally, these hormone levels should be equal in each mouse to assess the effects of chemosignals on sexual behavior. Researchers need to consider the individual differences in sexual behavior in their experiments.

3 Methods

3.1 Standard Procedures in Male Aggression

Generally, one resident male is assigned as a subject, and its behavior toward the opponent is analyzed. Socially grouped male aggressiveness is not so common in studies on the neurobiology of aggression, but are sometimes conducted from an ethological point of view [29]. One study found that attack by dominant male colony mice on intruders included chasing and lateral attack behaviors, whereas the corresponding intruder behaviors were flight, boxing, and checking [30]. Mice did not show a significant constraint on bites to ventral areas, and the rat defensive behavior of lying on the back was rare; the corresponding "on-top" behavior of attackers was almost absent in mice.

There are two major procedures for assessing aggression. One is evaluating the development of aggression in naïve mice, and the other is analyzing aggressive behavior in trained resident mice. In both procedures, the opponent of an intruder is not used repeatedly, and the same intruder is not used twice in the same day because of ethical issues. It is also preferable to change the intruder for every testing session in subject mice because familiar intruders are attacked less vigorously than unfamiliar mice. It seems likely that the defeated mouse becomes less potent as a stimulus eliciting attack as the resident becomes habituated to it.

3.1.1 Behavioral Parameters in Male Aggression

The behavioral inventory included agonistic and nonagonistic elements [31]: (1) attack bites and leaps, sideways movements and tail rattles, anogenital contacts, nosing, and pursuit recorded for the resident; (2) defensive upright posture, escape leaps, and supine posture recorded for the intruder; (3) locomotion, rearing, self-grooming, and mutual upright postures were recorded for both animals. Tail rattling, which is fast lateral rattling of the tail, and sideways movements, which is a lateral rotation of the body, accompanied by piloerection and short steps, are a prominent feature of aggressive encounters between mice and is thought to be a threatening behavior. Measuring locomotion, rearing and self-grooming can exclude other physiological or neurobiological factors that modulate aggression; for example, muscle relaxant effects can decrease aggressive behavior, but these effects also can be detectable in locomotor activity during the behavioral test.

3.1.2 Aggression Development

The subject male mouse is placed in his home cage because anxiety would inhibit aggression (*see* **Note 1**). An intruder mouse is introduced in the cage and the behavior of the resident is analyzed. Usually, a test session lasts 5–10 min if the resident does not show attack biting toward the intruder. Once the resident mouse shows attacking behaviors, the test is ended 3–5 min after the first attack or until 10–20 attacking bites, due to ethical issues. This procedure is repeated 2 or 3 days apart and 2–5 times. The repetition of the test session is necessary because sometimes the initial aggressive behavior would also be related to anxiety or adapting to the new environment. The comparable parameters are dependent on the procedures. If the researchers end the test sessions by the duration time, the number of attacking bites is the most important parameter, which usually increases by repetition of the session. When the bite numbers ends the test session, the time to the end of testing is the index to compare, and it decreases by the repetition of the sessions.

3.1.3 Aggression Analysis in Trained Residents

As a result of repletion of aggression experience, aggression in the resident mice becomes stable after approximately 4–5 repeated tests (*see* **Note 1**). Under conditions of pair housing and experiencing aggression, the resident male can discriminate their paired female from an unfamiliar one, and they are more sensitive to individual differences, such as showing vigorous aggression toward an unfamiliar castrated intruder [32] (*see* **Note 8**). This discriminative ability is based on the individual signature in urine, and this signature is acquired during the developmental period, although the responsible molecules have yet to be determined [33].

Using the trained resident, the durations of aggressive (sideways threats, attack bite, and tail rattles) and nonaggressive (anogenital contact, locomotion, rearing, digging, self grooming, and sniffing) behaviors are measured. To measure the resident's aggressiveness, the number of attack bites and the latency to the first attack bites are the most important parameters [33]. In this test procedure, all intruders avoided bloodshed, indicating that the intruders were not fiercely attacked.

3.2 Standard Procedures in Male Sexual Behavior

A test chamber is usually the home cage of the subject male mice and is placed on a shelf to allow for ventral viewing. Each of the males was individually housed in the test cage 1–24 h prior to the courtship testing. After 24 h, they were paired with estrogen- and progesterone-treated ovariectomized females, and their behavior was video-recorded for 1 h during the dark phase under a dim red light.

3.2.1 Behavioral Parameters in Male Sexual Behavior

The total numbers of male mounts, female lordosis behavior towards the mounts, intromission, and ejaculation behavior of males towards the females are indispensable parameters to score. Latency, frequency, duration, and other measures have been used as indices for these behaviors. Mount was defined as a male using both forepaws to climb onto a female from behind for copulation. Lordosis response, which is a great indicator of female receptivity, is defined as a female with all four paws grounded, the hind region elevated from the floor of the test chamber, no evidence of attempt to escape or exhibition of a defensive upright posture, and the back slightly arched (*see* **Note 5**). Intromission was defined as a male pelvic thrust with a stable frequency continuously for more than 6 s and demonstration of the female's anogenital area elevated over the ground when finished. The total duration of anogenital sniffing of the subject males toward females, the approach and sniffing behavior of the females toward males, the grooming behavior of the subject males, the grooming behavior of the females, and the rearing behavior of the females are also scored according to the following definitions, and these parameters reflect the ability of locomotion, exploring, recognition, and motivational state of the subject mice. Anogenital sniffing is defined as a male mouse actively approaching, touching, and sniffing the female's anogenital area. Approach and sniffing is defined as a female stretching out, approaching and sniffing a stud male's body, including the head and anogenital area. Grooming is defined as a male self-grooming its body, including the face, anogenital area, and body trunk.

3.3 Ultrasound Analysis of Male Ultrasonic Song Vocalization

The following is an example of the sound analysis methods, and the details of the settings can be modified. Spectrograms were generated with an FFT-length of 1,024 points and a time-window overlap of 75 % (100 % frame, Hamming window). The spectrogram was produced at a frequency resolution of 488 Hz and a time resolution of 1 ms. A lower cut-off frequency of 20–30 kHz was used to reduce background noise outside the relevant frequency band. Parameters analyzed for each subject included the number of syllables, duration of syllables, and qualitative and quantitative analyses of sound frequencies, measured in terms of frequency at the maximum of the spectrum (*see* **Note 9**).

Waveform patterns of calls collected from the subject mice can be analyzed in detail. Each syllable is identified as belonging to 1 of 10 distinct categories, based on internal pitch change, length, and shape. The classification of the ten categories of ultrasonic vocalization syllables is described elsewhere [34]. To date, there is no automatic categorization program available; therefore, researchers blind to the experimental treatment need to categorize the syllables using their own discretion. In order to confirm the categorization, a likelihood ratio test, which examines whether there is a systematic difference between the blinded experimenters, is performed by a generalized linear model that consists of response variables.

3.4 Methods for Chemosignal Inputs and Male Aggressive Behavior and Sexual Behavior

As described above, chemosignals of male mice play a critical role in the induction of male aggressive behavior, whereas female chemosignals induce male sexual behavior. In particular, urine contains maleness and femaleness signals and stimulates male aggression and sexual behavior, respectively (*see* **Note 10**). To assess the ability of urine to induce male behavior, the following behavioral assay is indispensable. Urine, collected by massaging the bladder region to induce urination of gently scruffed mice, is combined from multiple animals, in order to minimize the individual signatures and social ranks that are found in urine. If the researchers test female urine, they need to collect urine from more than ten females, in order to standardize the estrous cycle. Alternatively, they can monitor the estrous cycle by vaginal smears and collect the urine from a specific estrous stage of females. To swab urine or other materials onto the opponent, a cotton fluff (3×3 mm) is soaked in 60 μL of aqueous sample and applied to the back and anogenital area of the opponent before the behavioral testing. As an alternative, samples are dropped onto a cotton swab and presented to the subject mice before the behavioral testing [35].

4 Notes

1. Generally, not all male mice are aggressive, with 70–80 % of males displaying typical aggression. The reason is unknown, but because younger mice show more problems, adults of more than 10-weeks-old are preferable. Social contact experience, consisting of 5- to 10-min encounters with an unfamiliar male and female once a day for 3–5 days also provides stability to the results.
2. The optimal testing procedure of the transgenic mice study is to breed heterogeneous males and females and to assign all the

male siblings as the subjects. Therefore, the ratio of the testing mice becomes WT 1:Hetero 2:Knockout 1. This procedure can minimize so-called "litter effects" [36].

3. In the event that the opponent intruder attacks the subject resident, the data of this pair can be excluded because the aggressiveness of the subject mouse is clearly inhibited by the fight.
4. From our experience, some socially isolated males show very quick and reflex-like aggression towards the intruder. The latency of the first attack is less than a few seconds. In this case, the aggressor would not discriminate the intruder's sex and individuality but simply attack what they encounter. It is preferable to exclude these aggressors because their behavior is mostly one of being conditioned to attack the intruder and is not dependent on chemical discrimination.
5. If the female opponent shows escaping and upright posturing towards the subject male, the data of this pair can be excluded because the sexual behavior of the subject mice is clearly inhibited by rejection from the female.
6. A hormone-primed female vaginal smear is a good indicator of the success of the treatment. The smear reflects the stage of estrus, which is characterized by marked cornification of the cells and the disappearance of leukocytes.
7. Minimization of the noise level is important. The bedding should be filter paper-type, which does not make excessive noise when the mice are walking or digging. Further, filtering the sound frequency, cut-off at 20 kHz, is a useful method for increasing the sound/noise ratio.
8. The average of the number of trained aggressors is approximately 70–80 %; that is, 20–30 % of mice do not develop aggression. These mice, of course, can be excluded from the study.
9. Audition software, such as Avisoft Pro (Avisoft Vioacoustics) and Adobe Audition (Adobe), have an automatic measurement of these parameters. See the manufacturers' guidelines.
10. Fresh urine is the most effective, but urine that has been frozen at −20 °C is still bioactive. Do not store the urine at higher temperatures because volatile components will disappear.

References

1. Mugford RA, Nowell NW (1970) Pheromones and their effect on aggression in mice. Nature 524:967–968
2. Scott JP (1966) Agonistic behavior of mice and rats: a review. Am Zool 6:683–701
3. Nyby J, Wysocki CJ, Whitney G, Dizinno G (1977) Pheromonal regulation of male mouse ultrasonic courtship (*Mus musculus*). Anim Behav 2:333–341
4. Holy TE, Guo Z (2005) Ultrasonic songs of male mice. PLoS Biol 1:e386
5. Scott J (1942) Genetic differences in the social behavior of inbred strains of mice. J Hered 33:11–15

6. Crawley JN, Belknap JK, Collins A, Crabbe JC, Frankel W, Henderson N, Hitzemann RJ, Maxson SC, Miner LL, Silva AJ et al (1997) Behavioral phenotypes of inbred mouse strains: implications and recommendations for molecular studies. Psychopharmacology (Berl) 132:107–124
7. Simon NG (1979) The genetics of intermale aggressive behavior in mice: recent research and alternative strategies. Neurosci Biobehav Rev 2:97–106
8. Jones SE, Brain PF (1987) Performances of inbred and outbred laboratory mice in putative tests of aggression. Behav Genet 1:87–96
9. Selmanoff MK, Abreu E, Goldman BD, Ginsburg BE (1977) Manipulation of aggressive behavior in adult DBA/2/Bg and C57BL/10/Bg male mice implanted with testosterone in Silastic tubing. Horm Behav 3: 377–390
10. Maxson SC, Canastar A (2003) Conceptual and methodological issues in the genetics of mouse agonistic behavior. Horm Behav 3: 258–262
11. Abramov U, Puussaar T, Raud S, Kurrikoff K, Vasar E (2008) Behavioural differences between C57BL/6 and 129S6/SvEv strains are reinforced by environmental enrichment. Neurosci Lett 3:223–227
12. Shrenker P, Maxson SC (1983) The genetics of hormonal influences on male sexual behavior of mice and rats. Neurosci Biobehav Rev 3: 349–359
13. Brain PF, Benton D, Childs G, Parmigiani S (1981) The effect of the type of opponent in tests of murine aggression. Behav Processes 4:319–327
14. Novotny M, Harvey S, Jemiolo B, Alberts J (1985) Synthetic pheromones that promote inter-male aggression in mice. Proc Natl Acad Sci U S A 82:2059–2061
15. Ropartz P (1968) The relation between olfactory stimulation and aggressive behaviour in mice. Anim Behav 1:97–100
16. Rowe FA, Edwards DA (1971) Olfactory bulb removal: influences on the aggressive behaviors of male mice. Physiol Behav 6:889–892
17. Brodkin ES, Goforth SA, Keene AH, Fossella JA, Silver LM (2002) Identification of quantitative trait Loci that affect aggressive behavior in mice. J Neurosci 3:1165–1170
18. Miczek KA, O'Donnell JM (1978) Intruder-evoked aggression in isolated and nonisolated mice: effects of psychomotor stimulants and L-dopa. Psychopharmacology (Berl) 57:47–55
19. Goldsmith JF, Brain PF, Benton D (1976) Effects of age at differential housing and the duration of individual housing/grouping on Intermale fighting behavior and adrenocortical activity in TO strain mice. Aggressive Behav 4:307–323
20. Valzelli L (1973) The "isolation syndrome" in mice. Psychopharmacologia 31:305–320
21. Sato T, Matsumoto T, Kawano H, Watanabe T, Uematsu Y, Sekine K, Fukuda T, Aihara K, Krust A, Yamada T (2004) Brain masculinization requires androgen receptor function. Proc Natl Acad Sci U S A 6:1673
22. Kippin TE, Pfaus JG (2001) The nature of the conditioned response mediating olfactory conditioned ejaculatory preference in the male rat. Behav Brain Res 1:11–24
23. Burns-Cusato M, Scordalakes EM, Rissman EF (2004) Of mice and missing data: what we know (and need to learn) about male sexual behavior. Physiol Behav 2:217–232
24. De Vries GJ, Rissman EF, Simerly RB, Yang LY, Scordalakes EM, Auger CJ, Swain A, Lovell-Badge R, Burgoyne PS, Arnold AP (2002) A model system for study of sex chromosome effects on sexually dimorphic neural and behavioral traits. J Neurosci 20:9005–9014
25. McGill TE, Manning A (1976) Genotype and retention of the ejaculatory reflex in castrated male mice. Anim Behav 3:507–518
26. Green EL (1975) Biology of the laboratory mouse. Dover Publications, New York, NY
27. Pomerantz SM, Nunez AA, Bean NJ (1983) Female behavior is affected by male ultrasonic vocalizations in house mice. Physiol Behav 1:91–96
28. Wolff RJ (1985) Mating behaviour and female choice: their relation to social structure in wild caught House mice (Mus musculus) housed in a semi-natural environment. J Zool 1:43–51
29. Ginsburg B, Allee W (1942) Some effects of conditioning on social dominance and subordination in inbred strains of mice. Physiol Zool 15:484–506
30. Brain P, McAllister KH, Walmsley S (1989) Drug effects on social behavior. Neuromethods 13:687–739
31. Kletzkin M (1969) An experimental analysis of aggressive-defensive behaviour in mice. In: Garattini S, Sigg EB (eds) Aggressive behavior. Wiley, New York, NY, pp 253–262
32. Nakamura K, Kikusui T, Takeuchi Y, Mori Y (2007) The critical role of familiar urine odor in diminishing territorial aggression toward a castrated intruder in mice. Physiol Behav 2–3: 512–517
33. Nakamura K, Kikusui T, Takeuchi Y, Mori Y (2008) Influences of pre- and postnatal early life environments on the inhibitory properties

of familiar urine odors in male mouse aggression. Chem Senses 6:541–551

34. Kikusui T, Nakanishi K, Nakagawa R, Nagasawa M, Mogi K, Okanoya K (2011) Cross fostering experiments suggest that mice songs are innate. PLoS One 3:e17721

35. Haga S, Hattori T, Sato T, Sato K, Matsuda S, Kobayakawa R, Sakano H, Yoshihara Y, Kikusui T, Touhara K (2010) The male mouse pheromone ESP1 enhances female sexual receptive behaviour through a specific vomeronasal receptor. Nature 7302:118

36. Mendl M, Paul E (1991) Litter composition affects parental care, offspring growth and the development of aggressive behaviour in wild house mice. Behaviour 116:90–108

Chapter 24

Assessment of Urinary Pheromone Discrimination, Partner Preference, and Mating Behaviors in Female Mice

Olivier Brock, Julie Bakker, and Michael J. Baum

Abstract

Behavioral testing methods are described for determining whether female mice can discriminate between volatile urinary pheromones of conspecifics of the same vs. opposite sex and/or in different endocrine conditions, for determining sexual partner preference, for quantifying receptive (lordosis) behavior, and for monitoring the expression of male-typical mounting behavior in female mice.

Key words Estradiol, Progesterone, Main olfactory system, Vomeronasal organ, Lordosis, Mounting behavior

1 Introduction

Modern studies of the roles of genes, hormones, and pheromones in both the development and the adult expression of courtship behaviors in female mice rely on behavioral testing methods that produce results which are replicable among laboratories around the world. Studies conducted in both the laboratories of co-authors, Baum and Bakker [1, 2], over the past decade have assessed the capacity of female mice to discriminate urinary odors from male vs. female (or from testes-intact males vs. castrated males) conspecifics. The results obtained have been used to assess sex differences and/or gonadal hormone effects on pheromone detection and to assess the effects of lesioning either the main olfactory epithelium and/or the vomeronasal organ on sex discrimination capacity. We have also assessed the preference of female mice to approach and investigate volatile body (or urinary) odors and/or the sound and sight of male vs. female (or testes intact vs. castrated male) [3, 4]. The results obtained have been used to assess aspects of brain sexual differentiation in female mice, including the development of circuits that underlie females' normal preference to seek out a male reproductive partner. After a female approaches a male, she must be receptive to the male's mounts and intromissions in order both to receive

Kazushige Touhara (ed.), *Pheromone Signaling: Methods and Protocols*, Methods in Molecular Biology, vol. 1068, DOI 10.1007/978-1-62703-619-1_24, © Springer Science+Business Media, LLC 2013

sperm from the male and the vaginal-cervical stimulation needed to establish pregnancy. We [4, 5], and others [6–9], have developed a sequence of tests for lordosis behavior to assess females' receptivity. Finally, a full assessment of courtship capacity in female mice includes their display of male-like mounting and pelvic thrusting behaviors directed towards female conspecifics. Proestrous female mice show considerable mounting behavior directed towards other females; the functional significance of this mounting behavior is not known. We have studied female mounting behavior in studies designed to assess the role of perinatal estradiol signaling in male-typical sexual differentiation [4] and the possible role of VNO/pheromone vs. sex hormone signaling in the expression of this male-typical behavior in females [10].

2 Materials

2.1 Urinary Odor Discrimination Tests

1. Silastic capsules for chronic administration of estradiol to ovariectomized females are prepared using SILASTIC tubing (Silclear™ Tubing, Medical Grade Silicone Tubing, length 15 m, Ref 2110150949, Degania Silicone, Israel) by cutting 1-cm-long capsules (inner diameter, 1.57 mm; outer diameter, 2.41 mm). Close one end of the capsule with SILASTIC medical adhesive (Type A, Dow Corning Corporation, USA) for no more than 2 mm. Let the adhesive dry for 24 h. Fill the capsule with a mixture (1:1) of crystalline 17β-estradiol (cat. 101656, MP Biomedicals) and cholesterol (C8667, Sigma) for 5 mm. Close the remaining open end of the capsule with SILASTIC medical adhesive. Let the capsule dry for 24 h, and pre-incubate the capsule in saline at 37 °C for 24 h before s.c. implantation into subjects. The amount of steroid released from a SILASTIC implant is directly proportional to the surface of the capsule and inversely related to its thickness [11]. Our capsules produce circulating levels of estradiol around 150 pg/ml in ovariectomized female mice. This concentration is within the range of values observed during proestrus in ovary-intact, cycling female mice [12]. The capsules can be stored for several months in the dark. They have to be protected from the light to prevent degradation. Once implanted, these Silastic implants release a constant amount of estradiol for up to 2 months. It is preferable to remove the Silastic implant after this period to avoid side effects like vaginal inflammation.
2. In order to induce full behavioral estrus, progesterone must be administered s.c. to ovariectomized females previously given estradiol. Prepare a solution of progesterone by mixing progesterone (P0130, Sigma) in sesame oil to get a final concentration of 500 μg/100 μl. Let the solution stir overnight. The progesterone solution can be kept several months in the dark.

3. These tests are carried out in standard clear plastic colony cages (29 × 18 × 13 cm) in which the subject has lived alone for 48 h prior to the test without a change in bedding. The food is removed from the food hopper and the water bottle is removed a few minutes prior to the behavioral test.
4. Urine is collected from testes-intact male mice of the same strain as the female subjects and from castrated males as well as cycling females when they are in vaginal proestrus or estrus. Estrous female urine can also be collected from ovariectomized females that are primed with estradiol followed by progesterone to induce estrus (see details below). Urine can be collected by holding the mice by the scruff of the neck over a funnel and by pushing gently on their belly. Alternatively, a metabolic cage can be used to collect urine. Urine from four mice of the same sex/endocrine condition is pooled and mixed by vortexing. Aliquots of the pools of urine are then frozen at −80 °C in separate vials for later use.
5. Square pieces of filter paper (5 × 5 mm) are taped to plastic weigh boats (4.5 × 4.5 cm); a sufficient number of these urinary odor presentation weigh boats are prepared in advance to accommodate the total number of presentations of water or urine in the actual tests.

2.2 Partner Preference Tests

To assess partner preference using either visual, auditory, and/or olfactory stimuli, we use a Plexiglas box (60 cm long × 30 cm high × 13 cm wide) that is divided into three compartments by placing either opaque (black) or transparent partitions and has a perforated top. Each compartment is thus 20 cm in length. The partitions contain perforated holes (0.5 cm diameter) at a height of 8 cm to facilitate the diffusion of odors from the two side compartments to the middle compartment. Backside, floor, left and right sides are made of black Plexiglas; front side and top are made of transparent Plexiglas (*see* **Note 1**).

2.3 Lordosis Tests

Lordosis tests are usually conducted in a Plexiglas aquarium (35 cm long × 25 cm high × 19 cm wide) with a perforated top and whose floor is covered with fresh sawdust. We also have used the home cage (with food and water removed) of individual stimulus males in which to assess females' lordosis capacity after they are introduced.

2.4 Female Mounting Behavior Tests

1. These tests are carried out in standard colony cages in which the subject has lived alone for 48 h prior to testing.
2. Female stimulus mice of the same strain as the female subjects are ovariectomized at least 2 weeks prior to use. These stimulus females are given a s.c. silastic implant of diluted estradiol which is left in place for the duration of the study. Stimulus females are given a s.c. injection of progesterone 3–4 h prior to the onset of behavioral testing.

3 Methods

All behavioral testing is carried out under either dim yellow lighting or under red light in a separate room from the mouse colony during the dark portion of the 12 light/12 dark day/night cycle. Mice are singly housed beginning at least 1 week prior to any of the behavioral tests. A video camera with infrared night vision detection is used to record all behavioral tests. We either score behavior "on line" as it occurs or by analyzing videos of the behavior after the fact. We use Pocket PC and Noldus Observer Software to record the frequencies and durations of the particular behaviors under study. We use Sigma Stat software to carry out statistical analyses of the results. Whenever possible, the treatment condition of different subjects is concealed from the investigator who scores the behavior under study.

3.1 Urinary Odor Discrimination Tests

1. Female subjects are typically ovariectomized several weeks prior to assessing their capacity to detect/discriminate different volatile urinary odors. Ovariectomy is performed under general anesthesia using either a mixture of ketamine (80 mg/kg per mouse) and medetomidine (Domitor, Pfizer, 1 mg/kg) or continuous inhalation of 1 % isoflurane. Place the female gently on her side and make a small incision, first in the skin, then through the muscle layer. Lift the ovary and make a small suture to clamp of the uterus from the ovary and to avoid any bleeding. Then remove the ovary and suture the muscle layer, followed by the skin. Then place the female on her other side to repeat the same procedure. If females are going to be treated with estradiol by means of a SILASTIC capsule, make a small incision in the skin right below the neck, insert the capsule s.c. and then suture the skin (*see* **Note 2**). Mice receive atipamezole (Antisedan, Pfizer, 4 mg/kg s.c.) at the end of surgery in order to antagonize medetomidine-induced effects, thereby accelerating their recovery, an analgesic (Temgesic, Schering-Plough, 0.05 mg/kg s.c.) and are placed on a heating pad until they are mobile. Females should be allowed to recover from the surgery for at least 2 weeks before starting behavioral testing.
2. Beginning 2 weeks after ovariectomy, subjects can be tested in the absence of hormone replacement if the investigator is not interested in the activational effects of sex hormones on females' urinary odor discrimination ability. Otherwise, the female subjects should be brought into behavioral estrus at the time of odor discrimination assessment. Subjects that received a s.c. Silastic implant of estradiol at the time of ovariectomy can be given a single s.c. injection of progesterone (500 μg/mouse) 3 h prior to the onset of testing to induce full behavioral estrus.

3. Aliquots of urine are thawed whereupon 10 μL of a particular kind of urine is pipetted onto the filter paper attached to a weigh boat just prior to presentation. The weigh boat is placed behind the wire mesh of the cage's food chamber so that the subject is able to approach the stimulus with its nose coming within 5 mm without direct contact. No direct contact with the urinary stimulus with either snout or paws is possible. As a result, the subject is only exposed to volatiles emitted from the respective urinary stimuli; presumably only the main olfactory epithelium (and not the VNO) is involved in the detection of these odorants.
4. Tests of urinary odor discrimination begin with three consecutive 2-min presentations (1-min intervals between each stimulus) of deionized water. Beginning 1-min after presentation of the third water stimulus the subject is presented with three consecutive 2-min presentations of testes-intact male urine (at 1-min intervals; urine #1) followed by three consecutive 2-min presentations of either castrated male urine or estrous female urine (depending on the study; urine #2). This presentation sequence of type of urine can be varied either across subjects in a particular group during one test day. Alternatively, subjects can be retested with the reverse sequence of urine types on a separate day.
5. An observer uses a computer and Noldus software to record the number of seconds that each subject spends with its nose against the wire mesh grid abutting the odor stimulus during each 2-min stimulus presentation. Nonparametric one-tailed within-groups Wilcoxon tests are used to compare the time spent investigating the third presentation of water vs. the first presentation of urine #1 and to compare time investigating the third presentation of urine #1 vs. the first presentation of urine #2. Significant differences in each of these comparisons indicate whether or not subjects could perceive the newly introduced odorant from the previous one (*see* **Note 3**).
6. Subjects' performance in habituation/discrimination tests relies on the intrinsic motivation that mice have to explore urinary odors from conspecifics. As such, observed group differences in detection performance in this type of test may reflect a complex interaction between social motivation and odor perception (*see* **Note 4** for further discussion).

3.2 Partner Preference Tests

1. Allow each subject to adapt to the three compartment box once on the day before the onset of preference testing by placing them in the middle compartment for 10 min (with no stimulus animals placed in the two side compartments).
2. On the day of testing, place stimulus animals (e.g., an estrous female vs. a testes-intact male or, alternatively, a testes-intact

male vs. a castrated male) along with their own bedding to make the stimulus as odorous as possible into the two respective side compartments of the apparatus (*see* **Note 5**).

3. Then introduce the experimental subject into the middle compartment containing no sawdust, and observe it for 9 min. Record the time the subject spends poking its nose through the holes of each side partition or actively sniffing the bottom of the partition. Each test session is divided into three 3-min intervals to determine whether investigation times decrease during the test (*see* **Note 6**).
4. Prior to testing each new subject, clean the middle compartment with Norvanol while leaving the soiled bedding in the two side compartments. Place a new experimental subject in the middle compartment, and repeat the procedure until all experimental animals are tested (*see* **Note 7**).
5. For each female subject, calculate a preference score (PS) for the male by dividing the time spent investigating the testes-intact male compartment minus the time spent investigating the compartment housing either the estrous female or the castrated male (depending on which of these alternative choices to the testes-intact male is given to the female subject) divided by the total time spent investigating both compartments.

3.3 Lordosis Tests

1. At least 1 week before testing experimental female mice, make sure that a sufficient number of high quality stimulus males are available (*see* **Note 8**). Stimulus males are given mating experience, as follows: Place a stimulus male in the Plexiglas testing aquarium for 30 min. Then introduce a stimulus female, which has been implanted with an estradiol capsule for at least 1 week and primed with progesterone (500 μg/mouse; injected i.p.) 3 h before testing, into the aquarium. In order to optimize its utility as a stimulus male, he should mount the stimulus female within 5 min and also display intromissions with the female. After 10 min, or a maximum of 10 intromissions, remove the stimulus male and place him back into his homecage (*see* **Note 9**). Repeat this procedure on 2 or 3 consecutive days; males that mate each day will later be useful for lordosis testing.
2. At the beginning of each lordosis test, place a sexually experienced male mouse alone in the test aquarium for 15 min of habituation. No habituation is needed if tests are to be conducted in the male's home cage.
3. Beginning 3 h after receiving a s.c. progesterone injection (500 μg ip), place a female subject in the test aquarium and record her lordosis responses displayed in response to the mounts (with pelvic thrusting) received from the stimulus male. The test lasts until the female receives 10 mounts or

10 min have elapsed (*see* **Note 10**) after which the subject is replaced in her homecage (*see* **Note 11**). Repeat this process, using fresh stimulus males, as needed in order to sustain the mounting attempts needed to assess females' lordosis responsiveness (*see* **Note 12**).

4. After each test, calculate a subject's lordosis quotient (LQ) by dividing the number of lordosis responses displayed by the female subject by the number of mounts received (100×). A minimum of four tests is normally required to observe significant levels of lordosis behavior and to reach plateau values (LQ = 60–70 %; [5]). Unlike rats, female mice rarely achieve LQs of 100 %. Lordosis testing should be performed at 3–4 day intervals (*see* **Note 13**).

3.4 Female Mounting Behavior Tests

1. Ovariectomized female subjects are given testosterone either via daily s.c. injections (testosterone propionate dissolved in sesame oil; TP, 3 mg/kg) or via s.c. implantation of a Silastic capsule (inner D 1.57 mm; outer D 2.41 mm) containing a 5 mm length of undiluted powdered testosterone (*see* **Note 14**) for at least 7 days prior to the onset of tests for mounting behavior.
2. An ovariectomized stimulus female that is brought into behavioral estrus via estradiol and progesterone treatment is placed into the home cage of the testosterone-primed female subject for 30 min. An observer scores the occurrence of ano-genital investigations as well as mounts accompanied by pelvic thrusting movements directed towards the receptive stimulus female (both the frequency and duration in sec. of each of these behaviors is recorded). The number of pelvic thrusting movements associated with each mount is also recorded (*see* **Note 15**).
3. The following variables are summarized and subjected to statistical analyses: the percentage of female subjects in each treatment group that show mounts directed towards a stimulus mouse during any test, the total frequency and total duration (in sec.) of mounts directed towards a stimulus mouse during a 30-min test (individual means are computed, followed by grand group means if more than one behavioral test is given), and the total number of pelvic thrusts displayed by the subject over a 30 min test.

4 Notes

1. Another device commonly used in the literature to assess odor preferences is an enclosed, Plexiglas Y-Maze [13, 14]. It consists of the stem of the Y (55 cm long) and two arms (65 cm long) that diverge at 60° angle from the stem. All parts of the

maze are 9 cm high and 9 cm wide. A removable perforated Plexiglas door at the distal end of the each arm separates the goal box from the rest of the maze. We use either an opaque (to prevent visual cues) or a clear (to allow visual cues) Plexiglas door. A perforated Plexiglas door is placed at the other side of the goal box. The start box with a removable perforated Plexiglas door is located at the base of the stem. An electric fan is placed behind the start box, from which it is separated by a wire screen. The perforated doors and the fan make it possible to pull air over odor stimuli placed in the distal goal boxes through the entire maze into the start box. We decided not to use this device anymore for one major reason: the behavior observed in the Y-maze can be confounded by differences in exploratory activity. For example, after gonadectomy, C57Bl6 male and female mice become anxious in the Y-Maze and show a strong decrease in exploratory activity (most of the time, mice even do not approach the goal boxes). By contrast, gonadectomized CD1 mice are hyperactive in the Y-Maze and show very little interest in remaining in proximity to a specific stimulus.

2. After surgery, house female mice individually for a few days in order to avoid that they will nibble each other's sutures which will increase the risk of prematurely losing their capsule.

3. Subjects' detection thresholds for different urinary odors can be established using a variation on this method. On successive days subjects' ability to detect volatile odorants emitted from progressively lower dilutions of one type of urine (e.g., testes-intact male urine) following three presentations of deionized water can be assessed. In two studies [1, 2] long-term gonadectomized female subjects (no hormone replacement) successfully detected the urinary volatiles emitted from a 1:80 dilution of male urine (but not from a 1:160 dilution of male urine) whereas long-term castrated males (no hormone replacement) successfully detected volatiles emitted from a 1:20 (but not 1:40) dilution of male urine.

4. If the investigator is concerned that the outcome of a habituation/dishabituation study reflects group differences in sexual motivation as much as differences in odor perception capacity, studies may conducted in which hunger serves as the motivation for subjects' performance in an olfactory discrimination task. In one such two-choice odor detection task [15] the ability of food-deprived mice to use decreasing concentrations of male urine as a discriminative stimulus to find a food pellet buried in sand was assessed. In another instance [16], a go/no-go task (urinary volatiles were presented consecutively using an 8-channel liquid dilution olfactometer) was used to compare the ability of mice to discriminate volatiles emitted

from the urine of testes-intact males, castrated males, and ovariectomized females given either estradiol alone or estradiol + progesterone.

5. To assess partner preference based on subjects' visual, auditory, and olfactory senses, use transparent partitions. To assess preference based on auditory and/or olfactory stimuli, use opaque partitions. In the majority of published experiments female subjects are given a choice between a testes-intact male and an estrous female. Depending of the experimental question, a choice between a testes-intact male and a castrated male or an estrous female and an ovariectomized female can be offered. Anesthetized stimulus animals can also be used to present the subject with only olfactory cues that signal a partner preference.
6. Experimental subjects of some strains tend to investigate more during the first few minutes of the test. However, if no decrease in investigation time is observed across the entire test, the total test time spent investigating the two stimuli is calculated.
7. After being used in several test sessions, some stimulus animals will sleep in the corner of their respective side compartments. These stimuli should be replaced with new, active (awake) stimulus animals.
8. Note that not all male mice are "good breeders." Some make no effort to mount females or are very aggressive (attack) the female subjects. Thus before starting lordosis tests, make sure that you have a large batch of sexually active male mice which are not aggressive towards females.
9. Once male mice have experienced sexual behavior with a female, house them individually, otherwise they will fight to the death.
10. The receptivity of a female mouse can be scored as follows: (a) the female runs away and/or responds aggressively in response to a male's mounting attempt; (b) the female does not run away but rejects (kicks at the male) the mount; (c) the female accepts the mount but does not respond by showing lordosis (concave curvature of the back); (d) female responds by showing full lordosis without any resistance directed towards the stimulus male. In some strains (or transgenic models), female mice can require several mounting attempts in order to reach stage "d." In these particular cases, you can extend the test to 20 mounts or 15 min. The LQ should be based on the number of times the female subject displays full lordosis (d).
11. During the course of lordosis testing, gently separate the male from the female after the first intromission occurs in order to retard the occurrence of ejaculation. Once a male has ejaculated, he will not show any interest towards the subsequent

female subjects. If a male ejaculates with a female, replace him immediately with a new stimulus male in order to obtain the requisite number of mounts needed to compute the LQ for that test.

12. Do not change the bedding on the floor of the test aquarium between female subjects. Change it only before each daily test session.
13. You can easily score two females at the same time. The optimal procedure is to use three aquariums, each containing one male. Introduce one female into each of the first two aquariums. If one of these females is not mounted by the stimulus male, switch her into the third aquarium.
14. We have used both daily injections of TP [10] and a s.c. Silastic capsule [4] of testosterone to elevate circulating plasma levels of testosterone up to that characteristic of testes-intact males. Depending on whether subjects are to be tested under other endocrine conditions, investigators may choose either of these methods to administer testosterone.
15. Depending on the aim of our studies, we have administered 1–6 tests of mounting behavior in female subjects that were ovary intact or ovariectomized and treated with estradiol alone, estradiol + progesterone, or testosterone. In some instances we have also recorded mounting behavior that female subjects direct towards stimulus males that had been castrated several weeks previously. These males had urine from testes-intact males smeared on their backsides. These castrated males rarely attempted to mount our female subjects.

Acknowledgment

Preparation of this review was supported, in part, by FRSM grant 3.4571.10 to J.B.

References

1. Baum MJ, Keverne EB (2002) Sex difference in attraction thresholds for volatile odors from male and estrous female mouse urine. Horm Behav 41:213–219
2. Pierman S, Douhard Q, Balthazart J, Baum MJ, Bakker J (2006) Attraction thresholds and sex discrimination of urinary odorants in male and female aromatase knockout (ArKO) mice. Horm Behav 49:96–104
3. Brock O, Bakker J (2011) Potential contribution of prenatal estrogens to the sexual differentiation of mate preferences in mice. Horm Behav 59:83–89
4. Brock O, Baum MJ, Bakker J (2011) The development of female sexual behavior requires prepubertal estradiol. J Neurosci 31:5574–5578
5. Bakker J, Honda S, Harada N, Balthazart J (2002) The aromatase knock-out mouse provides new evidence that estradiol is required during development in the female for the expression of sociosexual behaviors in adulthood. J Neurosci 22:9104–9112
6. Mani SK, Reyna AM, Chen JZ, Mulac-Jericevic B, Conneely OM (2006) Differential response of progesterone receptor isoforms in

hormone-dependent and -independent facilitation of female sexual receptivity. Mol Endocrinol 20:1322–1332

7. Ogawa S, Chan J, Chester AE, Gustafsson JA, Korach KS, Pfaff DW (1999) Survival of reproductive behaviors in estrogen receptor beta gene- deficient (betaERKO) male and female mice. Proc Natl Acad Sci U S A 96: 12887–12892
8. Ogawa S, Eng V, Taylor J, Lubahn DB, Korach KS, Pfaff DW (1998) Roles of estrogen receptor-alpha gene expression in reproduction- related behaviors in female mice. Endocrinology 139:5070–5081
9. Rissman EF, Early AH, Taylor JA, Korach KS, Lubahn DB (1997) Estrogen receptors are essential for female sexual receptivity. Endocrinology 138:507–510
10. Martel KL, Baum MJ (2009) Adult testosterone treatment but not surgical disruption of vomeronasal function augments male-typical sexual behavior in female mice. J Neurosci 29:7658–7666
11. Smith ER, Damassa DA, Davidson JM (1977) Plasma testosterone and sexual behavior following intracerebral implantation of testosterone propionate in the castrated male rat. Horm Behav 8:77–87
12. Wersinger SR, Haisenleder DJ, Lubahn DB, Rissman EF (1999) Steroid feedback on gonadotropin release and pituitary gonadotropin subunit mRNA in mice lacking a functional estrogen receptor alpha. Endocrine 11: 137–143
13. Keller M, Douhard Q, Baum MJ, Bakker J (2006) Destruction of the main olfactory epithelium reduces female sexual behavior and olfactory investigation in female mice. Chem Senses 31:315–323
14. Pankevich DE, Cherry JA, Baum MJ (2006) Effect of vomeronasal organ removal from male mice on their preference for and neural Fos responses to female urinary odors. Behav Neurosci 120:925–936
15. Sorwell KG, Wesson DW, Baum MJ (2008) Sexually dimorphic enhancement by estradiol of male urinary odor detection thresholds in mice. Behav Neurosci 122:788–793
16. Wesson DW, Keller M, Douhard Q, Baum MJ, Bakker J (2006) Enhanced urinary odor discrimination in female aromatase knockout (ArKO) mice. Horm Behav 49:580–586

Chapter 25

Assessing Postpartum Maternal Care, Alloparental Behavior, and Infanticide in Mice: With Notes on Chemosensory Influences

Kumi O. Kuroda and Yousuke Tsuneoka

Abstract

Chemosensory signaling influences maternal care and other innate behaviors toward conspecific young animals in rodents. In this chapter, we describe basic protocols for assessment of postpartum maternal behavior and other pup-directed behaviors in laboratory mice. The specific aim of this protocol is to screen out the abnormal phenotypes in parenting of genetic mutant mice under the standard housing condition. The possible underlying mechanisms for a given abnormality in the mother-young interaction are briefly suggested as well.

Key words Parental care, Paternal behavior, Infanticide, *Mus musculus*, Olfaction, Pup retrieval assay, Nest building

1 Introduction

1.1 Parental Behavior and Related Pup-Directed Behaviors in Mice

Maternal behavior is defined as the collection of behaviors by the mother that can increase offspring survival [1, 2]. Similar nurturing behaviors as maternal behaviors, called as *paternal behavior* by fathers and *alloparental behavior* by older conspecifics, are widely seen in mammals. In this review, we will collectively refer to these maternal, paternal and alloparental behaviors as "parental behavior." In addition to these nurturing behaviors, mice also perform negative pup-directed behaviors, such as *nonresponding* (the subject mouse sniffs the pup at first, but stay away for most of the time), *avoidance* (moving away from the pup or burying the pup in the bedding) and *infanticide* (pup biting/killing, often but not necessarily combined with cannibalism). Infanticide is most frequent in virgin (sexually naïve) male mice. In C57BL/6J, a standard inbred strain with high sociability, the 70–80 % of virgin males commit infanticide even after repeated pup exposure (*see* **Note 1**). Once the same male has mated with a female and cohabitates with the pregnant mate, however, he eventually stops infanticide by the

Kazushige Touhara (ed.), *Pheromone Signaling: Methods and Protocols*, Methods in Molecular Biology, vol. 1068, DOI 10.1007/978-1-62703-619-1_25, © Springer Science+Business Media, LLC 2013

time of delivery of his biological offspring. And at this time, the father mice provide paternal care even toward nonbiological offspring [3, 4] (*see* **Note 2**).

1.2 Components of Mouse Parental Behaviors

Among the components of rodent parental behavior listed in Table 1, *pup-retrieval* behavior is widely used as an index of parental responsiveness in both rats and mice (e.g., [5–7]). The latency to retrieve each pup is easily and unambiguously measurable, and can be assessed not only in postpartum mothers but also in non-lactating females and males, as discussed in Subheading 3.2. Assessment of the *nest quality* is also a preferred measure in mice. For postpartum mothers, their ability of *placentophagia* and *provision of maternal milk* should be assessed as well. Therefore in this chapter, the basic protocols for evaluation of postpartum maternal behaviors and pup-retrieval behavior of mon-mothers are introduced. The presented protocol is compatible with the common practices of mutant mouse husbandry in the SPF condition, so that it may be useful for the initial screening of parental responsiveness in mutant mouse strains. For more detailed information on parental behaviors in rats and other mammalian species, please refer the previous literature [8–12], and for further discussions and for listing gene mutant strains implicated in this topic, refer [13].

1.3 Effects of Chemosensory Signaling on the Pup-Directed Behaviors

Pup-directed behaviors described above are largely dependent on olfaction, in many mammalian species including rodents [14]. In particular, the mouse parental behavior is totally dependent on the *main olfactory function* (*see* **Notes 3** and **4**). Anosmia caused either by postnatal surgery or by genetic mutation strongly inhibits most of parental behaviors in both virgin and postpartum mother mice. For the best studied example, the *Adcy3* gene encodes the type III adenylyl cyclase, which coupled with G_{olf}, both of which are required for sensory transduction of the main olfactory epithelium. *Adcy3* homozygous mutant (−/−) mice were anosmic, although some odorants could be detected through the VNO [15]. *Adcy3* (−/−) mutants are initially smaller than the wild-type littermates but catch up after 3 months of age. *Adcy3* (−/−) females are fertile and show normal placentophagia after parturition, but also exhibit severe deficits in pup retrieval and nest building, causing the majority of their pups' death within 2 days [16]. Maternal aggression of *Adcy3* (−/−) mothers is severely disturbed too. Furthermore, *Adcy3* (−/−) virgin females as well are almost devoid of a pup retrieval response. Consistently, the anosmic G_{olf} homozygous mutant mothers fail to retrieve pups and to crouch over the pups, so that all pups of four litters died without milk in their stomachs by postnatal day 2 [17]. The authors did not report the infanticide in these studies.

The *accessory olfactory function* is not required for pup retrieval behavior or nest building, but is required for maternal aggression

Table 1
Components of maternal behaviors in laboratory mice and rats

Description		General references	References for non-mothers
Pup-directed behaviors			
Nursing	Provision of the opportunity to suckle by exposing the nipples and being immobile in various nursing postures, such as high crouch, low crouch or lying positions.	[60]	Rat female [61].
Retrieval	Picking up the pup gently by a part of the body (most commonly, the dorsal skin) with the incisors and carries it to the nest site.	[35, 62]	Rat [5, 63]. Mouse female [38]; male [3].
Grouping	Gathering the pups together into one quadrant so that they touched one another.	[64, 65]	
Anogenital licking	Licking the anogenital region of the pup, which is stimulatory for pups' urination and defecation.	[66]	Rat female [67].
Body licking	Licking the pup's body generically except for the anogenital region.		
Tactile stimulation	Any contact with pups, such as stepping on pups or resting in contact with pups.	[28, 68]	Mouse pups with rat aunt [69].
Non-pup directed behaviors			
Placentophagia	Elimination and ingestion of placenta, umbilical cords, amniotic membrane and fluids attached to pups' body, by vigorous licking after delivery. This makes the pups' body dry and clean.	[70, 71]	Rat female [72]. Mouse female [73].
Nest building	Transporting the nesting materials toward the nest or manipulating the material to shape the enclosed nest edge.[a]	[74, 75]	Mouse female [51].
Defense of the young	Protection of the pups from intruders, predators and environmental hazards. It is called as "maternal aggression," if the target of maternal defense is unfamiliar conspecifics.	[76, 77]	Mouse female [78].

[a]Nonpregnant mice produces a small "sleeping nest," while females at late pregnancy make a bigger and more complex nest, termed a "brood nest." Brood nest building starts from one to a few days before parturition, continues for the first 2 weeks of lactation and declines

against intruders as shown by surgical VNO removal in mice [18]. This finding is supported by the reports of two genetic mutant mouse strains, *Trpc2* (transient receptor potential cation channel, subfamily C, member 2) mutant [19] and *Del(6)1*^{Mom} mutant (lacked a cluster of vomeronasal pheromone V1 receptor family) [20]; these mutants were fertile and showed normal pup retrieval behavior, but exhibited less maternal aggression. In addition, the

VNO dysfunction abolished the infanticide in virgin male mice (Tachikawa, manuscript in preparation), as well as the aversive responses toward pups in rats [21, 22]. Surgical ablation of VNO and the *Trpc2* mutant mice also caused the reduction of inter-male aggression as well, although *Del(6)1^{Mom}* mutants display normal inter-male aggression, suggesting the possible correlation of maternal aggression and infanticide (pup-directed aggression) with the inter-male aggression. Clearly, all of these aggressive behaviors toward conspecific animals are largely dependent on the pheromone signaling.

2 Materials

2.1 Cage System, Bedding and Nest Materials

The standard shoebox breeding cages (approx. size of 265 mm × 205 mm, 140 mm high) with automated ventilation and water-supply in the SPF condition can be used for the behavioral assay. For breeding the subject mice, normal cage bedding materials can be used. For the assay of the pup-directed behaviors, however, we use paper chips made from purified pulp paper-pulp (e.g., alpha-dri, Shepherd) as cage beddings. Wood-chip bedding may affect the outcome of parental behavior, as reported for the mice lacking Fyn tyrosine kinase [23, 24] and FosB transcription factor [25]. Hexanal, a volatile substance contained in plants and causing a grassy odor, was the responsible chemical component for this effect at least in fyn (−/−) [24]. The woods used to make the same wood-chip bedding product often vary by season. Contents of chemicals such as hexanal significantly vary between the types of wood used, and between the treatment of chips (autoclaving, addition of pesticides and so on). As mouse pup-directed behaviors are very sensitive to chemosensory signals, the wood chips are less suitable than quality-controlled, purified paper chips.

The addition of nest material makes it easy to identify the location and quality of the nest [26]. Normal cotton pads, balls, or thin paper strips can be used, but the compressed cotton piece (e.g., Nestlet, Ancare, Bellmore, NY) is ideal. Adult mice normally bite and tear this densely packed cotton sheet extensively into fluffier pieces to make their nest within a couple of days. If this square piece of Nestlet has not been torn and remains in its original form, it is highly suspected that the subject mouse should have some sort of health problems, or has serious defects with nest building behavior.

2.2 Genetic Background

The confounding effects of ***genetic background*** on behavioral phenotypes have been acknowledged [27], complicating the interpretation of findings. Also for maternal behavior, strain difference among different substrains has been reported [28, 29]. If the original mutant strain was constructed using 129Sv-derived embryonic stem cells, it is often preferable to *backcross* the mutant strain into C57BL/6, a standard congenic strain for neuroscience research (*see* **Note 5**). Backcrossing is usually performed by mating a heterozygous female with a C57BL/6 male, and the heterozygous female offspring is selected to mate with a C57BL/6 male again. After five generations of backcross, the 96.875 % of the whole genome is C57BL/6 background. However, genetic complications due the *flanking-gene effect*; that is, the closely linked genes surrounding the targeted locus tend to remain even after 20 generations of backcrossing, and affects the apparent phenotypic differences between the mutant (the flanking genes are of the original background) and the wild-type (the flanking genes are of C57BL/6 derived). This problem can be properly addressed by the specific breeding strategies proposed previously [30].

2.3 Production of Subject Animals

Here we describe the procedure for detecting the autosomal recessive phenotypes in adult parental behaviors of conventional targeted disruption (knockout, KO) of single genes as an example. The protocol should be appropriately modified for mutations that follow the other patterns of inheritance, such as X-linked or dominant mutations, conditional KOs, or parent-of-origin specific gene expression (genomic imprinting).

1. Cohabitate one to three heterozygous (+/−) female mice of 10–12 weeks old with one (+/−) male of 12–24 weeks old in one breeding cage (*see* **Note 6**).
2. According to the reproductive success, make a few to several breeding cages at a time, aiming at production of 6 ~ 8 homozygous (−/−) and the wild-type (+/−) subject females of similar ages (*see* **Note** 7), to be tested at a time (*see* **Note 8**).
3. Remove visibly pregnant females into a separate cage and check for delivery every morning.
4. The pups should be weaned at a determined time window after the birth (*see* **Note 9**) and group-housed until the behavioral experiments.
5. (a) To test adult virgin mice for their pup-directed behaviors, single-house subject females and males at 10 weeks and 12 weeks of age, respectively, in a new cage containing paper bedding with a cotton square as nest material described in Subheading 2.1. Measure the body weight of each animal at this point would be informative.

   (b) To test postpartum maternal behaviors, follow the next Subheading 2.4 to make the females pregnant, and single-house the pregnant females as above.

2.4 Breeding Strategy for Postpartum Maternal Behavior Tests

1. To compare the effects of maternal genotype on maternal behaviors, crosses between (+/+) males with (−/−) subject females and between (−/−) males with (+/+) subject females should be conducted, resulting in all the pups having the same genotype (+/−). If this is not possible because of the infertility of (−/−) males, then crosses between (+/+) females with (+/−) males can be conducted. The principle is to avoid the use of (−/−) pups as stimuli for testing maternal behavior, and also trying to equalize the pups' genotype for (−/−) and the control (+/+) subject mothers. Mating should preferably start with the 10–12 weeks of age for primiparous maternal behavior testing. Checking the females for a vaginal plug early in the morning following mating provides information about the possible delivery date and also the sexual behavior of this female, although this is not essential for maternal behavior assessment itself.
2. Once the female gets visibly pregnant, it is isolated in a new cage containing paper bedding with a cotton square as described above.
3. After the female has been moved into this cage, the bedding should not be changed during the first week of lactation, so as not to disturb the mother and pups. In case the bedding gets too dirty during this first week, one can remove the dirty part of the bedding manually, and add the same amount of new bedding, but one should avoid doing this during the peripartum period.

3 Methods

3.1 Assessing Postpartum Maternal Behavior on the First Morning After Delivery

The postpartum examinations should be done in this presented order, to minimize the stress of the mother. Throughout these observations, we try not to take mouse mothers out of the cage (*see* also [11]). To collect pups for retrieval test or for pup examination, if she is in the nest over pups, we wait for her to move away or gently push her away. Nor do we suspend her by the tail completely in the air, but let her forelimbs attached on the ground or the cage cover. Suspending a peripartum female by the tail may induce a stress reaction, which sometimes result in the mother destroying her nest or attacking her pups.

1. *Parturition check*: Around the estimated day of delivery, the cage should be checked for parturition every morning (*see* **Note 10**). With our animal husbandry of a 12:12-h light/dark cycle with lights on from 08:00 to 20:00, we check for delivery between 9 and 10 am every morning. This is because, after this time, the pups born in the previous dark phase will get weaker if the milk intake is not sufficient by any reason. If the pups get too

weak and hypothermic, the maternal responses will decrease and sometimes the mother may cannibalize the dying pups, obscuring the cause of pups' death. For the same reason, if one wants to cross foster the pups to rescue them, the success rate decreases after this time. In the afternoon it would be difficult to make a foster mother accept these weakened and chilled pups.

On the other hand, if the mother is still in delivery, one should not disturb the cage. It is normal for the nest to get destroyed during delivery, as the mother circulates and moves restlessly in the small cage during delivery of each pup. It is better to wait for about an hour after the delivery of the last pup, until the mother settles down and finishes placentophagia, rebuilding the nest, pup retrieval, and stays with the pups in the nest to nurse them, before starting the ratings described below.

2. *Nest grading*:(*see* **Note 11**) If the parturition has been ended, first evaluate the nest quality from outside of the cage, so as not to disturb the mother. The nest is rated as *0* when there is *no nest*, or the nest location is unclear because the nest material is distributed randomly in the cage. The nest is rated as *1* when the nest is *flat* and not well focused. Still the grade-1 nest should be able to be identified in the cage. The nest is *2* when the nest is similar to a *shallow soup bowl* [31]. The nest is graded as *3* when the nest is shaped into a hollow surrounded by a continuous bank (designated as incomplete dome or *half of a sphere* in [31]). With the optimal nest material such as fluffy paper strips, a completely enclosed nest, rated as *4* can be achieved (*full dome* [31]). In such nest, the pups are scarcely visible from any directions. The bedding is gathered toward the corner of the nest site, so that the nest floor is higher than the floor of the dirty corner of the cage, which the mother uses as an area for defecation and urination. With about 250 mL (one cup) of the paper-chip bedding and one piece of 5 cm × 5 cm Nestlet per a shoebox mouse cage as described here, however, this grade-4 nest is rather rare. Increasing the amount of nest material may cause mechanical troubles of automated water-supply and individual cage ventilation systems. If there are two clear nests in one cage, the grading can be made separately for each nest.
3. *Pup grouping evaluation*: Pups' location in relation to the nest and other pups (=if they are in body contact with other pups) should be briefly recorded, by examining them from outside of the cage (e.g., "three alive pups in the nest and grouped, two dead? pups buried under the bank of the nest, one alive pup outside of the nest") (status of each pup will be examined later in more detail, so that a brief note is sufficient at this moment). Ideally all the pups should be in the nest and tightly grouped (huddled) with each other. When any the pups are out of nest or

buried in the bottom or the bank of the nest (better observable from the bottom of the clear cage), the pups' condition (e.g., healthy and pinky, pale, or dead) should be examined thoroughly later as in the "pup examination" paragraph. Again, it is not abnormal maternal behavior to leave dead or dying pups out of the nest, and/or cannibalize them.

4. *Pup retrieval observation*: To quantify pup retrieval latency, remove the cage top, and gently take three healthy pups from the nest, and place one pup in each corner of the cage outside the nest (Fig. 1a). Then return the cage cover. The female and pups are observed continuously for 10 min and the following measures were recorded: latency to sniff a pup for the first time (*see* **Note 12**), to retrieve each pup into the nest, group all pups, and crouch over the pups continuously for >1 min. Pup carrying to the other place than the nest should be recorded separately. When the mother has finished retrieval and grouping, and has crouched over all the pups in the nest for more than 1 min, it is called as "full maternal behavior" (note that the criteria are slightly different in each literature [32–34]). The latency to show nest building or pup licking can also be recorded, although because of the small size, these behaviors in mice are more difficult to be distinguished from pup regrouping, pup sniffing or self-grooming compared with those in rats.

   If most of the pups are found in the nest at the initial check from outside, one can assume that the mother has already exhibited pup retrieval, because newborn pups are not able to group themselves together. When the pup are found grouped but the mother does not retrieve any pup during the pup retrieval observation, it is suspected that the mother is stressed by the current experimental maneuver. In that case, gently

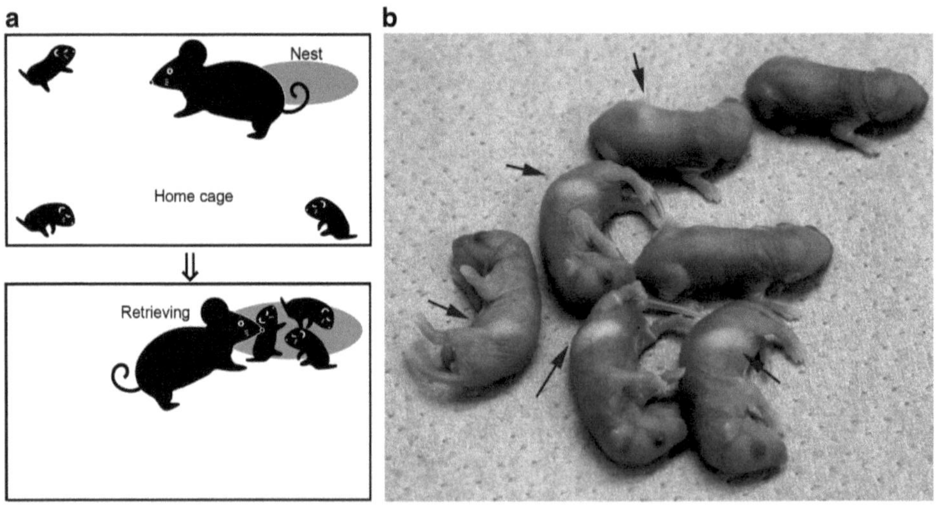

Fig. 1 Pup retrieval assay (**a**) and the milk band (**b**)

placing the cage back to the husbandry shelf, where it tends to be darker, and observe another 15 min for retrieval and full maternal behavior. If the mother shows retrieval, the delay of retrieval may have been caused by stress hypersensitivity rather than by lack of maternal responsivity itself.

In case where pups are visibly unhealthy because of inappropriate placentophagia, lack of milk intake, low body temperature or bleeding, these pups cannot be used for the stimulus pups in the pup retrieval test. In this case, the test can be made using the healthy pups (properly cleaned, milk in stomach and warm, preferably of the same genetic background) of the similar age from another mother (*see* **Note 13**). It is often seen that the mutant mother with insufficient milk secretion appears to abandon her weakened pups, but when fresh healthy donor pups are introduced, the same mother shows a quick retrieval response. Formally in this case, the control mothers should also be exposed with donor pups for retrieval scoring, to equalize the experimental conditions between mutant and control.

5. *Pup examination*: Finally, all live or dead pups are taken out of the cage, and are investigated and classified either as "alive with milk in the stomach" (Fig. 1b), "alive without milk," or "dead." In addition, it is recorded whether there are any remaining amniotic membranes, umbilical cord, or placenta attached to the body. When such sticky fetal tissue is remaining, the pup's skin may be covered by bedding materials (*see* [13]). Also the body of the pups should be briefly observed for possible bite marks or injuries. If cannibalization has occurred, any remaining pup bodies are carefully sought for in the bedding and are recorded.
6. *Maternal body examination*: If the pups are without milk, the *maternal nipple* should be examined briefly by holding its tail and lifting its hindlimbs slightly in the air, with forelimbs still on the cage top. At the same time the vaginal opening is checked for bleeding and any obstruction that could be caused by a dead fetus or other remaining tissue, which are the signs of parturition problems. If the mother has nursed the young and the pups can properly suckle, the nipples should be somewhat protruded from the ventral skin surface covered by short hair (easily identified by comparing the ventral surface of a lactating female with another female that is definitely not nursing). If the pups have been suckling vigorously because milk letdown is limited, the tip of the nipple is overly elongated and may show bleeding. Together with observations of maternal crouching over the pups in the nest, these nipple observations help to confirm that maternal nursing behavior has been performed.
7. *Interpretation of the results.* Interpret the above observatory data using the flowchart presented in Fig. 2. Analyze the latencies of each behavior during pup retrieval observations, and results from pup and maternal body examinations.

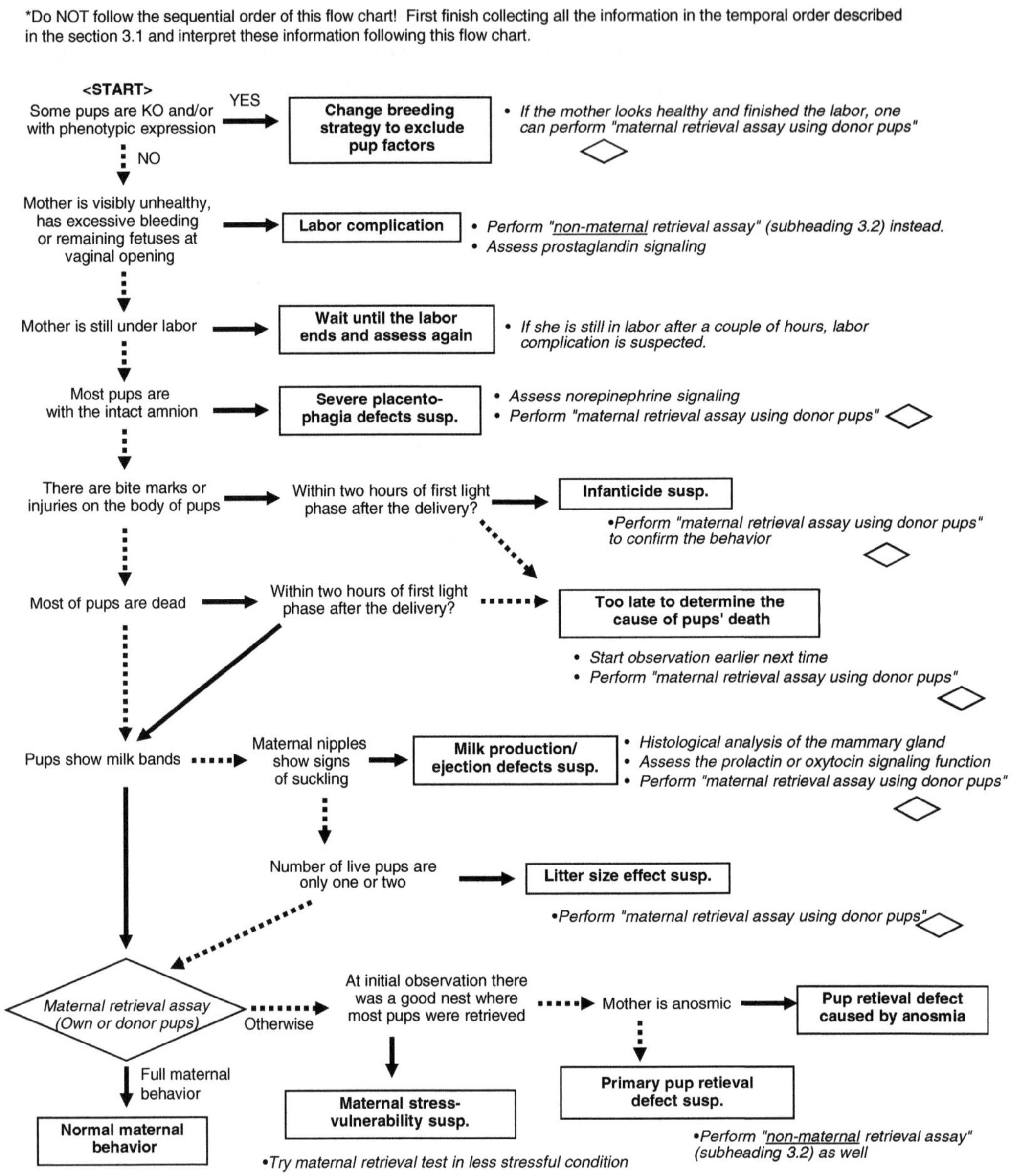

Fig. 2 Flowchart of maternal behavior assessment after the delivery

3.2 Parental Retrieval Behavior and Infanticide Toward Donor Pups by Non-maternal Mice

The pup retrieval in virgin female mice can be tested much more easily and quickly than that of postpartum mothers in mice or that of virgin female rats, since the virgin female mice are parental within half an hour at the first pup exposure [35, 36]. It is minimally 2-day procedure from single housing to testing for 30 min next day, if the donor pups can be provided from breeding colony. In addition the pup retrieval in virgin females is free of any physical or endocrine complications by parturition. It should also be noted that the virgin females' oparental behaviors are independent of hypophyseal hormones both in mice [37, 38] and in rats [5].

1. To quantify parental responsiveness of non-parturient animals, male or female subject mice are individually housed for 2 days (at least 24 h) prior of an experiment, in a new cage containing paper bedding with a cotton square as nest material as described above (*see* **Note 14**).
2. On the test day, the nest site and quality should be recorded as described above.
3. Then each animal is exposed to three 1- to 6-day-old donor pups (*see* **Note 15**). One pup is placed in each corner of the cage distant from the nest.
4. The cages are continually observed for the next 30 min and the following measures are recorded: latency to sniff a pup for the first time, to retrieve each pup into the nest, and to crouch over all the pups continuously for more than 1 min (*see* **Note 16**). If the subject mouse has performed all of these behaviors within 30 min, it is labeled as *fully parental*. If only part of these behaviors are seen by the end of the 30-min session, for example only one of the three pups is retrieved to the nest, or the subject does not retrieve at all but instead collects nest material to one of the pups and crouches over it, the subject is designated as *partially parental*. If the subject only sniffs the pups but does not show any retrieving or crouching responses throughout, it is labeled as *nonresponding*. If the subject does not approach and sniff any of the pups, it should be examined whether some kind of health problem or sensory dysfunction exists. If any of the pups is attacked during the test, which is normally observed during the first few minutes after the pup exposure, stop the observation and rescue all the pups immediately. The severely wounded pup is euthanized as described previously [39]. This subject is deemed as *infanticidal* (*see* **Note 17**). Otherwise, pups are left in the cage for 30 min.
5. After 30 min of observation, the pups are removed and, if suspected, examined for any bite marks.
6. If many of the subjects are nonresponding or partially parental, the same 30-min pup exposure session can be repeated for several days to examine whether changes in response subsequently develop by pup sensitization [7, 16]. On the other hand, once an animal becomes fully parental for 2 successive days, it seldom goes back to partially parental or infanticidal.

4 Notes

1. We found that the proportion of infanticidal virgin males is highly dependent on the rearing environment of the males as well as the testing conditions [25]. For example, the C57BL/6

males weaned at 21 days rather than 28 days are less infanticidal in our experimental conditions.

2. Evidence suggests that such infanticide seen in many wild mammalian species is adaptive in terms of inclusive fitness; that is, it is beneficial for the survival of their own biological offspring at the expense of nonbiological offspring that are potential competitors for environmental resources [40]. Especially for species forming a harem (polygyny) of a small number of males with multiple females such as langurs, lions, and mice, a new male tends to kill all the existing young upon take over a harem; such infanticide brings the females' lactation to an end, which hastens the occurrence of ovulation [41]. The new alpha male stops infanticide by the time that its biological offspring are born. This timing is clearly correlated with the gestation period of the particular species [42]. These findings support the idea that infanticide of non-offspring young is a valid and adaptive reproductive strategy selected through evolution.
3. Care should be taken for the species differences of the olfactory function in parental behaviors. Unlike in mice, maternal behavior in rats is under *multisensory control*, in terms that the maternal behavior is not dependent on any single sensory modalities (olfactory, visual, auditory, and somatosensory) [43, 44]. Elimination of any two, or all three of these sensory systems did not completely abolish the behavior, although some deficits in maternal retrieving were observed. More importantly, as in other mammalian species such as hamsters, rabbits and sheep, olfactory information from pups is necessary for the typical pup avoidant reaction that occurs in virgin female rats (*see* [14]). Olfactory bulbectomy, complete vomeronasal nerve cuts, pharmacological blockade of transmission in the accessory olfactory bulb, or destruction of the olfactory epithelium by intranasal application of zinc sulfate, all reduce the latency to the onset of parental responsiveness in virgin female rats. Therefore the high dependence of the mouse parental behaviors on olfaction may be rather atypical in mammalian species.
4. It should be noted that anosmic neonatal pups show defects in nipple finding and subsequent suckling, resulting in neonatal lethality [15, 45]. This means that the mother-pup interactions can be disturbed bidirectionally by anosmia. Therefore the proper breeding strategies are needed to segregate pup factors from maternal factors: that is, the observed mother-pup dyad should contain the homozygous mutant only in one side, either in the mother or in pups. Of additional note, the vomeronasal organ does not seem required for suckling in pups [46].

5. Milk-production deficit can be exaggerated by backcrossing the mutant lines from 129Sv into C57BL/6 as discussed [13].
6. Whenever appropriate, the subject experimental mice and the control mice should be the littermates (reared by the same mother) in general. That is, for production of experimental homozygous (−/−), heterozygous (+/−) and control (+/+) mice should be produced by crossing (+/−) females and (+/−) males, in order to minimize the potential confounding influences of background genes from breeder parents and rearing environmental factors [47]. Moreover, maternal behavior is known to transmit across generations nongenomically (future maternal behavior is affected by the maternal behavior received as neonates) [48].
7. If the preliminary analyses confirms that the given genetic mutation is complete recessive (i.e., (+/−) mothers and pups are essentially not different from (+/+)), use of the control females of (+/−) versus (−/−) subject females can be conducted for time/cost savings. In this case, produce (−/−) and (+/−) littermate females by crossing a (−/−) male with (+/−) females.
8. $N = 10$–20 mice per genotype are commonly needed to detect moderate behavioral differences using standard multivariate statistical analyses such as multiple and repeated-measures analysis of variance, followed by appropriate post hoc tests [49]. It is acceptable to start with about $N = 4$–6 animals/group for the initial screening, and then repeat at the same size or increased size for replications. These data can be combined, as long as the data are not significantly different across cohorts. As in any other research area, confirmation of the initially found behavioral phenotype across independent cohorts of mice provides compelling evidence for the functional outcome of the mutation [49].
9. The wild type and mutants on C57BL/6J background are weaned at 4 weeks of age in our laboratory, as well as in the Jackson Laboratory (*see* "Breeding strategies for maintaining colonies of laboratory mice," The Jackson Laboratory Resource Manual, 2009). This is because at 3 weeks of age, C57BL/6J pups appears still small.
10. The pregnant mouse mothers deliver pups most typically during the dark phase of 20 day after copulation, some variations (typically a half or 1 day delay) may occur [50] (*see* also "Biology of the laboratory mouse," The Jackson Laboratory, Adapted for the Web at http://www.informatics.jax.org/greenbook/index.shtml).

11. The method described here is according to the previous publications [31], with some simplifications to be compatible to other measures of maternal behavior assessments in mice [13]. *See* also [26, 51–54].
12. This measure of sniffing latency is useful to show the possible insensitivity or retardation of reaction to the pup introduction in the experimental animals.
13. It should be noted that laboratory mice do not selectively care their own young, but also retrieve alien young, provided their age is comparable. This nonselective caring may be related to the trait of communal nesting and nursing in feral *Mus musculus* species [55].
14. Because 1-month of social isolation has been shown to be stressful and may increase aggressive behavior in male mice [47, 56], they are group housed prior to the experiment, and they are not isolated for more than 2 weeks. The cage bedding should not be changed before the testing.
15. The pups with fur are not optimal stimuli for elicitation of retrieving response.
16. If one wishes to gain more information of other components of parental behavior than retrieval, the observer can simultaneously score the behavior of each mouse once every 15 s during the 30 min test for a total 120 observations, as either crouching, sniffing and/or licking, nest building, or other behaviors. Similar methods were described in [57].
17. In case that infanticide is expected as dominant response, there is a technique to minimize harming the stimulus pups [58, 59]; in essence, instead of placing pups into the subject's home cage directly, first place a pup contained in a wire-mesh tube or a small metal tea strainer. If the subject mouse is observed to start biting the container aggressively with eyes squinted and tail rattled, it is highly probable to be infanticidal. If the subject mouse does not show any signs of attack, the wire-covered pup is taken away and there naked pups are introduced in the cage as described above to further test pup retrieval behavior.

Acknowledgments

We thank Kashiko Tachikawa and Yoshihiro Yoshihara for helpful discussion and comments on this topic. This work was supported by RIKEN, the Uehara Memorial Foundation, and the JSPS Grant-in-Aid programs. We would appreciate if the readers kindly let us know if any corrections or comments on this article at: oyako@brain.riken.jp.

References

1. Krasnegor NA, Bridges RS (1990) Mammalian parenting: biochemical, neurobiological, and behavioral determinants. Oxford University Press, New York, NY
2. Numan M, Insel TR (2003) The neurobiology of parental behavior. Springer, New York, NY
3. vom Saal FS, Howard LS (1982) The regulation of infanticide and parental behavior: implications for reproductive success in male mice. Science 215(4537):1270–1272
4. Priestnall R, Young S (1978) An observational study of caretaking behavior of male and female mice housed together. Dev Psychobiol 11(1): 23–30
5. Rosenblatt JS (1967) Nonhormonal basis of maternal behavior in the rat. Science 156(781):1512–1514
6. Numan M, McSparren J, Numan MJ (1990) Dorsolateral connections of the medial preoptic area and maternal behavior in rats. Behav Neurosci 104(6):964–979
7. Brown JR, Ye H, Bronson RT, Dikkes P, Greenberg ME (1996) A defect in nurturing in mice lacking the immediate early gene fosB. Cell 86(2):297–309
8. Elwood RW (1983) Parental behaviour of rodents. Wiley, New York, NY
9. Sluckin W, Herbert M (1986) Parental Behaviour. Basil Blackwell, Oxford
10. Rosenblatt JS, Snowdon CT (eds) (1996) Parental care: evolution, mechanism, and adaptive significance, vol 25, Advances in the study of behavior. Academic, San Diego, CA
11. Lonstein JS, Fleming AS (2002) Parental behaviors in rats and mice. Curr Protoc Neurosci Chapter 8:8.15.11–18.15.26
12. Capone F, Bonsignore L, Cirulli F (2005) UNIT 13.9 methods in the analysis of maternal behavior in the rodent. Curr Protoc Toxicol 13(9):1–16
13. Kuroda KO, Tachikawa K, Yoshida S, Tsuneoka Y, Numan M (2011) Neuromolecular basis of parental behavior in laboratory mice and rats: with special emphasis on technical issues of using mouse genetics. Prog Neuropsychopharmacol Biol Psychiatry 35(5):1205–1231. doi: 10.1016/j.pnpbp.2011.02.008, S0278-5846(11)00062-5 [pii]
14. Levy F, Keller M (2009) Olfactory mediation of maternal behavior in selected mammalian species. Behav Brain Res 200(2):336–345
15. Wong ST, Trinh K, Hacker B, Chan GC, Lowe G, Gaggar A, Xia Z, Gold GH, Storm DR (2000) Disruption of the type III adenylyl cyclase gene leads to peripheral and behavioral anosmia in transgenic mice. Neuron 27(3):487–497, S0896-6273(00)00060-X [pii]
16. Wang Z, Storm DR (2011) Maternal behavior is impaired in female mice lacking type 3 adenylyl cyclase. Neuropsychopharmacology 36(4):772–781
17. Belluscio L, Gold GH, Nemes A, Axel R (1998) Mice deficient in G(olf) are anosmic. Neuron 20(1):69–81
18. Bean NJ, Wysocki CJ (1989) Vomeronasal organ removal and female mouse aggression: the role of experience. Physiol Behav 45(5): 875–882
19. Leypold BG, Yu CR, Leinders-Zufall T, Kim MM, Zufall F, Axel R (2002) Altered sexual and social behaviors in trp2 mutant mice. Proc Natl Acad Sci U S A 99(9):6376–6381. doi:10.1073/pnas.082127599, 082127599 [pii]
20. Del Punta K, Leinders-Zufall T, Rodriguez I, Jukam D, Wysocki CJ, Ogawa S, Zufall F, Mombaerts P (2002) Deficient pheromone responses in mice lacking a cluster of vomeronasal receptor genes. Nature 419(6902):70–74
21. Mennella JA, Moltz H (1988) Infanticide in the male rat: the role of the vomeronasal organ. Physiol Behav 42(3):303–306
22. Fleming AS, Gavarth K, Sarker J (1992) Effects of transections to the vomeronasal nerves or to the main olfactory bulbs on the initiation and long-term retention of maternal behavior in primiparous rats. Behav Neural Biol 57(3): 177–188
23. Yagi T, Aizawa S, Tokunaga T, Shigetani Y, Takeda N, Ikawa Y (1993) A role for Fyn tyrosine kinase in the suckling behaviour of neonatal mice. Nature 366(6457):742–745. doi:10.1038/366742a0
24. Hamaguchi-Hamada K, Sanbo C, Hamada S, Yagi T (2004) Exposure to hexanal odor influences maternal behavior and induces neonatal death in Fyn tyrosine kinase-deficient mice. Neurosci Res 48(3):259–267
25. Kuroda KO, Meaney MJ, Uetani N, Kato T (2008) Neurobehavioral basis of the impaired nurturing in mice lacking the immediate early gene FosB. Brain Res 1211:57–71
26. Deacon RM (2006) Assessing nest building in mice. Nat Protoc 1(3):1117–1119
27. Doetschman T (2009) Influence of genetic background on genetically engineered mouse phenotypes. Methods Mol Biol 530:423–433
28. Brown RE, Mathieson WB, Stapleton J, Neumann PE (1999) Maternal behavior in female C57BL/6J and DBA/2J inbred mice. Physiol Behav 67(4):599–605

29. Shoji H, Kato K (2006) Maternal behavior of primiparous females in inbred strains of mice: a detailed descriptive analysis. Physiol Behav 89(3):320–328
30. Wolfer DP, Crusio WE, Lipp HP (2002) Knockout mice: simple solutions to the problems of genetic background and flanking genes. Trends Neurosci 25(7):336–340
31. Hess SE, Rohr S, Dufour BD, Gaskill BN, Pajor EA, Garner JP (2008) Home improvement: C57BL/6J mice given more naturalistic nesting materials build better nests. J Am Assoc Lab Anim Sci 47(6):25–31
32. Moltz H, Lubin M, Leon M, Numan M (1970) Hormonal induction of maternal behavior in the ovariectomized nulliparous rat. Physiol Behav 5(12):1373–1377
33. Pedersen CA, Prange AJ Jr (1979) Induction of maternal behavior in virgin rats after intracerebroventricular administration of oxytocin. Proc Natl Acad Sci U S A 76(12):6661–6665
34. Bridges RS, DiBiase R, Loundes DD, Doherty PC (1985) Prolactin stimulation of maternal behavior in female rats. Science 227(4688): 782–784
35. Noirot E (1972) The onset of maternal behavior in rats, hamsters, and mice a selective review. In: Lehrman DS, Hinde RA, Shaw E (eds) Advances in the study of behavior, vol 4. Elsevier, New York, NY, pp 107–145
36. Lonstein JS, De Vries GJ (2000) Sex differences in the parental behavior of rodents. Neurosci Biobehav Rev 24(6):669–686
37. Leblond CP (1940) Extra-hormonal factors in maternal behavior. Proc Soc Exp Biol Med 38:66–70
38. Leblond CP, Nelson WO (1937) Maternal behavior in hypophysectomized male and female mice. Am J Physiol 120:167–172
39. Perrigo G, Belvin L, Quindry P, Kadir T, Becker J, van Look C, Niewoehner J, vom Saal FS (1993) Genetic mediation of infanticide and parental behavior in male and female domestic and wild stock house mice. Behav Genet 23(6):525–531
40. Trivers RL (1972) Parental investment and sexual selection. In: Campbell B (ed) Sexual selection and the descent of man 1871-1971. Aldine-Atherton, Chicago, IL, pp 136–172
41. Hrdy SB (1974) Male-male competition and infanticide among the langurs (Presbytis entellus) of Abu, Rajasthan. Folia Primatol (Basel) 22(1):19–58
42. Parmigiani S, Vom Saal FS (eds) (1990) Infanticide and parental care. Harwood academic publishers, London
43. Beach FA, Jaynes J (1956) Studies of Maternal Retrieving in Rats: III. Sensory cues involved in the lactating female's response to her young. Behavior 10:104–125
44. Herrenkohl LR, Rosenberg PA (1972) Exteroceptive stimulation of maternal behavior in the naive rat. Physiol Behav 8(4):595–598
45. Tasaki T, Sohr R, Xia Z, Hellweg R, Hortnagl H, Varshavsky A, Kwon YT (2007) Biochemical and genetic studies of UBR3, a ubiquitin ligase with a function in olfactory and other sensory systems. J Biol Chem 282(25):18510–18520
46. Kimchi T, Xu J, Dulac C (2007) A functional circuit underlying male sexual behaviour in the female mouse brain. Nature 448(7157): 1009–1014. doi:10.1038/nature06089, nature06089 [pii]
47. Crawley JN (2007) Whats wrong with my mouse? Behavioral phenotyping of trangenic and knockout mice, 2nd edn. Wiley-Liss, New York, NY
48. Francis D, Diorio J, Liu D, Meaney MJ (1999) Nongenomic transmission across generations of maternal behavior and stress responses in the rat. Science 286(5442):1155–1158
49. Crawley JN (2008) Behavioral phenotyping strategies for mutant mice. Neuron 57(6): 809–818
50. Hedrich H (ed) (2004) The laboratory mouse, The hand book of experimental animals. Elsevier, Amsterdam
51. Gandelman R (1973) Induction of maternal nest building in virgin female mice by the presentation of young. Horm Behav 4(3): 191–197
52. Slotnick BM, Nigrosh BJ (1975) Maternal behavior of mice with cingulate cortical, amygdala, or septal lesions. J Comp Physiol Psychol 88(1):118–127
53. Numan M, Callahan EC (1980) The connections of the medial preoptic region and maternal behavior in the rat. Physiol Behav 25(5): 653–665
54. Mann PE (1993) Measurement of maternal behavior. Methods Neurosci 14:343–358
55. Manning CJ, Wakeland EK, Potts WK (1992) Communal nesting patterns in mice implicate MHC genes in kin recognition. Nature 360(6404):581–583
56. Valzelli L, Bernasconi S, Gomba P (1974) Effect of isolation on some behavioral characteristics in three strains of mice. Biol Psychiatry 9(3):329–334
57. Liu D, Diorio J, Tannenbaum B, Caldji C, Francis D, Freedman A, Sharma S, Pearson D, Plotsky PM, Meaney MJ (1997) Maternal care,

hippocampal glucocorticoid receptors, and hypothalamic-pituitary-adrenal responses to stress. Science 277(5332):1659–1662
58. Perrigo G, Belvin L, vom Saal F (1990) Humane behavioral assays in the laboratory: an alternative test for assessing how rodents behave toward young. Humane Innov Alternat Anim Experiment 4:208–209
59. Perrigo G, Bryant W, vom Saal F (1989) The use of live pups in a humane, injury-free test for infanticidal behaviour in male mice. Anim Behav 38:897–904
60. Stern JM, Lonstein JS (2001) Neural mediation of nursing and related maternal behaviors. Prog Brain Res 133(64):263–278
61. Lonstein JS, Wagner CK, De Vries GJ (1999) Comparison of the "nursing" and other parental behaviors of nulliparous and lactating female rats. Horm Behav 36(3):242–251
62. Wiesner BP, Sheard NM (1933) Maternal behaviour in the Rat. Oliver and Boyd, London
63. Bridges RS, Zarrow MX, Goldman BD, Denenberg VH (1974) A developmental study of maternal responsiveness in the rat. Physiol Behav 12(1):149–151
64. Fahrbach SE, Morrell JI, Pfaff DW (1984) Oxytocin induction of short-latency maternal behavior in nulliparous, estrogen-primed female rats. Horm Behav 18(3):267–286, 0018-506X(84)90016-3 [pii]
65. Kinsley CH, Bridges RS (1988) Parity-associated reductions in behavioral sensitivity to opiates. Biol Reprod 39(2):270–278
66. Moore CL (1992) The role of maternal stimulation in the development of sexual behavior and its neural basis. Ann N Y Acad Sci 662: 160–177
67. Gubernick DJ, Alberts JR (1985) Maternal licking by virgin and lactating rats: water transfer from pups. Physiol Behav 34(4):501–506
68. Ambrose A (ed) (1969) Stimulation in early infancy. Academic, New York, NY
69. Rosenberg KM, Denenberg VH, Zarrow MX (1970) Mice (Mus musculus) reared with rat aunts: the role of rat-mouse contact in mediating behavioural and physiological changes in the mouse. Anim Behav 18(1):138–143
70. Kristal MB (1980) Placentophagia: a biobehavioral enigma (or De gustibus non disputandum est). Neurosci Biobehav Rev 4(2):141–150
71. Kristal MB (1991) Enhancement of opioid-mediated analgesia: a solution to the enigma of placentophagia. Neurosci Biobehav Rev 15(3):425–435, S0149-7634(05)80035-1 [pii]
72. Kristal MB, Graber GC (1976) Placentophagia in nonpregnant rats: influence of estrous cycle stage and birthplace. Physiol Behav 17(4):599–605
73. Kristal MB, Eleftheriou BE (1975) Placentophagia in nonpregnant nulliparous mice: a genetic investigation. Behav Biol 13(1): 113–119
74. Koller G (1952) Der Nestbau der Weissen Maus und seine hormonale Auslosung. Zool Anz Suppl 17:160–168
75. Denenberg VH, Taylor RE, Zarrow MX (1969) Maternal behavior in the rat: an investigation and quantification of nest building. Behaviour 34(1):1–16
76. Lonstein JS, Gammie SC (2002) Sensory, hormonal, and neural control of maternal aggression in laboratory rodents. Neurosci Biobehav Rev 26(8):869–888
77. Gammie SC (2005) Current models and future directions for understanding the neural circuitries of maternal behaviors in rodents. Behav Cogn Neurosci Rev 4(2):119–135
78. Svare B, Gandelman R (1976) Suckling stimulation induces aggression in virgin female mice. Nature 260(5552):606–608

Chapter 26

Testing Smell When It Is Really Vital: Behavioral Assays of Social Odors in the Neonatal Mouse

Benoist Schaal, Syrina Al Aïn, and Bruno Patris

Abstract

The initial interactions of mouse newborns with their mother are crucial for their survival. These interactions rapidly end in the pups reaching nipples and getting milk. While we realize that olfaction is clearly prevailing in the success of these first suckling episodes, we still understand little about the nature and range of the natural odorants involved. Here we non-exhaustively describe some experimental principles and methods to assay the behavior of newly born and infant mice exposed to different odor stimuli from conspecifics. Testing neonatal and young mice with chemostimuli which they are evolutionarily or developmentally canalized to detect may be a productive way to trace unanticipated odor signals. Moreover, testing neonates also may also lead to characterize unsuspected strategies of murine females to produce and release odor messages.

Key words Mouse (*Mus musculus*), Newborn, Infant, Olfaction, Social odor, Pheromones, Behavior

1 Introduction

The survival of newborns mammals depends on the success of their first suckling of colostrum and milk, ensuring continued supply in hydration, energy, nutrients, passive immunization, as well as the transfer of growth factors, hormones and regulatory peptides, and nonpathogenic bacteria. Through milk, mothers also convey sensory information that impinges on the current and future adaptive competence of their offspring. Thus, after successful birth, it is imperative for neonates to perceive their mother and locate her mammae. The corollary of this inescapable bottleneck of the mammalian condition is that neonates must have been evolutionarily tailored to use any sensory cue susceptible to speedily bring them to those goals. Thus, testing neonatal and infant animals with stimuli which they are canalized to detect may be a productive way to trace unanticipated odor signals. Testing neonates also leads to characterize strategies of females to produce and release odor messages that evolved to match perceptual biases in their newborns.

Kazushige Touhara (ed.), *Pheromone Signaling: Methods and Protocols*, Methods in Molecular Biology, vol. 1068,
DOI 10.1007/978-1-62703-619-1_26,

Despite we know that these interactions with the mother are crucial in the viability of newborns [1–3], we have still much to understand about the nature and range of the stimuli that direct their initial behavior. Here, we will address first general advantages and difficulties in investigating early chemosensory development in mammals, emphasizing a species that has become a privileged mammalian model in biology, the mouse. Then, we will provide selected methods and protocols that have been devised to quantify chemosensory abilities of mouse newborns and infants toward natural odors conveyed by biological secretions of conspecifics.

1.1 Studying Chemoreception in Newborn and Young Animals: Advantages and Problems

A range of psychobiological features make mammalian neonates and infants particularly profitable to investigate in the context of olfactory exchanges in general, and of pheromonal communication in particular:

1. Pronounced taxonomic differences in neonatal development are associated with qualitative and quantitative differences in initial capacities. While some species give birth to newborns of the *altricial* type, others deliver *precocial* newborns, and still others engender newborn of an *intermediate* type. The most obvious criteria underlying this altriciality–precociality "scale" relate to neonatal thermoregulatory, locomotor, and sensory abilities. Regarding sensory abilities, precocial newborns have a complete (although still maturing) set of active sensory systems, while those of altricial newborns are only partially functional. In mammals, altriciality means that somesthesis and chemoreception are the only sensory means to control neonatal behavior. In contrast, precocial newborns additionally dispose of a good level of visual and auditory awareness. Thus, relying predominantly on chemosensory processes to survive, *altricial newborns are outstanding models to assess the earliest olfaction-guided responses.*
2. Mammalian newborns, especially those of the altricial type, are obviously limited in their behavior, but they are far from being incomplete, incompetent or hindered adults. In much contrast, they are endowed with specialized responses that are best expressed in their natural/usual ecologies. Neonatal abilities to move and orient, explore and choose, grasp and suck are indeed fully adjusted to the sequence of operations these inexperienced organisms have to achieve in their species-typical environment. Each developmental phase has its typical ways to generate behavior adaptively, and the specialized responses of neonates have once been considered as automatic and reflex-like, conceptualized as "fixed action patterns" prompted by specific "releasing stimuli." Although these conceptions have become more flexible with advancing psychobiological research (e.g., [4, 5]), *neonatal sensorimotor abilities are first organized*

around stimulus-response loops that are ready to operate right at birth, and much remains to be understood about the sensory triggers of these initial functional loops.

3. Neonatal mammals rapidly learn and escape sensorimotor automatisms in the regulation of their behavior. Newborn rodents are especially skilled to learn about the odor environment [6–8]: (1) they acquire any odorant which they are merely exposed to or which is associated with the mother, and such acquisitions can take only minutes to establish; (2) such learning is functional from birth onwards, although its efficacy increases with maturation and experience; (3) neonatal olfactory sensitivity and preferences are partially induced prenatally [9–11], and thus, at first presentation of an odorant, neonatal reactions may trace back into fetal acquisition [10, 12, 13]; (4) multiple, redundant reinforcing processes and agents afforded by the mother's body or her milk promote rapid odor learning in neonates (e.g., [7]). In sum, these neonatal learning abilities are of great empirical interest to investigate early plasticity of the senses, brain and behavior, but on the other side the *fast odor learning of neonates may obscure the salience of odor signals that do not depend on exposure effects.* This is especially inconvenient to track behaviorally active odorants such as pheromones which, by definition, imply no or minimal induction by exposure effects or learning (cf. below).
4. It ensues from the previous point that newborn rodents are practical to investigate because their olfactory experience can be somehow titrated. Their odor memory stores can be pre-shaped by controlled odor exposure during prenatal development. They can be gestated by a female transferring them experimental odorants through the placenta and then cross-fostered to a lactating dam transferring the same or a different odorants into her milk. They can be prevented from any postnatal contact with their mother, and tested for odorant stimuli before they had any direct exposure to them postnatally. *Neonatal rodents are thus suitable to appreciate whether a natural odorant can be behaviorally active in absence of previous exposure.*
5. In many altricial species, newborns and juvenile individuals are easily tractable and amenable to approach an "attachment" figure (e.g., the mother or sibling) or salient sensory traits "disembodied" from her (odorants carried in biological secretions). However, *optimal species-specific responsiveness of neonate animals will be obtained only if a range of basic ethological principles are applied* (e.g., [14]) that take into account: (1) age-dependency of test conditions (in terms of perceptual, motor and cognitive abilities of young organisms; existence of sensitive periods), (2) state-dependency of responses (separation stress; thermoregulation; food deprivation; circadian rhythms),

and (3) the social-ecological validity of the stimuli being assayed, of the testing context and of target responses. Although such principles may be general in the investigation of any animal, they appear especially sensitive with more physiologically unstable and psychologically susceptible newborns and infants.

1.2 Mammalian Newborns and the Pheromone Concept: Methodological Consequences

Newborn mammals are canalized into adaptive responses by two types of *nonexclusive* mechanisms, predisposed processes which do (apparently) not depend on previous exposure effects, and highly efficient learning processes which fine-tune them to the current or prospective environment. Associative or non-associative learning mechanisms work in extracting arbitrary and incidental information from the current odor environment to convert them into meaningful *odor cues.* By contrast, predisposed mechanisms rely on hardwired stimulus-response loops that can operate at the very first stimulus presentation and do apparently not need induction by previous contact with it. Such *predisposed stimuli* have been selected because they bear an inherent signal value to the neonatal receiver [15]. While cues tend to be lastingly "on," signals can be switched "on-off" according to the emitter's behavior or condition [16]. Any biological secretion may convey numerous cues, including informative olfactants derived from the environment or current physiological or psychological states of a conspecific. In contrast, signals are less frequent. Therefore, in the complex mixtures given off in body secretions or excretions, *rare odor signals will unavoidably be embedded in an abundance of odor cues.* Thus, it is always a challenging endeavor to separate rare or unique odor signals from affluent odor cues in a chemically complex body secretion.

Before sorting odor signals from odor cues, a number of steps have to be fulfilled. First, an age- and state-adapted behavioral assay has to be validated. Second, behaviorally active biological substrates must be identified, and the emitter's state in which this substrate bears maximal activity characterized. Then, the active substrate has to be reduced in complexity to target fractions that remain active on the target response, e.g., extricating composite substrates; separating volatile from nonvolatile compounds, or hydrosoluble from liposoluble compounds. Then, comes the difficult task of designing techniques that combine chemical separation and functional responses to pinpoint (the) active compound(s) in these fractions. Once chemically identified, candidate compounds have then to be related in activity to the original odor substrate, and then screened through a set of criteria to assess whether they are "mere" cues or if they qualify as signal(s), i.e., pheromone(s) (see below). Finally, once a pheromone is identified, its communicative function and adaptive meaning between emitting and receiving organisms need to be ascertained. The successful attainment of each of these steps depends on the elaboration of reliable behavioral assays.

Regarding the pheromone concept, its initial definition [17] now appears minimalist, encompassing almost any olfactant exchanged between conspecifics, be it a cue or a predisposed signal. To avoid this latent confusion and in an attempt to critically resize the concept in relation with the cognitive complexity of mammalian behavior, Beauchamp, Doty, Moulton and Mugford [18] (see also [19–22]) proposed criteria implying that any candidate pheromone should: (1) be chemically "simple," in the sense that the active fraction is composed of one chemical or a small set of chemicals in a given ratio; (2) release in a conspecific receiver unambiguous behavioral/physiological responses which are morphologically invariant in a same context, and (3) obvious in terms of adaptive function; (4) elicit these responses in a selective way (thus tested against a set of reference compounds), and (5) be tested in various species to establish specificity at taxonomic level (genus, species, subspecies); finally, (6) the candidate stimulus-response link should minimally or not at all depend on previous exposure effects and learning. Thus, in principle, to qualify a candidate compound as a pheromone one should make sure that the responsiveness to it cannot be ascribed to prenatal induction by direct exposure, to facilitated learning during the natal process, or to rapid learning immediately after birth. This latter criterion is generally overlooked in most cases of mammalian pheromones described in adult animals [18–20, 23]. However, investigating young organisms should make its assessment feasible.

The above methodological steps and concepts have been successfully applied to mammalian neonates. In fact, the most complete (although yet incomplete!) demonstrations to date of mammalian pheromones according to Beauchamp et al.'s operational criteria [18] have made use of newborns. Let us develop two examples which are borrowed beyond the mouse literature. First, in the newborn rat, pup saliva and pup salivary gland extract were shown to be reliable elicitors of nipple grasping [24]. Subjected to gas chromatography-mass spectrometry (GC-MS), these substrates were shown to carry dimethyl disulfide (DMDS). Synthetic DMDS was indeed effective in eliciting the grasping of nipples that were washed before the test, although its releasing potency was about 50 % relative to that of intact nipples [24]. Thus, DMDS explained only in part the elicitation of nipple grasping in rat pups, unknown compounds from the original mixture coating nipples being probably involved. DMDS is special, however, as it also reduces aversive responses to perioral stimulation in the fetus. This effect, mediated by opioidergic processes, suggests unconditional reward properties of DMDS already established in late gestation [25]. As DMDS was reported to be undetectable by GC in amniotic fluid [26], its behavioral activity may not depend on prenatal exposure. In these studies, however, pups were aged postnatal days (PND) 3–5, implying that the activity of DMDS could derive from exposure to saliva prior to testing. Thus, it is difficult to decide whether DMDS is an odor cue acquired in the first suckling episodes or if it is an evolved signal to promote nipple attachment. Further, to be categorized as a pheromone, its species-specificity remains to be established and it needs to be assayed against reference compounds to eliminate the possibility of unspecific arousal effects.

The second example also goes beyond the murine focus of the present chapter, but remains in scope as it involves newborns. The extensive corpus of behavioral investigation by P Schley and R Hudson on odor cues that direct rabbit pups to the nipple and milk [27–29] paved the way for the chemical characterization of the behaviorally active volatiles involved in milk. A GC-MS-olfactometry analysis of rabbit milk effluvium resulted in identifying 2-methyl-but-2-enal (2MB2) as accounting for the bioactivity of milk, as it released a similar response at the highest possible rate in very different contexts of stimulation [30]. 2MB2 was then assayed to verify whether it qualified as a pheromone. In chemical terms, it is the "simplest" (viz., monomolecular) possible stimulus bearing a strong potency to elicit nipple grasping actions in pups, although other, unidentified compounds may act in the same direction. Further, 2MB2 triggers the criterion response regardless of mode of presentation (at least during the first day 10 days postbirth) in a highly selective manner (tested against 40 reference odorants, [31]). Its activity is optimal within a concentration range (10^{-9} to 10^{-5} g/mL, [32]), giving flexibility for intensity fluctuations of the signal. The species-level bioactivity of 2MB2 was established through its independence from maternal diet and genetic background, and its inertness for neonatal rats, mice, cats, and even brown hares [30]. Finally, to be behaviorally active, 2MB2 does not appear to require contingent labor-related arousal, suckling, or contact with the mother, and it is resistant to exposure deprivation during the first 6 days of life. Further, it is active in pups delivered 1–2 days before term, and GC-MS failed detecting it in amniotic fluid and blood plasma of lactating does [30]. These results allowed qualifying 2MB2 from rabbit milk as a pheromone. As it was detected in ejected milk it was called "mammary pheromone."

Regarding its functional involvement, 2MB2 was shown to control mutually beneficial functions as it contributes to the rabbit doe-offspring reciprocity in eliciting pups' contagious directional actions toward the female, and guidance to nipples; on the doe's side, it boosts tactile stimulations from pups that trigger, synchronize and sustain lactational physiology. Those pups that do not react to 2MB2 on PND 1 have low survival chance at weaning [34]. Otherwise, maximal activity of 2MB2 closely matches the period of strict milk-dependence of pups (up to PND 10) and declines when they begin eating solid food [33]. 2MB2 elicits then an automatic drive to grasp a nipple at any time during the first critical postnatal days, and it is only after PND 10 that its activity "relaxes" and comes under prandial and circadian control [35]. Finally, 2MB2 also works as a primary reinforcing stimulus transferring its behavioral activity to any (initially neutral) odorant after a single, brief association [36].

The above examples designate mammalian female-neonate units as promising model systems to uncover odor cues and signals that regulate their vital exchanges, and then to characterize molecular and neurobehavioral processes underlying these ways of communication. Similar advances are needed to understand the system of cues and signals that rule early relationships between murine mothers and their newborns.

2 The Mouse Pup as a Model to Track Behaviorally Significant Odorants

2.1 Nasal Chemoreception in the Context of Murine Developmental Milestones

Mouse pups are born altricial. Accordingly, they are raised in a nest in quasi-permanent contact with the mother (and, sometimes, father) and, in the case of wild mice, with other related females, as communal nursing is frequent in this species [37]. Newborn mice are essentially guided by somesthesis and chemoreception in their early orientation responses. Audition comes into play by PND 9, and vision by PND 14–15 [38, 39]. Before PND 9, mouse pups are thus only governed by tactile, thermal and chemical cues (nasal and oral chemoreception). Murine nasal chemoreception is based on multiple receptive systems. These include the main olfactory epithelium, the vomeronasal organ, which are part of the main and accessory systems, respectively, the trigeminal system, the Grueneberg ganglion, and the septal organ of Masera [40, 41]. Much remains to be learned about the functional onset of these distinct nasal systems, especially the last three of them. The main olfactory system sets on prenatally [42], thus contributing very early to the experiential induction of behavioral phenotypes [43]. But the vomeronasal system does not appear to be functional before birth and seems to remain immature until puberty ([44, 45]; but see [46]). Regarding motor development, mouse pups are crawlers during the first week of like, and begin to walk by PND 7 and to run and jump by PND 15 [47, 48]. Inventing behavioral assays will obviously depend on these states of motor development.

The general role of olfaction in neonatal rodents (i.e., rat and mouse) was ascertained through its abolishment by bulbar lesion [49–51], by $ZnSO_4$ peripheral denervation [52], or by genetic disruption inducing early dysfunction in olfactory transduction (e.g., [53]). Neonatal rodents are also disturbed when the normal odor cuing system of the dam is disrupted. For example, washing the abdomen of a lactating dam results in a drastic reduction of nipple seizing performance in mouse pups [55, 56] as well as in rat pups [54].

2.2 Assaying Social Odor-Based Behavior in Infant Mice

The mouse being a privileged model species in various subdisciplines of biology it has received extensive coverage regarding methods and protocols to assay behavior in adult (e.g., [57–63]) and in young individuals (e.g., [47, 64, 65]). Here, we will focus on behavioral assays that have been devised to measure responsiveness of newborn and infant mice when exposed to biological substrates collected from conspecifics. After presenting some general conditions for testing young mice, we provide a brief survey of bioassays adapted to three age-periods, newborn mice (PND 0–2), previsual infant mice (from PND 3 to 14), and visual preweaning infant mice (PND 14–21).

2.2.1 General Conditions

1. *Temperature*: Mouse pups have poor thermoregulatory abilities up to PND 12–28 [66, 67], and they then mainly rely on thermotaxis and huddling to maintain body temperature [68]. In general, breeding and experimental rooms are kept between 20 and 24 °C. When pups have to be separated from the dam before testing, hypothermia can be prevented by keeping the male with the litter and/or putting the home cage on a heating plate (Gestigkeit, Germany; set at the temperature that ensures that the nest is stabilized at 30–32 °C). All tests should be realized in conditions that reproduce those prevailing in the nest (30–32 °C), which can be obtained by placing the test device on a heating plate.
2. *Lighting*: Experimental rooms are generally lit up following a given reversed light–dark cycle (e.g., 12:12 h). Before pups have opened eyes, they can be assayed under artificial light. But, when vision gets functional it is best to test them under red light during the dark phase.
3. *Noise*: As the auditory system is functional on PND 9 and as pups emit and likely detect ultrasonic vocalizations during the first two weeks [69–72], experiments should be conducted in a dedicated, quiet room.
4. *Parasite odors*: Odor cues given off by murine conspecifics, the experimental material or experimenters themselves may interfere with target odor stimuli and thereby negatively influence the responses of neonates. Thus, to avoid interference with such unwanted odors, any behavioral testing should be conducted in a dedicated room. Experimenters are recommended to limit their body odors in wearing a scentless clean blouse that remains in the testing room, avoiding perfume, coffee or tobacco, and handling animals with gloved hands (do not use latex gloves). Finally, the experimental material should be thoroughly washed twice, with (70 %) ethanol and then with water, and finally dried at each testing session. It is often necessary to visually mark animals after a treatment, a test or whatever: in all cases, take care to select a scentless marker.

5. *Stress associated with separation, cold, novelty and handling*: All sorts of stressors related to social isolation can adversely affect newborn and infant mice. When the protocol necessitates separation from the dam, keeping the male with the litter in their home cage can limit social as well as cold stress. Cold stress is controlled in maintaining the home cage or the experimental device on a heating plate [73, 74, 76, 77] or under 60-W red bulbs suspended 20 cm above the device [78, 79]. The handling method is essential to preserve mice from fear [80], and it is thus recommended to handle pups with extreme delicacy and consistency across experimenters to limit artifactual response variability. A protocol to familiarize pups with the experimental device can have positive effects on stress responses. For example, pups can be allowed to explore the device for 1–3 min to reduce neophobia before to the actual test.
6. *Motivation of animals*: To optimize arousal and motivation, pups are often deprived of contact with the mother and, hence, of suckling, prior behavioral testing. Several separation schedules were used without apparent immediate harm to pups or dams (i.e., 1 h [81]; 2 h [76]; 4 h [73, 74]). During the first 2 postnatal weeks, mouse pups stay in the nest and huddle and exhibit low levels of arousal and activity. In an attempt to standardize the pups' arousal level right before the tests, some experimenters apply a stimulation protocol, for example by stroking them along the back (e.g., ten strokes) with a soft paintbrush [73].
7. *Litter effects*. Neonatal mice come in litters, and littermates are more similar to one another than to members of other litters because of differences in genetic and environmental background [82]. The first way to reduce litter effects is thus through rigorous husbandry methods, such as using inbred strains, diminishing the number of breeding males, and standardizing housing conditions and pup development in equalizing litter size (to control for competition between pups). Litters effects are further limited when testing. Not only should every pup be assayed only once, but the number of littermates per test condition should be reduced to preserve independence of data. The optimum is one littermate per condition, or one male and one female (e.g., [83]), but most studies are constrained to be less conservative because of limited number of litters in their breeding colony (e.g., 4-littermates/condition [73, 74]; pups from three litters distributed in two or more test conditions [56]). Finally, data should be assessed statistically for potential litter effects in treating each litter as the statistical unit [77, 83] or in using the litter as a covariable [43, 73, 74].

2.2.2 Sampling Odorant Stimuli from Conspecifics: Collection and Conservation

Many different types of stimuli have been used to assess behavioral responses in newborn and infant mice. Before entering into more detailed analyses, experimenters often use first entire animals as stimuli to verify whether these are olfactorily differentiable to pups along a selected dimension. Then, more analytic experiments evaluate which biological substrates from distinct body regions of conspecifics are efficient in reproducing the responses elicited by olfactorily intact entire animals. Some of these substrates are exemplified below. In certain cases, artificial arbitrary odorants are used as controls.

1. *Live animals as stimuli.* Entire animals can be used to assess whether neonates detect olfactory correlates of age, sex, physiological states, genetic relatedness or merely familiarity. Pups are then exposed to two animals simultaneously (a female vs. a male, a lactating vs. a non-lactating female, the mother vs. the father, etc.). For example, Breen and Leshner [84] presented mouse pups with live lactating vs. non-lactating dams in a two-choice device which allowed to hide them while only blowing their body odor toward the tested pup. In similar tests in recent studies, two stimulus-females were sedated (intraperitoneal injection of 45 mg/kg body weight of ketamine and 9 mg/kg Xylazine; [75]) to be simultaneously exposed to pups in a choice-arena. Entire animals can also be used to analyze intra-individual discrimination by pups, scrutinizing whether limited body regions emit volatiles. For example, nipples were especially reactogenic for newly born pups and were the target of numerous studies in the mouse [55, 56, 75]. In this case also, the stimulus females were anesthetized (e.g., 100 mg/kg ketamine, 10 mg/kg xylazine, and 0.2 mg/kg acepromazine [56]; 50 mg/kg of pentobarbital sodium [55]). Other sources of salient odorous substrates are the face, mouth, pedal, or ano-genital regions. If some regions are not of direct interest for the bioassay, care should be taken to avoid their inclusion in the presentation device.
2. *Odor-active biological substrates.* To achieve normal life-sustenance activities, any living organism produces innumerable chemosensorily active compounds conveyed in excretions, secretions and wastes. *Mus musculus* carries an extremely diverse network of scent producing sources that varies along all dimensions of biological functioning (e.g., individual, sex, age, kin, social status, physiological state, stress-anxiety, pathology and parasite load) (e.g., [85–87, 104]). Here, we highlight some of the biological substrates that are of interest in the context of infants nosing to their dams. But many other biological substrates from the dam, but also from the father or littermates, will be interesting to screen for behavioral activity of newborns.

(a) *Skin secretions.* The skin surface of adult mice emits a range of unspecified volatile profiles that pups can regionally discriminate. For example, they olfactorily differentiate back from ventrum. Smears of dorsal and abdominal skin secretions can be collected by rubbing one edge of a small petri dish on the mouse's abdomen [77], or by rubbing a humidified cotton-tip over dorsal fur or nipples [73].

(b) *Amniotic fluid.* At parturition, dams self-lick and thereby spread traces of amniotic fluid, blood, saliva on their abdominal fur and around nipples. This fluid is then avidly attended by newly born pups [56, 75, 88]. Its collection necessitates anesthesia and rapid euthanization of the female [75] before externalizing uterine horns and amniotic sacs using established surgical techniques. Amniotic fluid can then be aspired through the amnion (untainted with blood) with a 2-mL glass Pasteur pipette.

(c) *Saliva.* For the same reasons as amniotic fluid, saliva is omnipresent over the nursing dams' fur, and in addition is of both maternal and infant origin. Separate sampling of maternal and pup saliva consists in delicately introducing the tip of a glass stick or a humidified small cotton-tip into the animals' mouth (adult or pup; glass stick diameter: 0.4 cm and 0.1 cm, respectively) and let it there for 10–15 s [73, 74]. Saliva can also be collected from anesthetized dams injected with 0.1 mg pilocarpine to stimulate salivation [56].

(d) *Milk* is an obvious odorivector substrate in lactating females. What is less obvious is that milk is in constant change with advancing lactation, and with genotypes, parity, sex of offspring, mother's diet, etc. Thus, its collection should be done as a function of this compositional variability. It may be noted that pups respond preferentially to the odor of a milk collected in dams which lactation age matches their own age (i.e., PND-2 pups prefer milk of lactation day 2 rather milk from lactation day 15 [75]). To optimize the milking procedure (see also [90]), it is better to separate dams from their litter for 2 h. Donor dams are then anesthetized and injected intraperitoneally with 0.15 mL oxytocin, to stimulate milk ejection in association with gentle massages on the mammary areas (anesthesia does not appear to notably affect the odor of milk [74]). Milk is aspired from each nipple with a glass Pasteur pipette (2 mL) to be aliquoted into glass vessels kept on ice. Milk is then either immediately deep-frozen (−80 °C) or (much better) used fresh [73–75]. Milking should not be repeated on a same donor female during a same lactation cycle.

(e) *Urine, feces.* Females can be put on a clean glass surface (at ambient temperature) to stimulate spontaneous urination. Abdominal massage can ease micturition [56]. Fresh urine is then pipetted into glass vessels. Urine and feces are also often collected in a less controlled fashion, in using the soiled bedding material from the nest or after letting the animals defecate/urinate in a test box [84]. Soiled bedding is then often used to test olfactory discrimination, in contrasting own litter's vs. alien litter's beddings (e.g., [72, 83, 91]).

(f) *Sample conservation issues.* To limit alteration (by oxidization, evaporation) of the odor of the above substrates, collection operations should be as swift as possible and recipient vessels kept on ice during collection. Ideally, the substrates should be tested immediately after sampling. If not, immediate storing at –20 °C, or better at –80 °C, is mandatory (but even at this temperature samples remain instable and they should be used as soon as possible). It is advisable to aliquot the samples in smallest possible glass vessels in volumes necessary for a single test. To reduce individual variability of samples, and if individual variation is not an issue under test, pools of them should be constituted.

3. *Artificial odorants.* These odorants are of interest to understand olfaction in neonates for multiple reasons. First, artificial odorants are often used in social contexts to assess the effects of experience on emerging preferences, with the assumption that the learning processes involved with artificial odorants simulate those that operate in the natural situation. Such odor "probes" contributed, for example, to highlight that fetuses learn odors in utero [42, 92], that neonatal behavior depends on fetal learning [3, 12], and that neonates learn any odor associated with the mother as subsequently positive (e.g., [7, 93]). Numerous pure odorant compounds were used with mouse pups in such learning experiments associating them with nursing or other aspects of maternal care (ano-genital licking or its experimental simulation with soft paintbrush) (for example, *n*-hexanoic acid and citronellal [76], peppermint and lemon [79], +carvone and citral [72], benzaldehyde and limonene [94], or acetophenone and isopropyl tiglate [43]). Care should be taken when selecting such odorants that they bear little trigeminal properties and that they are as much as possible matched in intensity (at least for a panel of human noses). Second, artificial odorants constitute useful controls to test whether a response is specifically caused by a candidate chemosignal and does not result from nonspecific arousal or novelty effects that would be elicited by any odorant.

2.2.3 Perinatal and Neonatal Mice (PND 0–2)

1. *Test paradigms*

   Behavioral assays generally aim to examine an organism's discriminative responsiveness to stimuli that differ along qualitative or quantitative dimensions controlled by the experimenter. Optimally, they intend to elicit specific responses that the animals display spontaneously in a given natural situation, rather than only a nonspecific response (such as nasal flaring or sniffing). A diversity of methods has been achieved to investigate olfaction-related abilities in neonatal and infant mice. The most used paradigms rely on differential orientation responses of the body or head, or on responses that are more specific to neonates, involving oral reactivity (oral seizing of a nipple or a nipple-shaped object, sucking).

   (a) *Differential attraction.* In their initial stage of limited motility, newly born pups can be put in assays requiring head orientation without additional motor demand. Such assays, first used in neonatal rats [95] were applied to mouse pups on PND 0, several minutes or hours after birth, or on PND 1–2 [74, 77]. Their principle is that the pup is set between two odorant stimuli and ensuing head-turning responses are timed in latency and duration. These paired-choice tests can be run in two modes, either in contrasting two odorant stimuli (O1 vs. O2) to assess *relative orientation*, or in contrasting an odorant and a control stimulus (O1 vs. Control) to assess *absolute orientation*. To make sure that the outcome of a relative orientation assay results from attraction to O1 and not from avoidance of O2, it should always be arranged with two absolute orientation assays testing either odorants against a scentless control. Thus, absolute attraction assays evaluate in the same time detection and intrinsic attraction/avoidance of an odorant against the control, and help interpreting the results of relative attraction assays.

   (b) *Oral responses.* Beyond head orientation, newborns display oral behavior toward certain odor cues, interpreted as reflecting attraction and appetence. Following a short sequence of bilateral rooting motions, these oral responses involve jaw extension, grasping and initiating rhythmic sucking actions. An oral response assay initially developed in the rat newborn consisted in the presentation of a natural nipple to elicit oral grasping and sucking [54, 89]. This paradigm was also applied to the mouse newborn [55, 56, 75].

2. *Devices and procedures*

   (a) *Head-orientation assays.* Several variants of head-orientation assays were developed. One variant consisted in a heating plate (Gestigkeit, Germany) covered with a blotting paper on which two scentless auto-adhesive labels were stuck symmetrically, at a distance of 2 mm (for PND-0 pups) or 3 mm (for PND-2 pups) on each side of a midline. A certain volume (5–10 μL) of each stimulus was pipetted on either label (Fig. 1) [74]. Then, the assay began when the body and head of the pup were aligned along the midline of the device. The assays lasted 3 min for PND-0 pups and or 2 min for PND-2 pups, during which they were videotaped. In subsequent analyses of these videotapes, the pup orientation toward each odor stimulus was timed using a PC-based recording software (e.g., The Observer, Wageningen, The Netherlands). The tested pup was considered to be oriented toward a stimulus when both of its nares crossed the midline of the device in direction of that stimulus [74]. In the other variant, two scented dishes were placed 2.5 cm apart at equidistance from a line drawn on clean tissue paper [77]. The pup was immobilized between the tips of the experimenter's thumb and index finger and aligned on the line at a distance of 4 cm from the odor sources. It was held in this way until its head and shoulders moved enough that their nose touched one of the dishes, signifying the choice [77]. These assays came out with latency to orient, total duration spent oriented to each odor stimulus, or the proportion of litters in which pups chose one odor stimulus.

(b) *Oral response assays.* Nipple attachment and suckling assays were adapted from tests on rat pups by Teicher and Blass [54, 89] to assess neonatal mice on PND 1–2 and earlier when exposed to a nipple in the context of a female's abdomen. Prior to the assay, nipples of stimulus-females were treated to suppress (washing, masking) their natural odor, as indicated by the abolishment of the typical oral

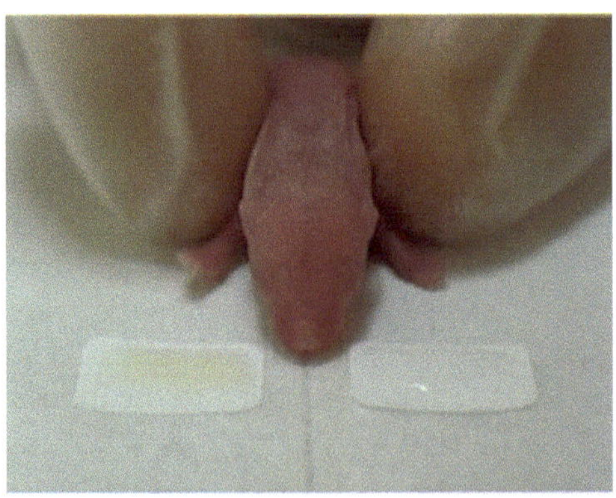

Fig. 1 Head-orientation assay adapted for mouse pups aged postnatal day 0–2. For the testing, the device is set on a heating plate to keep the pup warm. From [74]; reproduced by authorization from Public Library of Science

grasping and suckling responses of pups. In studies using lactating dams, the washing procedure consisted in rubbing nipples for 30 s with 50 μL distilled water and drying them with a cotton swab [56]. Other studies avoided the nipple washing step in using non-lactating females whose nipples were shown to be naturally behaviorally inert to newborn pups [75]. When neutral nipples are achieved by either method, different target substrates can be painted onto them by rotating a glass stick soaked with the substrate for 10 s [75] or by painting them with a soft brush [56]. Whether the substrates of interest reinstate the typical form and level of response is then directly tested in presenting the treated nipples to the pups.

Two options were used to present pups to nipples. First, stimulus-females were anesthetized and laid supine, and the pups were vertically held by an experimenter at 1 cm from the female's nipple for 2 min to let them approach and orally seize the nipple [56]. In the second option, the stimulus-females were sedated and hand-held by an experimenter A by pinching the inter-scapular skin between thumb and index to vertically expose their abdomen (Fig. 2) [75]. The pups to be tested were held in the same way by an experimenter B. The 2-min test began when a pup's muzzle was set in slight contact with the nipple of the stimulus-female. Both assays were videotaped for subsequent blind coding of latency to

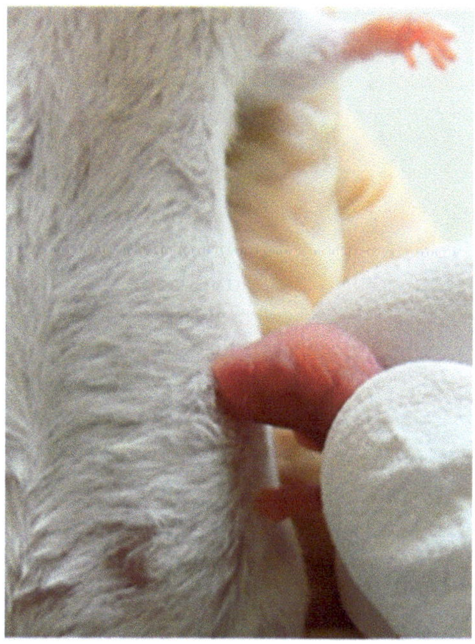

Fig. 2 Nipple seizing assay for newly born mouse pups. A 1-day-old pup is presented to a nipple in the context of an unfamiliar, behaviorally neutral non-lactating female's abdomen. From [75]; reproduced by authorization from Wiley

the first nipple grasping [56], total duration of nipple grasping and frequency of mouth actions (sucking) during grasping [75]. A device eliciting oral grasping of an artificial (plastic) nipple was developed to investigate fetal and neonatal rats [96–98], and should be feasible in mouse newborns as well.

2.2.4 Previsual Mouse Pups (PND 3–14)

1. *Test paradigms*
   During the first week, mouse pups become rapidly mobile. By PND 3 they are sufficiently active to be introduced into test arenas for short assays measuring differential attraction towards stimuli. By that age, they also emit increasingly frequent ultrasonic vocalizations in response to isolation from the dam/litter/nest. These vocalizations are attenuated or stopped when returned to the normal situation or when familiar odor stimuli are restituted.
   (a) *Differential attraction.* Several studies have set up double-choice tests to evaluate differential exploration of social odorants or of artificial odorants with which pups had distinctive experience (e.g., on PND 6 [73, 74]; on PND 2–P12 [91]).
   (b) *Detection—differential stress buffering effect.* The relief effect familiar odors have on ultrasonic vocalizations expressed in stressed pups was used as an index of detection and differential stress buffering potency of odors in several recent studies (e.g., on PND 8 [72], PND 2–P12 [91]).
2. *Devices and procedure*
   (a) *Differential investigation: paired-odor choice test.* The most usual test device used to investigate attraction/preference behavior in rodents is the double-choice arena. Several such arenas have been successfully applied in infant mice. The principle is based on a rectangular polypropylene box with a mesh floor above the bottom, under which both stimuli are presented (Fig. 3). Several dimensions of the box were used, some designed for younger pups (e.g., Length × width × height: 7.1 × 5 × 7 cm [73, 74]) and some for older pups (e.g., 20 × 36 × 10 cm [78]; 19 × 32 × 13 cm [76]). Under the mesh floor, two equal compartments are separated by a barrier to avoid mixing of odors. It may be practical to delimit a neutral zone below the floor to increase the separation of both odor stimuli (as, e.g., in [78, 79]). Also, to limit the mixing of odors that vertically convect above the mesh floor, it is better not to cover the test arena. In certain cases (when only minute quantities of the stimulus are available), both stimuli can be presented in the corners of the arena above the mesh floor to take advantage of pups' tendency to explore corners [73]. The stimuli can be composed of live animals (Fig. 3) to assess

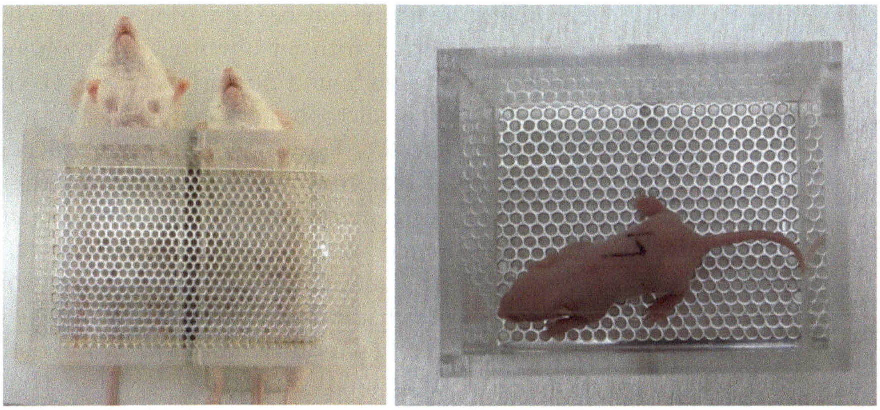

Fig. 3 Two-odor choice-arenas used to assess differential investigation of paired odor stimuli in mouse pups aged 6 days old. During testing, the device is set on a heating plate to keep the pup warm. *Left*: On the bottom of the device, below the grid, two live (anesthetized) mice are presented simultaneously to assess pups' differential investigation of adult conspecifics in different physiological stages. From Al Aïn et al. [73]; reproduced by authorization from Wiley. *Right*: Mouse pup roaming in the device on the bottom of which two stimuli are presented below the grid, placed symmetrically from the midline. From [74]; reproduced by authorization from Public Library of Science

their relative attraction as a function of physiological state, or of biological substrates (disposed in one or several petri dish(es) set symmetrically relative to the midline).

Once stimuli are placed beneath the mesh, the pup is placed on the midline of the arena, and its roaming in the two halves of the arena can be videotaped for 2 min [73], 3 min [76, 78], 6 min [91], or 5 × 1 min [79]. Subsequently the duration of investigative roaming in either scented compartment of the arena is blindly analyzed. A clear, reproducible criterion of orientation should be decided: an animal was considered oriented over one half of the arena when its entire muzzle including both eyes crossed the midline [73, 74] or when the anterior snout of the animal was in that half of the apparatus [91]. When the stimuli were smeared on corners of the test arena, the orienting criterion toward a stimulus corner was when the tip of the pup's muzzle was within the 1-cm radius to it (a line being drawn on the TV screen for analysis) [73].

(b) *Ultrasonic vocalization test*. Each tested pup is placed in a chamber where odorants are delivered in a time-controlled fashion. The stimuli can be conveyed by an olfactometer or, more simply, by petri dishes placed under the mesh set above the bottom of the chamber [72, 91]. The test consists of an initial isolation period of 90 s for baseline recording of ultrasonic vocalization (USV) followed by 60-s odor exposure [72, 91]. The baseline level of USV is recorded some time after introducing the pup in the test chamber to avoid the enhanced emission related to handling. USV are detected with an ultrasonic microphone (SM2 Microphone, Ultra Sound Advice) located at 10 cm above of the pup. This microphone is coupled to a bat detector (S25 Bat Detector, Ultrasound advice; frequency range 10–180 kHz) that converts ultrasounds into the audible frequency range. Afterwards, USV are extracted and automatically counted using an in-house software (cf. [85]). The rate of USV emitted during the last 30 s of isolation is compared with that emitted during the first 30 s of odor exposure [72, 91]. Another procedure [71] consists in a 2-min odor-free familiarization period, followed by a 1-min odor exposure in placing a petri dish containing an odor stimulus under the mesh, and next followed by replacing this first dish by a dish containing the other odor for 1 min. The total duration of USVs during each 1-min odor exposure was measured.

2.2.5 Visual Preweaning Mouse Pups (PND 14–21)

After PND 14, mouse pups have mobility and reactiveness levels that compare to those of adults. They begin to gnaw at nest materials and solid foods, to be fully weaned by PND 21. As early as PND14, and increasingly so around weaning, they can thus be submitted to a variety of bioassays devised for adult mice. Pups can then investigate their environment in moving around, in approaching a target area or object, in digging, in feeding or drinking. Different kinds of situations include a*ttraction-preferences assays* evaluating differential investigation of social odors sources presented at opposed sides of two-choice arenas or Y-mazes, or in water or feeding trays. In such assays, when one odor stimulus is not differentiated from the other, explanations may pertain to non-preference because the stimuli were not detected, because they were not discriminated, or finally because they had equivalent reinforcing value despite they were discriminated. Thus, in addition to differential attraction or preference assays, d*iscrimination assays* are required to reach a conclusion on whether two odorants are differentiated although they bear similar reinforcing value (*see*, e.g., [99, 102]). At this age (and perhaps even earlier), mouse pups can also be subjected to assays evaluating conditioned preferences (*see*, e.g., [99, 102]).

Here, we will only summarize assays that evaluate differential attraction. Multiple versions of two-odor choice assays have been designed to test differential investigative responses in preweanling mice. One case is an arena made of a rectangular polycarbonate/ Plexiglas box (16 × 11 × 13.5 cm [74]; 29 × 22 × 10 cm [83]) with a stainless steel mesh placed 0.7 cm above the bottom. Under the mesh, two equal compartments (or three compartments of which one is a neutral area), are delimited by a polypropylene barrier (Fig. 4) [74, 83]. Both stimuli are either directly inserted into each compartment [83], or carried by rectangular polypropylene plates (7.8 × 11 cm, 0.2 cm thickness [74]). At the beginning of each test,

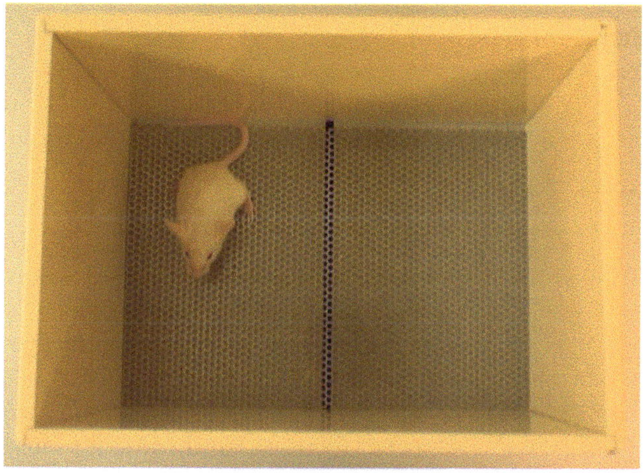

Fig. 4 The test box used to assess differential attraction to paired odor stimuli for mouse pups aged 15 days. From [75]; reproduced by authorization from Wiley

the pup is placed along the midline of the arena between the two test odors. Each mouse is allowed to move freely in the arena, while it is videotaped for 2 min [74] or 3 min [83] to code the duration of investigation in each compartment. A mouse pup is defined as investigating one of the odors when its entire muzzle has crossed the midline [74] or its snout and forepaws have entered the area [83]. A pup is excluded from further analyses when it performed less than three midline crossings or when it urinated [83].

In other studies, the choice device takes the form of a Y-maze [94]. Petri dishes containing odorized filter paper are placed at the end of each short arms of the Y-maze (height: 13 cm, width: 6.5 cm, long arm length: 21 cm, short arms' length: 15 cm). Both dishes have a porous cap in order to prevent any contact with the odorant. The tested pup is placed in the neutral zone of the Y-maze and left to roam for 3 min while videotaped. The amount of time the pup spent investigating each short arm is recorded. Another interesting device has been designed to automatically record the choice behavior of adult mice [99]. It is constituted of a hole-board (40 × 40 cm) in which five holes are opened to receive glass vessels in which odorized swabs are placed and covered with bedding material. Depending on the test 2–5 holes can be made accessible to the tested animal, so that different paradigms can be applied, including detection tasks (in varying the concentrations of the odorant), discrimination tasks (using a habituation–dishabituation procedure), memory tasks, and preference tasks. When the mouse introduces its muzzle into a hole, sensors connected to a computer are triggered to time duration of sniffing and digging activity [99]. So far, this apparatus has not been used with young mice, but it holds promise to investigate the development of differential responsiveness to social odors.

3 Conclusion

How limited our knowledge is on odor-regulated behaviors in mammalian infants is astounding in regard of their critical importance for survival and typical development. After the promising wave of data gathered by developmental psychobiologists in the 1970s and 1980s (e.g., [1, 6, 8, 26, 100]), the area is worthy of rekindled interest. Such effort promises indeed to be scientifically rewarding to understand proximate mechanisms of chemo-emission and chemoreception in murine rodents, especially in a time of explosive genetic and molecular neurogenetic research in relation with chemosensory processes in mice. The present methodological survey focusing on the infant mouse will undoubtedly be enriched by borrowing additional methods and ideas from similar research on other rodents (for reviews, e.g., [100–102]) and other animals at large (e.g., [103, 104]).

Acknowledgments

This work was supported by the Centre National de la Recherche Scientifique (INSB, INEE) and the Regional Council of Burgundy.

References

1. Blass EM, Teicher MH (1980) Suckling. Science 210:15–22
2. Schaal B (2010) Mammary odor cues and pheromones: mammalian infant-directed communication about maternal state, mammae, and milk. Vitam Horm 83:81–134
3. Alberts JR, Ronca AE (2012) The experience of being born: a natural context for learning to suckle. Int J Pediatr. doi: 10.1155/2012/129328
4. Blass EM (ed) (1986) Handbook of behavioral neurobiology. Developmental psychobiology and behavioral ecology, vol 9. Plenum, New York
5. Blumberg MS, Freeman JH, Robinson SR (eds) (2010) Oxford handbook of developmental behavioral neuroscience. Oxford University Press, New York
6. Rosenblatt JS (1983) Olfaction mediates developmental transitions in the altricial newborn of selected species of mammals. Dev Psychobiol 16:347–375
7. Brake SC, Shair H, Hofer MA (1986) Exploiting the nursing niche: the infant's sucking and feeding in the context of the mother-infant interaction. In: Blass EM (ed) Handbook of behavioral neurobiology, vol 9, Developmental psychobiology and behavioral ecology. Plenum, New York, pp 347–388
8. Alberts JR (1987) Early learning and ontogenetic adaptation. In: Krasnegor NA, Blass EM, Hofer MA, Smotherman WP (eds) Perinatal development: a psychobiological perspective. Academic, Orlando, FL, pp 11–37
9. Smotherman WP, Robinson SR (1987) Psychobiology of fetal experience in the rat. In: Krasnegor NA, Blass EM, Hofer MA, Smotherman WP (eds) Perinatal development: a psychobiological perspective. Academic, Orlando, FL, pp 39–60
10. Schaal B, Orgeur P (1992) Olfaction in utero: can the rodent model be generalized? Quart J Exp Psychol 44B:245–278
11. Youngentob SL, Kent PF, Sheehe PR et al (2007) Experience-induced fetal plasticity: the effect of gestational ethanol exposure on the behavioral and neurophysiologic olfactory response to ethanol odor in early postnatal and adult rays. Behav Neurosci 121: 1293–1305
12. Pedersen PA, Blass EM (1982) Prenatal and postnatal determinants of the 1st suckling

episode in albino rats. Dev Psychobiol 15: 349–355
13. Schaal B (2005) From amnion to colostrum to milk: odor bridging in early developmental transitions. In: Hopkins B, Johnson SP (eds) Prenatal development of postnatal functions. Praeger, London, pp 51–102
14. Calamandrei G (2004) Ethological and methodological considerations in the use of newborn rodents in biomedical research. Ann Ist Super Sanita 40:195–200
15. Maynard-Smith J, Harper D (2003) Animal signals. Oxford University Press, Oxford, UK
16. Hauser MD (ed) (1996) The evolution of communication. MIT Press, Cambridge, MA
17. Karlson P, Lüscher M (1959) "Pheromones": a new term for a class of biologically active substances. Nature 183:55–56
18. Beauchamp GK, Doty RL, Moulton DG et al (1976) The pheromone concept in mammals: a critique. In: Doty RL (ed) Mammalian olfaction, reproductive processes, and behavior. Academic, New York, pp 143–160
19. Johnston RE (2000) Chemical communication and pheromones: The types of chemical signals and the role of the vomeronasal system. In: Finger TE, Silver WL, Restrepo D (eds) The neurobiology of taste and smell. Wiley, New York, pp 101–127
20. Doty RL (2003) Mammalian pheromones: fact or fantasy? In: Doty RL (ed) Handbook of olfaction and gustation, 2nd edn. Marcel Dekker, New York, pp 345–383
21. Schaal B (2008) Social odors and pheromones in mammals. Biofutur 27:41–45
22. Wyatt TD (2010) Pheromones and signature mixtures: defining species-wide signals and variable cues for identity in both invertebrates and vertebrates. J Comp Physiol A. doi:10.1007/s00359-010-0564-y
23. Doty RL (2010) The great pheromone myth. John Hopkins University, Baltimore, MD
24. Pedersen PA, Blass EM (1981) Olfactory control over suckling in albino rats. In: Aslin RN, Alberts JR, Petersen MR (eds) Development of perception, psychobiological perspectives, vol 1, Audition somatic perception and the chemical senses. Academic, New York, pp 359–381
25. Smotherman WP, Robinson SR (1992) Dimethyl disulfide mimics the effect of milk on fetal behavior and responsiveness to cutaneous stimuli. Physiol Behav 52:761–765
26. Blass EM (1990) Suckling: determinants, changes, mechanisms, and lasting impressions. Dev Psychol 26:520–533
27. Schley P (1979) Olfaction and suckling behavior in young rabbits. In: Myers K, MacInnes CD (eds) Proceedings of the 1st world lagomorph conference. University of Guelph, Guelph, Canada, pp 291–294
28. Hudson R, Distel H (1983) Nipple location by newborn rabbits: Evidence for pheromonal guidance. Behaviour 82:260–275
29. Keil W, von Stralendorff F, Hudson R (1990) A behavioral bioassay for analysis of rabbit nipple-search pheromone. Physiol Behav 47: 525–529
30. Schaal B, Coureaud G, Langlois D et al (2003) The mammary pheromone of the rabbit: chemical and behavioural characterisation. Nature 424:68–72
31. Coureaud G, Langlois D, Perrier G et al (2003) A single key-odorant accounts for the pheromonal effect of rabbit milk: Further test of the mammary pheromone's activity against a wide sample of volatiles from milk. ChemoEcol 13:187–192
32. Coureaud G, Langlois D, Sicard G et al (2004) Newborn rabbit reactivity to the mammary pheromone: concentration-response relationship. Chem Senses 29:341–350
33. Coureaud G, Rödel H, Kurz CA et al (2008) Age dependent responsiveness to the mammary pheromone in domestic and wild rabbits. ChemoEcol 18:52–59
34. Coureaud G, Fortun-Lamothe L, Langlois D et al (2007) The reactivity of neonatal rabbits to the mammary pheromone as a probe for viability. Animal 1:1026–1032
35. Montigny D, Coureaud G, Schaal B (2006) Shift from automatism to prandial control in the response of newborn rabbits to the mammary pheromone. Physiol Behav 89:742–749
36. Coureaud G, Moncomble AS, Montigny D et al (2006) A pheromone that rapidly promotes learning in the newborn. Curr Biol 16:1956–1961
37. König B, Markl H (1987) Maternal care in house mice. I. The weaning strategy as a means for parental manipulation of offspring quality. Behav Ecol Sociobiol 20:1–9
38. Ehret G (1976) Development of absolute auditory thresholds in the house mouse (*Mus musculus*). J Am Audiol Soc 1:179–184
39. Weber EM, Olsson IA (2008) Maternal behaviour in Mus musculus sp. An ethological review. Appl Anim Behav Sci 114:1–22
40. Breer H, Fleischer J, Strotmann J (2006) The sense of smell: multiple olfactory subsystems. Cell Mol Life Sci 63:1465–1475
41. Rodriguez I, Boehm U (2008) Pheromone sensing in mice. Results Probl Cell Differ. doi:10.1007/400_2008_8
42. Coppola DM, Millar LC (1997) Olfaction in utero: behavioral studies of the mouse fetus. Behav Proc 39:53–68

43. Todrank J, Heth G, Restrepo D (2010) Effects of in utero odorant exposure on neuroanatomical development of the olfactory bulb and odor preferences. Proc Roy Soc B 278:1949–1955
44. Coppola DM, O'Connell RJ (1989) Stimulus access to olfactory vomeronasal receptors in utero. Neurosci Lett 106:241–248
45. Coppola DM, Budde J, Millar L (1993) The Vomeronasal duct has a protracted postnatal development in the mouse. J Morphol 218: 59–64
46. Tarozzo G, Cappello P, De Andre M et al (1998) Prenatal differentiation of mouse vomeronasal neurones. Eur J Neurosci 10: 392–396
47. Fox WM (1965) Reflex-ontogeny and behavioural development of the mouse. Anim Behav 13:234–241
48. Crawley JN (2000) What's wrong with my mouse? Behavioral phenotyping of transgenic and knockout mice. Wiley-Liss, New York
49. Kovach JK, Kling A (1967) Mechanisms of neonate sucking behaviour in the kitten. Anim Behav 15:91–101
50. Singh PJ, Tobach E (1975) Olfactory bulbectomy and nursing behavior in rat pups. Dev Psychobiol 8:151–164
51. Teicher MH, Flaum LE, Williams M et al (1978) Survival, growth and suckling behavior of neonatally bulbectomized rats. Physiol Behav 21:553–561
52. Hofer MA, Shair H, Singh P (1976) Evidence that maternal ventral skin substances promote suckling in infant rats. Physiol Behav 17: 131–136
53. Wong ST, Trinh K, Hacker B et al (2000) Disruption of the type III adenylyl cyclase gene leads to peripheral and behavioral anosmia in transgenic mice. Neuron 27:487–497
54. Teicher MH, Blass EM (1976) Suckling in newborn rats: eliminated by nipple lavage, reinstated by pup saliva. Science 193:422–425
55. Hongo T, Hakuba A, Shiota K et al (2000) Suckling dysfunction caused by defects in the olfactory system in genetic arhinencephaly mice. Biol Neonate 78:293–299
56. Logan DW, Brunet LJ, Webb WR et al (2012) Learned recognition of maternal signature odors mediates the first suckling episode in mice. Curr Biol. doi:10.1016/j.cub.2012.08.041
57. Butcher RE, Vorhees CV (1979) A preliminary test battery for the investigation of the behavioral teratology of selected psychotropic drugs. Neurobehav Toxicol 1:207–212
58. Adams J (1985) Collaborative behavioral teratology study: protocol design and testing procedures. Neurobehav Toxicol Teratol 7: 579–586
59. Crawley JN, Paylor R (1997) A proposed test battery and constellations of specific behavioral paradigms to investigate the behavioral phenotypes of transgenic and knockout mice. Horm Behav 31:197–211
60. Brown RE, Stanford L, Schellinck HM (2000) Developing standardized tests for knockout and mutant mice. ILAR J 41:163–174
61. Schellinck HM, Forrestel CA, LoLordo VM (2001) A simple and reliable test of olfactory learning and memory in mice. Chem Senses 26:663–672
62. van Meer P, Raber J (2005) Mouse behavioural analysis in systems biology. Biochem J 389:593–610
63. Youngentob SL (2005) A method for the rapid automated assessment of olfactory function. Chem Senses 30:219–229
64. Cuomo V, De Salvia MA, Petruzzi S et al (1996) Appropriate end points for the characterization of behavioral changes in developmental toxicology. Environ Health Perspect 104:307–315
65. Branchi I, Ricceri L (2002) Transgenic and knock-out mouse pups: the growing need for behavioral analysis. Genes Brain Behav 1: 135–141
66. Clough G (1982) Environmental effects on animals used in biomedical research. Biol Rev 57:487–523
67. Berry R, Bronson FH (1992) Life history and bioeconomy of the house mouse. Biol Rev 67:519–550
68. Smith JS (1981) Mouse model for abortive rabies infection in the central nervous system. Infect Immun 16:297–308
69. Elwood RW, Keeling F (1982) Temporal organization of ultrasonic vocalizations in infant mice. Dev Psychobiol 15:221–227
70. Branchi I, Santucci D, Vitale A et al (1998) Ultrasonic vocalizations by infant laboratory mice: a preliminary spectrographic characterization under different conditions. Dev Psychobiol 33:249–256
71. Branchi I, Santucci D, Alleva E (2001) Ultrasonic vocalisation emitted by infant rodents: a tool for assessment of neurobehavioural development. Behav Brain Res 125:49–56
72. Lemasson M, Delbé C, Gheusi G et al (2005) Use of ultrasonic vocalizations to assess olfactory detection in mouse pups treated with 3-methylindole. Behav Proc 68:13–23
73. Al Aïn S, Chraïti A, Schaal B et al (2011) Orientation of newborn mice to lactating females: biological substrates of semiochemical interest. Dev Psychobiol. doi:10.1002/dev.21003

74. Al Aïn S, Belin L, Patris B et al (2012) An odor timer in milk? Synchrony in the odor of milk effluvium and neonatal chemosensation in the mouse. PLoS One 7:e47228
75. Al Aïn S, Belin L, Schaal B et al (2012) How does a newly born mouse get to the nipple? Odor substrates eliciting first nipple grasping and sucking responses. Dev Psychobiol. doi:10.1002/dev.21082
76. Armstrong CM, DeVito LM, Cleland TA (2006) One-trial associative odor learning in neonatal mice. Chem Senses 31:343–349
77. Todrank J, Busquet N, Baudoin C et al (2005) Preferences of newborn mice for odours indicating closer genetic relatedness: is experience necessary? Proc Roy Soc B 272: 2083–2088
78. Alleva E, Calamandrei G (1986) Odor-aversion learning and retention span in neonatal mouse pups. Behav Neural Biol 46:348–357
79. Bouslama M, Durand E, Chauviere L et al (2005) Olfactory classical conditioning in newborn mice. Behav Brain Res 161: 102–106
80. Hurst JL, West RS (2010) Taming anxiety in laboratory mice. Nat Methods 7:825–826
81. Durand E, Dauger S, Vardon G et al (2003) Classical conditioning of breathing pattern after two acquisition trials in 2-day-old mice. J Appl Physiol 94:812–818
82. Zorrilla EP (1997) Multiparous species present problems (and possibilities) to developmentalists. Dev Psychobiol 30:141–150
83. Thomas NR, Fonken LK, LeBlanc M et al (2010) Maternal separation alters social odor preference development in infant mice (Mus musculus). J Comp Psychol 124:295–301
84. Breen MF, Leshner AI (1977) Maternal pheromone: a demonstration of its existence in the mouse (*Mus musculus*). Physiol Behav 18: 527–529
85. Hurst JL, Beynon RJ (2004) Scent wars: the chemobiology of competitive signalling in mice. Bioessays 26:1288–1298
86. Brennan PA, Zufall F (2006) Pheromonal communication in vertebrates. Nature 444:308–315
87. Ferrero D, Liberles SD (2010) The secret codes of mammalian scents. WIREs Syst Biol Med 2:23–33
88. Kodama N (1990) Preference for amniotic fluid in newborn mice. Annual Meeting of the Int Soc Dev Psychobiol, Cambridge, UK
89. Teicher MH, Blass EM (1977) First suckling response of the newborn albino rat: the roles of olfaction and amniotic fluid. Science 198: 635–636
90. DePeters E, Hovey R (2009) Methods for collecting milk from mice. J Mammary Gland Biol 14:397–400
91. de Chevigny A, Lemasson M, Saghatelyan A et al (2006) Delayed onset of odor detection in neonatal mice lacking tenascin-C. Mol Cell Neurosci 32:174–186
92. SmothermanW P (1982) Odor aversion learning by the rat fetus. Physiol Behav 29:769–771
93. Fillion TJ, Blass EM (1986) Infantile experience with suckling odors determines adult sexual behavior in male rats. Science 231:729–731
94. Hellier JL, Arevalo NL, Smith L et al (2012) α7-Nicotinic acetylcholine receptor: role in early odor learning preference in mice. PLoS ONE 7(4):e35251
95. Hepper PG (1988) Adaptive fetal learning: prenatal exposure to garlic affects postnatal preferences. Anim Behav 36:935–936
96. Petrov ES, Varlinskaya EI, Smotherman WP (1997) The newborn rat ingests fluids through a surrogate nipple: a new technique for the study of early suckling behavior. Physiol Behav 62:1155–1158
97. Cheslock S, Varlinskaya E, Petrov E et al (2000) Rapid and robust olfactory conditioning with milk before suckling experience: promotion of nipple attachment in the newborn rat. Behav Neurosci 114:484–495
98. Smotherman WP, Goffman D, Petrov E et al (1997) Oral grasping of a surrogate nipple by the newborn rat. Dev Psychobiol 31:3–17
99. Mandairon N, Sultan S, Rey N et al (2009) A computer-assisted odorized hole-board for testing olfactory perception in mice. J Neurosci Methods 180:296–303
100. Shair HN, Barr GA, Hofer MA (eds) (1991) Developmental psychobiology. New methods and changing concepts. Oxford University Press, New York
101. Slotnick BM, Schellinck HM (2002) Methods in olfactory research with rodents. In: Simon SA, Nicolelis M (eds) Frontiers and methods in chemosenses. CRC, New York, pp 21–61
102. Schelling HM, Price SR, Wong JM (2008) Using ethologically relevant tasks to study olfactory communication in rodents. In: Hurst JL, Beynon RJ, Roberts SC, Wyatt TD (eds) Chemical signals in vertebrates, vol 11. Springer, New York, pp 71–80
103. Haynes KF, Millar JG (eds) (1998) Methods in chemical ecology, vol 2, Bioassay methods. Klüwer Academic, Norwell, MA
104. Brown RE (1985) The rodents II: suborder Myomorpha. In: Brown RE, Macdonald DW (eds) Social odours in mammals, vol 1. Clarendon, Oxford, pp 345–457

Chapter 27

An Assay for Human Chemosignals

Idan Frumin and Noam Sobel

Abstract

Like all mammals, humans use chemosignals. Nevertheless, only few such chemosignals have been identified. Here we describe an experimental arrangement that casts a wide net for the possible chemosignaling functions of target molecules. This experimental arrangement can be used in concert with various methods for measuring human behavioral and brain responses, including psychophysiology and brain imaging. Moreover, many of the methodological issues we describe are relevant to any study with human chemosignals.

Key words Human, Chemosignals, Pheromones, fMRI, Psychophysiology, Psychophysics

1 Introduction

All mammals communicate using chemosignals, and humans are no different. Mammalian chemosignaling is especially prominent in reproduction-related behaviors, and this too is true for humans. For example, the clearest case of chemical communication in humans is the phenomenon of menstrual synchrony, whereby women who live in close proximity, such as roommates in dorms, synchronize their menstrual cycle over time [1]. This effect is mediated by an odor in sweat. This was verified in a series of studies where experimenters obtained underarm sweat extracts from *donor* women during either the ovulatory or follicular menstrual phase. These extracts were then deposited on the upper lips of *recipient* women, where follicular sweat accelerated ovulation, and ovulatory sweat delayed it [2, 3]. Moreover, variation in menstrual timing can be increased by the odor of other lactating women [4], or regulated by the odor of male hormones [5, 6].

A second human reproduction-related chemosignaling behavior relates to mate selection. The human genome includes a region called Human Leukocyte Antigen (HLA), or more broadly termed Major Histocompatibility Complex (MHC), which consists of many genes related to the immune system, in addition to olfactory

Kazushige Touhara (ed.), *Pheromone Signaling: Methods and Protocols*, Methods in Molecular Biology, vol. 1068, DOI 10.1007/978-1-62703-619-1_27, © Springer Science+Business Media, LLC 2013

receptor genes and pseudogenes. Several studies have found that women can use smell to discriminate between men as a function of similarity between their own, and the men's HLA alleles [7–11]. The "ideal" smell of genetic makeup remains controversial, yet most evidence suggests that women prefer an odor of a man with HLA alleles not identical to their own, but at the same time not too different [11, 12]. In turn, this preference may be for MHC heterozygosity rather than dissimilarity [13]. This chemosignaling dependent mate preference is plastic. For example, single women preferred odors of MHC-similar men, while women in relationships preferred odors of MHC-dissimilar men [14]. Moreover, olfactory mate preferences are influenced by the menstrual cycle [15–18] and by hormone-based contraceptives [7, 8, 19]. Finally, olfactory influences on mate preferences are not restricted to women. Men can detect an HLA odor different from their own when taken from either men or women odor donors, and rate the similar odor as more pleasant for both of the sexes [8, 13]. In addition, men preferred the scent of common over rare MHC alleles [13]. Moreover, unrelated to HLA similarity, male raters can detect the menstrual phase of female body odor donors. The follicular phase is rated as more pleasant and sexy than the luteal phase [18], an effect that is diminished when the women use hormonal contraceptives [13, 20].

These behavioral results are echoed in hormone expression. Men exposed to the scent of an ovulating woman subsequently displayed higher levels of testosterone than did men exposed to the scent of a non-ovulating woman or a control scent [21]. Moreover, a recent study on chemosignals in human tears revealed a host of influences on sexual arousal [22]. Sniffing negative-emotion-related odorless tears obtained from women donors induced reductions in sexual appeal attributed by men to pictures of women's faces. Sniffing tears also reduced self-rated sexual arousal, reduced physiological measures of arousal, and reduced levels of testosterone (recently also seen by Oh and colleagues [23]). Finally, functional magnetic resonance imaging revealed that sniffing women's tears selectively reduced activity in brain substrates of sexual arousal in men [22].

Human chemosignaling is not restricted to reproduction-related behavior. Although many types of social chemosignaling have been examined [24], here we will detail one particular case, and that is the ability of humans to smell fear. Fear or distress chemosignals are prevalent throughout animal species [25, 26]. In an initial study in humans, Chen and Haviland-Jones [27] collected underarm odors on gauze pads from young women and men after they watched funny or frightening movies. They later asked other women and men to determine by smell, which was the odor of people when they were "happy" or "afraid." Women correctly identified happiness in men and women, and fear in men. Men correctly identified happiness in women and fear in men. A similar result was later obtained in a study that examined women only [28]. Moreover, women had improved performance in a cognitive verbal task after smelling fear sweat versus neutral sweat [29], and the smell of fearful sweat biased women toward interpreting ambiguous expressions as more fearful, but had no effect when the facial emotion was more discernible [30]. Also, subjects had an increased startle reflex when exposed to anxiety-related sweat versus sports-related sweat [31]. Finally, imaging studies have revealed dissociable brain representations after smelling anxiety sweat versus sports-related sweat [32]. These differences are particularly pronounced in the amygdala, a brain substrate common to olfaction, fear responses, and emotional regulation of behavior [33]. Taken together, this body of research strongly suggests that humans can discriminate the scent of fear from other body odors, and it is not unlikely that this influences behavior.

How can we assay whether a given substance is, or contains, a human chemosignal? The rational for how to do this is simple: Once we have identified a chemosignal and a behavior we think it relates to, we can measure that behavior, or its neural substrates, with and without exposure to the chemosignal. Thus, if we have identified a fear-related chemosignal, we can measure fear in the presence of the chemosignal versus the presence of an unrelated control substance, or measure brain activity in the amygdala for example, again in the presence of the chemosignal versus the presence of an unrelated control substance. However, in many cases we may have a potential chemosignal in hand without a clear notion regarding its expected influence. With this in mind, we have developed a behavioral assay that provides a rather widely cast net. This is an experiment aimed at probing for a host of potential psychological, physiological, and brain responses. In this chapter we will describe this assay. The human responses within this assay can be measured with several standard methods, for example psychophysiology and brain imaging. The general application and analysis of psychophysiology and brain imaging has been detailed in various chapters of this series [34–37], and is indeed beyond the scope of one text. Therefore, here we will concentrate on the unique aspects of assaying human chemosignals with these methods. To reiterate, this chapter is not intended to teach psychophysiology and brain imaging, but rather how to bring human chemosignals into this environment. We also detail various alternatives regarding major design aspects of such experiments. For example, for stimuli one can use the full human-derived media that presumably contains the chemosignals (sweat, tears, etc.), or individual synthetic molecules that have been dubbed putative human chemosignals. Also, one can measure the direct impact of the stimuli, or the influence of the stimuli on some task, such as emotional appraisal, or startle response. An additional major design aspect with extreme alternatives has to do with stimulus delivery. If one wants to deliver the stimuli with millisecond temporal resolution, one needs an olfactometer. These devices, which can be self-built [38–40] or bought [41], are complicated and expensive, and are therefore in the hands of relatively few labs. With this in mind, in this chapter we will restrict our description to methods that do not call for an olfactometer. Finally, Subheading 4 contains various insights from our experience with apparently small decisions that sometimes make all the difference between a successful and unsuccessful experiment. We wish you good luck with yours.

2 Materials

2.1 General

1. Ethical approval for the procedures from appropriate authorities (Helsinki or IRB committee).
2. Human volunteers (~30 per study).
3. General questionnaires. These should include a comprehensive demographics questionnaire, and the Ekman mood questionnaire [42] (*see* **Note 1**). Questionnaires should be made executable on-screen using presentation software (Fig. 1).
4. Well-ventilated room, subserved by Carbon and HEPA filtration, ideally coated with odorant non-adherent material such as stainless steel (*see* **Note 2**). The room should be observable from a neighboring control room through one-way mirror and/or video monitors such that subjects can be left alone in the room during the experiments. An intercom between experimental and neighboring control room is helpful.
5. A subject-chair that is both comfortable and adjustable, ideally a dentist-type patient chair (Fig. 2). The chair should be equipped with a wide armrest that can be refitted for either the left or right arm. This armrest is for the non-dominant hand

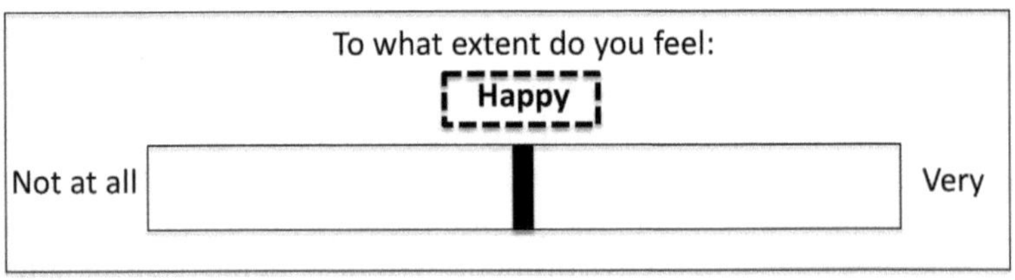

Fig. 1 Schematic of on-screen Visual Analog Scale (VAS). This graphic is presented on the screen in front of the subject. The subject uses the mouse to drag the marker horizontally to a position that reflects their self-assessment of the current mood in question (in this case "happy"). Once the marker is in the appropriate place, the subject clicks the mouse to enter their judgment, and the next mood question appears (e.g., "happy" is replaced with "sad"). This continues for the 17 mood questions (*see* **Note 1**). Although this may seem crude, it is in fact informative and reliable

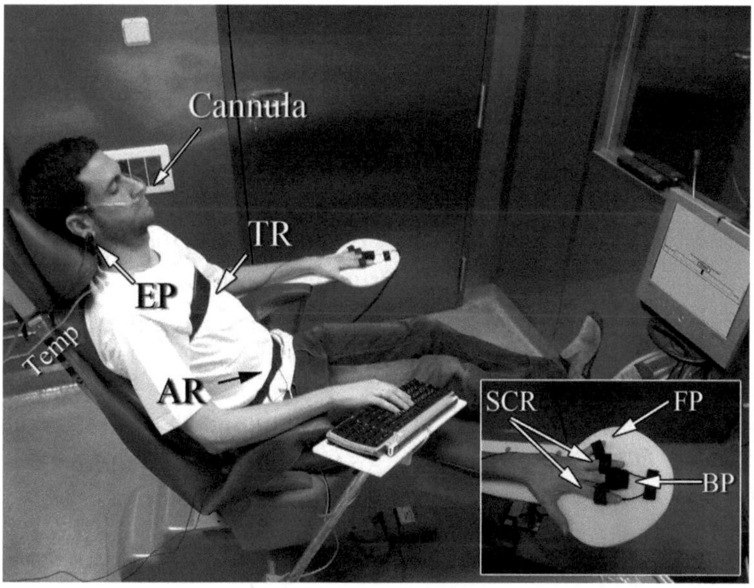

Fig. 2 Subject set-up in chair. Subject comfortably seated in stainless steel room. Visible transducers include body temperature (temp.), ear pulse (EP), nasal respiration (Cannula), thoracic respiration belt (TR), abdominal respiration belt (AR), and *inset* highlights skin conductance sensors (SCR), finger pulse (FP), and blood pressure (BP). Note monitor in easy viewing angle, and one-way mirror behind monitor, which allows viewing from neighboring experimenter control room

later fitted with physiological transducers. The opposing armrest should have a keyboard and mouse holder, to allow subject responses. Display hardware, e.g., computer monitor, should ideally be situated in comfortable viewing angle from the chair.

2.2 Collecting Body-Odor or Sweat

1. Scentless soap.
2. Cotton pads or cotton shirts.
3. Medical adhesive tape.
4. Sealable aluminum-lined plastic bags (sized to contain aforementioned pads/shirts) (*see* **Note 3**).
5. Refrigeration for the samples at 4 °C or below (*see* **Note 4**).
6. For sweat: An emotional setting for the active condition (*see* Subheading 3.1.2), and a treadmill/exercise bicycle for the control.

2.3 Collecting Tears

In all cases:

1. Vials/Tubes (preferably glass, wide opening (1–2 cm diameter)).
2. Saline/physiological solution, or ringer solution (*see* **Note 5**).

2.3.1 Emotional Tears

1. A computer or TV/DVD set for screening movies (headphones optional).
2. A sad movie (ideally chosen by the subject) (*see* **Note 6**).
3. A small make-up mirror, or similar.

2.3.2 Trigeminal/Reflexive Tears

1. Nasal endoscope (ideally <2 mm OD flexible) (*see* **Note 7**).
2. Glass Capillaries (50–200 μl) (e.g., Hirschmann Micropipettes or Drummond Microcaps).
3. A rubber bulb or micropipette pump to fit the aforementioned capillaries.

2.4 Preparing Synthetic Putative Chemosignals

1. Synthetic chemosignal, currently commonly used include:
   (a) Androstadienone (androsta-4,16,-dien-3-one; CAS# 4075-07-4) (*see* **Note 8**).
   (b) Estratetraenol (1,3,5(10),16-estratetraen-3-ol; CAS# 1150-90-9).
   (c) Androstenone (5α-androst-16-en-3-one ; CAS# 18339-16-7) (*see* **Note 9**).
2. Diluent—Most commonly propylene glycol (1,2 propanediol; CAS# 57-55-6) (*see* **Note 10**).
3. Clove oil/Eugenol (4-Allyl-2-methoxyphenol; CAS# 97-53-0) (*see* **Note 11**).
4. Analytical scale.
5. Chemical hood.
6. Glass vials and working tools such as spatula (*see* **Notes 2** and **17**).
7. Vortex and/or magnetic stirrer (*see* **Note 12**).

2.5 Delivering Chemosignals

1. Band-aids.
2. Pipette.
3. Opaque wide-mouthed jars (ideally glass) with both sealed and mesh covers.
4. Absorbent material, either cotton pads or PTFE beads (*see* **Note 13**).

2.6 Measuring Psychophysical and Psychophysiological Responses

1. A computer with stimulus-presentation software, either designated (e.g., Psyscope [43], E-Prime [44]) or programmable (e.g., Matlab [45], LabVIEW [46]). The computer should output to two separate monitors, one in front of the subject chair, and one in the experimenter control room.
2. Pre-rated sets of short emotional movie clips. We use 10-min segments of the nature film "Deep Blue" as "neutral," 5-min segments of the film "Nine and a half weeks" as "erotic," a 5-min segment from the movie "The Champ" as "sad," and 5-min segments from "Monty Python" as "funny" (*see* **Notes 6** and **14**).

3. A human approved psychophysiology rig (e.g., AD Instruments [47]) consisting of:
4. 16-channel (or more) biosignal amplifier.
5. Skin conductance (SCR) transducer with two bipolar finger Ag/AgCl Electrodes (surface: 1 cm^2).
6. Electrocardiogram transducer (ECG) with three circular Ag/AgCl conductive adhesive electrodes (0.9 cm diameter).
7. Finger and/or ear pulse transducer (FP/EP) using an IR plethysmograph (size: 15 mm × 15 mm × 6.3 mm).
8. Skin surface temperature transducer (ST) made of a small ceramic-encapsulated metal oxide semiconductor (9.5 mm in length, 2 mm in diameter), designed to operate from 0 to 50 °C.
9. Abdominal and thoracic respiration belt transducers (AR and TR) (30 cm rest length, 10 cm maximum elongation, 4.5 cm in width) mounted on Velcro belts. The transducers contain a piezoelectric device that responds linearly to changes in length (sensitivity, 4.5 ± 1 mV/mm).
10. Motion transducer attached to the subject chair (e.g., high-sensitivity (2,500 mV/g) (EGCS, Entran Devices, Fairfield, NJ [48])) (*see* **Note 15**).
11. Continuous blood-pressure (BP) finger-cuff monitor (e.g., Finapres Ohmeda 2300 [49]) (*see* **Note 16**).
12. Nasal airflow transducer/spirometer and nasal cannula to fit.
13. Biosignal presentation and analysis software (e.g., AD Instruments Chart [47]).

2.7 Measuring Endocrine Responses

1. Bioassy kits for measuring hormones in saliva (e.g., Salimetrics [50] Testosterone and Cortisol kits [51, 52]).
2. Collection tubes (e.g., 15 ml conical PP tubes). Optional—short straws.
3. Refrigerator + Freezer.
4. Polypropylene (PP) Microtubes (e.g., Eppendorf conical 1.5 ml microtubes).
5. Centrifuge (suitable for microtubes (1.5 ml)).
6. Precision pipettes for 20 μl range and for 200 μl range.
7. Multichannel pipette for 200 μl range.
8. Plate agitator/rotator.
9. Spectrophotometric plate reader, to match kit instructions (e.g., for Salimetrics kits—450 nm filter).
10. Computer software capable of 4-parameter sigmoid minus curve fit (4PL logistics non-linear regression) (e.g., SigmaPlot [53], Microsoft Excel Solver tool [54] or MyAssays add-in [55], Matlab [45]).

2.8 Imaging Neural Responses

1. An MRI machine (either 1.5 or 3 T) with gradient capability for functional MRI (fMRI) and appropriate head-coil for functional brain imaging.
2. MRI compatible projection, and MRI compatible earphones.
3. High-powered personal computer with extensive memory and imaging analysis software (e.g., SPM [56], BrainVoyager [57], MR Vista [58]).

3 Methods

First, one must select which human chemosignal one intends to study, and collect/prepare it as follows (*see* **Note 17**):

3.1 Collecting and Preparing the Chemosignal

3.1.1 Collecting Body-Odor

1. Supply the subject with scentless soap and either a shirt or pads in a sealed plastic bag (*see* **Note 3**).
2. Instruct the subject to avoid eating extremely odorous foods that may influence body odor, such as fenugreek, asparagus and garlic.
3. Instruct the subject to wash with the scentless soap before bedtime, not to wear any perfume or deodorant, and wear the supplied shirt over night.
4. In the morning, the subject is asked to put the shirt in the bag and reseal it, and put it in refrigeration for later collection by the experimenter (*see* **Note 4**).

3.1.2 Collecting Axillary Sweat

1. Fit the subject with cotton pads placed under the armpits and secured in place using medical adhesive tape. Alternatively, the subject wears a cotton shirt whose armpits are later cut out.
2. Place the subject in the sweat generating condition.

   Control sweat: Subject is placed on either a treadmill or exercise bicycle for ~30 min.

   Chemosignal sweat: For the condition of interest, the subject is placed in an appropriate setting. For example, for fear sweat we recommend either tandem skydiving or bungee jumping. Stress sweat can be typically obtained from college students during a statistics exam.
3. Immediately after the sweat is obtained (either control or chemosignal), the relevant emotion is assessed. For example, for fear, after the activity, subjects are requested to rate their level of fear on a 10-point scale ranging from "not afraid at all" to "the most afraid I have ever been."
4. Immediately after the activity, the pads or the cut armpit areas of the shirt are stored in a sealed plastic bag and kept under refrigeration (*see* **Note 4**).

Fig. 3 Collecting tears. To collect tears, the subject watches a sad movie of their own choice in isolation, and once tears start to trickle, they collect them directly into a vial, assisted by a small mirror. Observe the process from neighboring room via one-way mirror or video monitor to assure reliability.

3.1.3 Collecting Tears

Tear collection technique may influence tear fluid content [59, 60]. Ask donors to avoid use of creams, lotions or makeup. Instruct subjects to wash their face without using soap before the collection begins.

Emotional Tears and Control

1. Seat the subject in a comfortable room, equipped with a display device such as a computer, DVD or projector.
2. Instruct the subject on how to collect tears—by using a mirror and a collection vial, capturing the teardrops directly into the vial as they visibly trickle down the cheek (Fig. 3).
3. To demonstrate this to the subject, use a pipette to apply drops of saline or ringer solution under the subject's eye, and let the subject capture these drops into the collection vial. This trickled saline will later serve as the control compound.
4. Leave the subject alone in the room, and project the preselected sad movie (*see* **Note 6**). Continue to monitor the subject through one-way mirror or video.
5. Ask the subject to call for you once they have obtained tears in the vial, ideally up to the 1 ml mark.
6. Use the tears as soon as possible, but always before 3 h.

Trigeminal/Reflexive Tears

1. Seat the subject in a comfortable chair that allows tilting of the head backwards and sideways.
2. An experienced ENT physician should insert a sterilized flexible endoscope into the naris, and make contact with the septum near the inferior turbinate (*see* **Note 7**).
3. The contact of the endoscope tip with the trigeminal nerve endings-rich tissue should elicit a tearing reflex in the ipsilateral eye.

4. As soon as tears begin to collect in the eye, tilt the head of the subject gently sideways, while keeping the endoscope in place and avoiding harsh movements. The tilt should be in the direction of the naris—i.e., for a right naris and right eye tilt the head to the right (*see* **Note 18**).
5. Use a glass capillary tube to capture the tears as they drop out of the eye (*see* **Note 19**).
6. Immediately empty the capillary into a vial.

3.1.4 Synthetic Putative Chemosignals

The chosen chemosignal can be presented in either solid form or solution (*see* **Note 20**). Conduct all the following procedures inside a chemical hood to avoid dust and smell contamination of the surrounding area (*see* **Notes 2** and **17**).

Preparation in Solution

1. Weigh the desired amount of solid chemosignal (e.g., for a 2 mM concentration of 10 ml androstadienone solution take ~5 mg solid androstadienone crystals (m.w. 270.4 g/mol) and dissolve in 10 ml of propylene glycol).
2. To generate masked-odor solution (as in [61, 62]), use 1 % w/w eugenol in propylene glycol as the diluent (*see* **Note 11**).
3. Rigorously vortex or magnetically stir the solution until no crystals are visible. This might require a couple of hours. Mild heat (<35 °C) can be applied to facilitate dissolution (*see* **Note 12**).
4. The above stock solution can be further diluted to any desired concentration by using propylene glycol or propylene glycol + 1 % eugenol (e.g., for 10 ml 250 μM final concentration take 1.25 ml of the above stock solution and mix it with 9.75 ml diluent). For threshold concentrations *see* ref. [63].
5. Keep the stock and diluted solutions in an airtight glass vial, preferably under refrigeration and/or well-ventilated storage.

3.2 Narrative of Psychophysiology Experiment

The most powerful designs are "within-subjects," that is, the same subjects are tested in two conditions (chemosignal and control), counter-balanced for order, and double-blind as to compound identity. In other words, neither the participant, nor the experimenter interacting with them, should know which experiment is with the chemosignal or with the control. We recommend that for each subject, the two experiments be conducted day after day and at the same time (*see* **Note 21**), so as to minimize external sources of variance. In overview, for each experiment you will obtain a baseline for all measures, then you will conduct a stimulus exposure, and then continue to monitor the response over about an hour. Together, this makes for nearly a 2-h experiment, which you will repeat day after day.

3.3 Psycho-physiology Set-Up

1. A same-sex experimenter should greet the subject (i.e., women subject by women experimenter).
2. Obtain informed consent.
3. Escort the subject to the experimental room, and leave them there alone to complete an initial on-screen baseline mood questionnaire.
4. Return to the room, and provide the subject with a collection tube and collect a baseline saliva sample (*see* **Notes 22** and **26**).
5. Set the remaining collection tubes, all clearly consecutively numbered, within a holder in easy reach of the subject's dominant hand.
6. The same-sex experimenter now applies the various psychophysiology transducers to the subject:

   Skin Conductance Response (*SCR*): Place the SCR electrodes on the second phalanx of the index and the third digit of the non-dominant hand and attach with Velcro strap.

   Electrocardiogram (*ECG*): Paste the two signal electrodes on the left and right lower rib, above the abdomen. Paste the ground electrode on the left ankle.

   Finger pulse/ear pulse (*FP/EP*): Using a Velcro strap, place one plethysmograph on the pinky finger of the non-dominant hand. Using an ear-clip, place another plethysmograph on the ear lobe on the side of the non-dominant hand.

   Skin temperature (*ST*): Using medical adhesive tape, paste the thermistor directly below the non-dominant axilla.

   Abdominal and thoracic respiration (*AR and TR*): Place AR belt around the belt line, and the TR belt around the chest of the subject, just below the axilla.

   Continuous blood pressure (*BP*): Place the BP finger-cuff on the third finger of the non-dominant hand, i.e., the finger between the two fingers with SCR electrodes.

   Spirometer: Fit a nasal cannula to the subject.

   Motion: This is the only transducer not fitted to the subject, but rather to the armrest of the subject chair.
7. Leave the subject alone in the experimental room, instructing them that if an experimenter later enters, they are not to engage in conversation with them.
8. From the control room, observe that all physiological variables are reading properly at 1 kHz, and if any are particularly noisy, return to the subject room to reconnect sensors as necessary (*see* **Note 23**). Start the recording at 1 kHz (make sure you have sufficient memory for a 2-h recording).

3.4 Psychophysiology Experiment

1. Commence with 10-min baseline recording. During these 10 min project a baseline nature film.
2. Project the mood questionnaire.
3. Project instruction to spit into the second collection tube.
4. An opposite sex experimenter now enters the room with a jar containing the stimulus. The experimenter stands next to the subject, opens the jar such that it remains with mesh cover only. At this point, the presentation software should sound an auditory instruction as follows: "At the tone, sniff for the duration of the tone, three, two, one, TONE." Tone duration should be 1.6 s. At the "two count," the experimenter should bring the open jar to the nose of the subject, allowing them to take a sniff. The experimenter should pull the jar away at the end of the tone. At this point, consecutive on-screen VAS scales should appear, requesting the subject to rate intensity and pleasantness of the stimulus. The above constitutes one trial. The entire exposure should consist of ten trials, with 30-s inter-trial intervals.
5. After the ten sniffs, the experimenter removes the nasal cannula from the subject, and pastes the pre-prepared stimulus-containing (~100 μl) band-aid to the upper lip of the subject, to allow continued exposure throughout the experiment (Fig. 4). The experimenter then exits the room.
6. The subject receives an on-screen mood questionnaire.
7. The subject receives on-screen instructions to spit into the third collection tube.

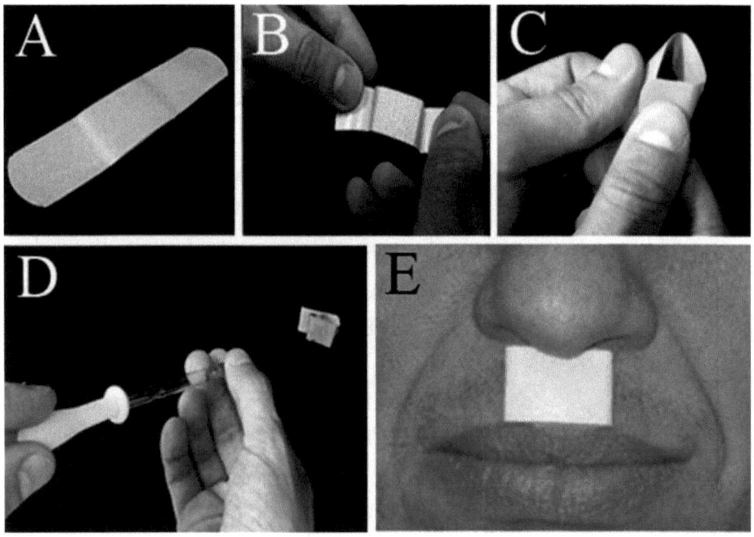

Fig. 4 Continued exposure to chemosignals. (**a**) Use a simple band-aid. (**b**, **c**) Fold the band-aid backwards to form a pad with adhesive on the back. (**d**) Use a pipette to deposit chemosignal onto the pad (~100 μl). (**e**) Paste the pad on the upper lip

8. Project one of three mood-induction movies (erotic/sad/funny, counterbalanced for order across subjects).
9. The subject receives an on-screen mood questionnaire.
10. The subject receives on-screen instructions to spit into the fourth collection tube.
11. Project a baseline nature film.
12. Project the second of three movies (erotic/sad/funny).
13. The subject receives an on-screen mood questionnaire.
14. The subject receives on-screen instructions to spit into the fifth collection tube.
15. Project a baseline nature film.
16. Project the third of three movies (erotic/sad/funny).
17. The subject receives an on-screen mood questionnaire.
18. The subject receives on-screen instructions to spit into the sixth collection tube.
19. A same-sex experimenter enters the room, disconnects the subject from the various transducers, and invites them to return at the same time tomorrow. The full timeline for this experiment is in Fig. 5.
20. Freeze saliva samples at or below −20 °C. Samples can be transferred to 1.5 ml microtubes at this stage to save space (*see* **Note 24**).
21. On the next day, the procedures are repeated exactly, yet using the second compound (chemosignal/control). The order of the three mood-induction film clips should be counterbalanced across subjects.

3.5 Narrative of Imaging Experiment

Those who have an fMRI-compatible olfactometer (likely less than ten labs in the world) can investigate the event-related response to the chemosignal alone. Most labs, however, lacking such an olfactometer, are primarily restricted to examining the brain response to a task or set of stimuli from either the auditory or visual domains, yet under two separate conditions: after exposure to chemosignals

M	S	Set-up	VN	M	S	Exp	M	S	VE1	M	S	VN	VE2	M	S	VN	VE3	M	S
~2	~1	~15	~10	~2	~1	~8	~2	~1	~5	~2	~1	~10	~5	~2	~1	~10	~5	~2	~1

~Duration (min)

Fig. 5 Schematic timeline of psychophysiology experiment. *M* mood questionnaire, *S* saliva collection, *Set-up* hook up all physiological transducers, *VN* emotionally neutral video clip taken from a nature film, *Exp* exposure to chemosignal or control, starting with ten sniffs, and then placing of band-aid on upper lip, *VE1–VE3* the three mood-induction video clips; sad, happy, and sexually arousing counter balanced for order across subjects

versus after exposure to control. We further recommend independent localizer tasks before exposure to either compound. The imaging task can include complex event-related designs, yet here we detail the simplest block-design approach.

3.6 Imaging Experiment

1. Screen subject for MRI compatibility.
2. Place subject in scanner.
3. Obtain full-head T1 weighted high-resolution (1 mm^3) 3D acquisition.
4. Obtain a T1 weighted reduced resolution full-brain acquisition at identical parameters (resolution and orientation) to those later used for functional scans. This is for later drawing of regions of interest (ROIs), and realignment.
5. Set parameters for full-brain functional scans (typically T2* weighted). Do not use TRs greater than 2 s, and use low TE to reduce loss in ventral temporal susceptibility regions.
6. Commence block-design localizer: project alternating 30-s movie segments containing either neutral or emotion-of-interest content. For example, to study sexual behavior related chemosignals, alternate "neutral" and "erotic" clips. To study fear responses, alternate "neutral" and "frightening" clips. Provide 6 alternations across conditions.
7. Extract scanner bed, but instruct subject to remain motionless.
8. With the subject supine in the scanner bed, conduct stimulus exposure as in the psychophysiology experiment. Specifically:
9. An opposite sex experimenter stands next to the subject, opens the jar such that it remains with mesh cover only. At this point, the presentation software should sound an auditory instruction as follows: "At the tone, sniff for the duration of the tone, three, two, one, TONE." Tone duration should be 1.6 s. At the "two count," the experimenter should bring the open jar to the nose of the subject, allowing them to take a sniff. The experimenter should pull the jar away at the end of the tone. At this point, consecutive on-screen VAS scales should appear, requesting the subject to rate intensity and pleasantness of the stimulus. The above constitutes one trial. The entire exposure should consist of ten trials, with 30-s inter-trial intervals.
10. After the ten sniffs, the experimenter pastes the pre-prepared stimulus-containing (~100 μl) band-aid to the upper lip of the subject, to allow continued exposure throughout the experiment (Fig. 4). The subject is then reinserted into the scanner.
11. Again obtain a T1 weighted reduced resolution full-brain acquisition exactly as in **step #4**.
12. Again set parameters for full-brain functional scans exactly as in **step #5**.

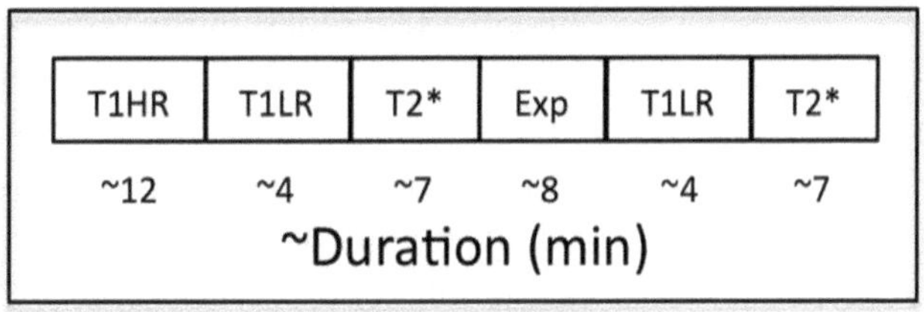

Fig. 6 Schematic timeline of imaging experiment. *T1HR* high-resolution T1 full head acquisition, *T1LR* low resolution T1 acquisition at same orientation as later functional scans, *T2** functional scans with alternating video clips, "neutral" and "emotion of interest," *Exp* extraction from scanner, and exposure to chemosignal or control, starting with ten sniffs, and then placing of band-aid on upper lip

13. Commence block-design experiment exactly as in **step #6**.
14. Extract subject, and invite them to return at the exact same time on the next day. The full time line for this experiment is in Fig. 6.
15. Repeat all of the above on the next day, yet using the other (chemosignal/control) compound.

3.7 Psychophysiology Analysis Pointers

For each subject, for each day, and for each physiological measure, mood question, and hormone in saliva:

1. Extract the baseline values obtained during the 10-min baseline.
2. Extract the values obtained during the ten-sniff stimulus exposure.
3. Extract the values obtained during each type of emotional setting.
4. For each subject, compute a percent change score from baseline for each measure on each day.
5. Compare the change from baseline in the chemosignal condition to the change from baseline in the control condition (*see* **Note 25**).

3.8 Endocrine Analysis Pointers

1. Thaw saliva, vortex for a few seconds, and centrifuge at 1,500 × *g* for 15 min to precipitate particulate material and proteins. Use only the clear supernatant, carefully avoiding disruption of the pellet.
2. Use your specific chosen kit instructions for reagent preparation and needed volumes for analysis.
3. Analysis is best performed in triplicates for each time point of each subject.
4. Arrange the layout of the plate in advance. Leave space as instructed by the kit protocol for standard curve measures and

controls, and design the rest of the space so that each triplicate spans two rows and columns, in an L-shaped pattern. This is done in order to avoid intra-column differences across the plate.

5. Upon completion of the kit procedure, you should have a read-out of the plate in relative optical density (OD) values. Use the readouts of the standard curve to fit a 4PL logistics line, using one of the suggested computer programs (*see* Subheading 2.7, **item 10**).
6. Average the triplicates and notice any aberrant results. Omit outliers as necessary.

3.9 Imaging Analysis Pointers

1. Conduct the standard pre-processing steps of your analysis scheme (i.e., motion correction, realignment, etc.).
2. Combine the localizer scans (**step #6**) from Day 1 and 2, and conduct a standard linear contrast to generate functional regions of interest (fROIs). For example, extract areas significantly more or less active in the contrast of "neutral" versus "erotic."
3. Extract the time course within these fROIs from the later conducted task (**step #13**).
4. Compute the difference in percent signal change across the two conditions (e.g., "neutral" and "erotic") within the fROI.
5. Repeat the above for the second day experiment.
6. Compare the change from baseline across the 2 days, chemosignal versus control.

4 Notes

1. The Ekman mood scale originally contains the following 16 variables: Amused; Content; Happy; Calm; Confident; Interested; Angry; Anxious; Annoyed; Bored; Contemptuous; Embarrassed; Stressed; Afraid; Disgusted and Sad. To these we have added a 17th variable: Sexually aroused. To obtain mood state, the 17 variables are displayed in succession with a visual-analogue scale (VAS) ranging between "very" and "not at all," as in Fig. 1 [22, 64, 65].
2. If you intend to conduct a single study with human chemosignals, then you can get passed without stainless steel-coated rooms. In contrast, if you are building a lab that will conduct many such studies over time (years), then experimental rooms coated with nonporous material (e.g., Stainless steel, PTFE) are important, as without them you will eventually have contamination across studies.
3. Plastic bags used to store shirts should be odorless, non-absorbing, non-diffusing and properly sealable. We have found

that aluminum-lined anti-static plastic bags (such as the ones used to store electronic components) work well for this purpose. These are usually sealable using a cheap heat-sealer. Alternatively, zip-lock style sealing is also optional, provided that the seal is hermetic. Amber/Opaque glass jars are also a viable option.

4. Methods for storing chemosignals are critical. In the case of tears, we used only fresh non-refrigerated tears obtained within 3 h or less of their use. Moreover, in a small pilot study we found that storing them for longer periods altered the effects. As to sweat and body-odor, the highest recommended temperature for keeping biological samples is generally 4 °C (regular home fridge temperature). This is a fairly bacteriostatic temperature (i.e., prevents bacterial growth), it slows enzymatic reactions and also reduces evaporation (and hence also loss of volatiles)—but does not allow long-term storage without degradation due to oxidation or eventual bacterial/fungal spoilage. Freezing samples at the usual −20 °C (regular home freezer, but avoid no-frost devices) or −80 °C (laboratory deep freezer) would result in better sample preservation, but care should be given to the rate of freezing: Slow freezing, by simply putting the samples into the freezer chamber, may result in the formation of large crystals of ice. The growing ice crystals might in turn disrupt the membranes of cells found in the sample, resulting in altered composition of the sample due to lost cellular compartmentalization and subsequent sample degradation. Therefore, unless otherwise specified, it is advised to flash-freeze samples prior to long term storage, by transferring the samples into cryogenic tubes and dropping them into liquid nitrogen, followed by storage in freezers. Having said all this, be advised that some plastic materials tend to adsorb lipid components in a way which is aggravated by flash-freezing. Therefore the composition of lipids in the sample might be altered.
5. We use medical grade sterile saline solution (0.9 % w/v Sodium Chloride for IV injection). Alternatives can be Ringer solution (different compositions are available) or a more specialized "artificial tears" (*see* [66, 67]).
6. To obtain tear donors, we first post an add asking for individuals who cry easily. When they call in, we ask them which film made/makes them cry, and we obtain this film from the video library. Our experience is that only about 10 % of those who think they cry easily, in fact do so. Such donors, however, can return to lab on a regular basis, and will typically generate 1 ml of tears in response to the same film segment again and again.
7. Flexible thin endoscopes allow the most comfortable means of eliciting reflexive tears. However, other more "low-tech" alternatives, such as a thin long shaft medical transfer swab (e.g., Copan Italia 160C) have also proven useful.

8. We have found that for obtaining good quality steroid chemosignal molecules, Steraloids (Newport, RI) is a good source.
9. Androstenone is a swine pheromone. Pheromones are presumably species-specific. Nevertheless, androstenone has been studied in various human chemosignaling contexts. Like with all steroids, be sure to check purity.
10. While Propylene Glycol (PG; 1,2 propanediol) is a widely used diluting solvent in the odor and fragrance world, Dipropylene Glycol (DPG; 4-Oxa-2,6-heptandiol and 4-Oxa-1,7-heptandiol mixture) is also suitable as an alternative. DPG is very similar to PG, although it has lower freezing and higher boiling points, and a lower vapor pressure. Both are almost odorless, and effectively dissolve most moderate to highly hydrophilic compounds and essential oils. Mineral oil can also be considered, as the solubility of steroidal compounds may be higher in hydrophobic solvents, but it is less commonly used.
11. McClintock and colleagues used clove oil to mask the odor of androstadienone [61, 62]. Clove oil is comprised mainly of eugenol, but with varying concentrations (60–95 %, depending on the natural source). To ensure consistent outcome, pure solid eugenol dissolved in the diluent is preferable. Use 1 % w/v eugenol dissolved in propylene glycol. The resulting solution has a recognizable yellow tint, so be sure to use opaque jars for delivery.
12. Dissolution of solid steroid chemosignal compounds might prove hard at higher concentrations (i.e., stock solutions). Possible aids are the use of a sonicator (either ultrasonic bath or probe) and the application of gentle heat (~35 °C) combined with a lengthy stir or vortex.
13. PTFE beads (we usually use 2–3 mm in diameter) are a cheap, inert and simple means of enlarging the effective surface area of solutions. Wet a bed of beads inside a jar to avoid using the liquid solution directly.
14. In different cultures, it may be worth validating these sources in a separate short validation study. For example, American subjects rated clips from "Mr. Bean" as funny, but Israeli subjects did not. In both cultures, however, "The Champ" induced sadness, and "Nine and a half weeks" induced sexual arousal in both sexes. Note that more explicit film clips often induce disgust rather than arousal.
15. The motion sensor is best fitted to the non-dominant hand armrest. This way, the motion trace can later be used to identify motion-artifacts in the physiological data.
16. Finapres Ohmeda 2300 are the best continuous finger-cuff PB monitors, yet they are no longer produced. Thus, each time one shows up on eBay (typically from hospital surplus equipment

dealers, as they were used in child anesthesiology), all the psychophysiologists fight over it. Keep your eyes out for one.

17. We strongly recommend that when working with chemosignals, one applies contamination prevention standards typical of work with radioactive material. Under our working hypothesis of extreme sensitivity to chemosignals, with functional threshold far below conscious detection threshold, the possibility of contaminating a study is ever present. With this notion in mind, it is also advised to work with single-use tools whenever possible (i.e., disposable plastic spatulas, plastic weighing "boats," etc.).
18. When collecting reflex tears, make sure to tilt the subject's head in the correct direction, so that tears won't flow toward the medial canthus and thus drain into the nasolacrimal duct.
19. Hold the glass capillary tube in parallel to the ground or in a small tilt so that the fluid flows freely into the tube aided by gravity and capillary force. Do not over-fill the capillary tube and use the marking to assess the amount of fluid withdrawn.
20. Although many studies have used solid androstadienone crystals as the stimulus, this results in a headspace concentration that is most likely non-biological.
21. Various hormones peak at various times of day. Thus, if you select particular hormones to follow, you should first investigate this phenomenon for the selected hormone, as it may imply particular times of day to avoid, e.g., early morning for testosterone.
22. You should leave a glass of water at reach of the dominant hand of the subject, and suggest that they take a sip after each saliva donation (i.e., ~15 min before their next donation). Otherwise, they might run out of saliva across the 2-h study.
23. A trick to know whether your SCR recording is working is to abruptly bang (once) on the door of the subject room. This should startle the subject, and you should see a clear SCR deviation.
24. Saliva samples should be frozen at −20 °C or below, and not only for long term storage. The freezing itself precipitates proteins (e.g., mucins) out of the saliva liquid, allowing for the separation by centrifugation of the potentially interfering factors for the analysis. However, multiple freeze–thaw cycles should be avoided [51, 52].
25. Note that this analysis scheme is susceptible to the risks associated with multiple comparisons. This is inherent to casting a wide net. A solution is to first study a small group of subject in a separate pilot study, in order to identify the impact of a given chemosignal. For example, a chemosignal may have profound

effects on SCR but not on BP. You should then continue the main study collecting only the measures of interest identified in the pilot. This way you will avoid multiple comparisons. A second alternative is to combine several of the measures into composite measures, as in references [22, 64, 65, 68].

26. Steroid hormones secretion is episodic in nature. Thus, a minimum of three samples per experiment is advised [51, 52]. We found that taking five samples is better, dispersed throughout the length of the experiment [22]. To achieve an average baseline, equal volumes of samples can be physically pooled, but to detect fluctuations stemming from experimental intervention take individual samples across different time points.

Acknowledgment

This work was supported by the James S. McDonnell Foundation.

References

1. McClintock MK (1971) Menstrual synchorony and suppression. Nature 229(5282):244–245
2. Stern K, McClintock MK (1998) Regulation of ovulation by human pheromones. Nature 392(6672):177–179. doi:10.1038/32408
3. Russell MJ, Switz GM, Thompson K (1980) Olfactory influences on the human menstrual cycle. Pharmacol Biochem Behav 13(5): 737–738
4. Jacob S, Spencer NA, Bullivant SB, Sellergren SA, Mennella JA, McClintock MK (2004) Effects of breastfeeding chemosignals on the human menstrual cycle. Hum Reprod 19(2): 422–429
5. Cutler WB, Preti G, Krieger A, Huggins GR, Garcia CR, Lawley HJ (1986) Human axillary secretions influence women's menstrual cycles: the role of donor extract from men. Horm Behav 20(4):463–473
6. Wysocki CJ, Preti G (2004) Facts, fallacies, fears, and frustrations with human pheromones. Anat Rec A Discov Mol Cell Evol Biol 281(1):1201–1211. doi:10.1002/ar.a.20125
7. Wedekind C, Seebeck T, Bettens F, Paepke AJ (1995) MHC-dependent mate preferences in humans. Proc Biol Sci 260(1359):245–249. doi:10.1098/rspb.1995.0087
8. Wedekind C, Furi S (1997) Body odour preferences in men and women: do they aim for specific MHC combinations or simply heterozygosity? Proc Biol Sci 264(1387):1471–1479. doi:10.1098/rspb.1997.0204
9. Ober C, Weitkamp LR, Cox N, Dytch H, Kostyu D, Elias S (1997) HLA and mate choice in humans. Am J Hum Genet 61(3): 497–504, doi:S0002-9297(07)64312-2 [pii] 10.1086/515511
10. Eggert F, Muller-Ruchholtz W, Ferstl R (1998) Olfactory cues associated with the major histocompatibility complex. Genetica 104(3):191–197
11. Jacob S, McClintock MK, Zelano B, Ober C (2002) Paternally inherited HLA alleles are associated with women's choice of male odor. Nat Genet 30(2):175–179, doi:10.1038/ng830 ng830 [pii]
12. Roberts T, Roiser JP (2010) In the nose of the beholder: are olfactory influences on human mate choice driven by variation in immune system genes or sex hormone levels? Exp Biol Med (Maywood) 235(11):1277–1281, doi:ebm.2010.010112 [pii] 10.1258/ebm.2010.010112
13. Thornhill R, Gangestad SW, Miller R, Scheyd G, McCollough JK, Franklin M (2003) Major histocompatibility complex genes, symmetry, and body scent attractiveness in men and women. Behav Ecol 14(5):668–678
14. Roberts SC, Little AC (2008) Good genes, complementary genes and human mate preferences. Genetica 132(3):309–321. doi:10.1007/s10709-007-9174-1
15. Gangestad SW, Cousins AJ (2001) Adaptive design, female mate preferences, and shifts across the menstrual cycle. Annu Rev Sex Res 12:145–185

16. Havlicek J, Roberts SC, Flegr J (2005) Women's preference for dominant male odour: effects of menstrual cycle and relationship status. Biol Lett 1(3):256–259, doi:A79R9E6B7CCHUEN7 [pii]10.1098/rsbl.2005.0332
17. Little AC, Jones BC, Burriss RP (2007) Preferences for masculinity in male bodies change across the menstrual cycle. Horm Behav 51(5):633–639
18. Singh D, Bronstad PM (2001) Female body odour is a potential cue to ovulation. Proc Biol Sci 268(1469):797–801. doi:10.1098/rspb.2001.1589
19. Roberts SC, Gosling LM, Carter V, Petrie M (2008) MHC-correlated odour preferences in humans and the use of oral contraceptives. Proc Biol Sci 275(1652):2715–2722, doi:F542428772R96X64 [pii]10.1098/rspb.2008.0825
20. Kuukasjarvi S, Eriksson CJP, Koskela E, Mappes T, Nissinen K, Rantala MJ (2004) Attractiveness of women's body odors over the menstrual cycle: the role of oral contraceptives and receiver sex. Behav Ecol 15(4):579–584
21. Miller SL, Maner JK (2010) Scent of a woman: men's testosterone responses to olfactory ovulation cues. Psychol Sci 21(2):276–283, doi:0956797609357733[pii]10.1177/0956797609357733
22. Gelstein S, Yeshurun Y, Rozenkrantz L, Shushan S, Frumin I, Roth Y, Sobel N (2011) Human tears contain a chemosignal. Science 331(6014):226–230,doi:science.1198331[pii]10.1126/science.1198331
23. Oh TJ, Kim MY, Park KS, Cho YM (2012) Effects of chemosignals from sad tears and postprandial plasma on appetite and food intake in humans. PLoS One 7(8):e42352, doi:10.1371/journal.pone.0042352PONE-D-11-08927 [pii]
24. Meredith M (2001) Human vomeronasal organ function: a critical review of best and worst cases. Chem Senses 26(4):433
25. Hauser R, Marczak M, Karaszewski B, Wiergowski M, Kaliszan M, Penkowski M, Kernbach-Wighton G, Jankowski Z, Namiesnik J (2008) A preliminary study for identifying olfactory markers of fear in the rat. Lab Anim (N Y) 37(2):76–80, doi:laban0208-76 [pii]10.1038/laban0208-76
26. Pageat P, Gaultier E (2003) Current research in canine and feline pheromones. Vet Clin North Am Small Anim Pract 33(2):187–211
27. Chen D, Haviland-Jones J (2000) Human olfactory communication of emotion. Percept Mot Skills 91(3 Pt 1):771
28. Ackerl K, Atzmueller M, Grammer K (2002) The scent of fear. Neuroendocrinol Lett 23(2):79–84
29. Chen D, Katdare A, Lucas N (2006) Chemosignals of fear enhance cognitive performance in humans. Chem Senses 31(5):415
30. Zhou W, Chen D (2009) Fear-related chemosignals modulate recognition of fear in ambiguous facial expressions. Psychol Sci 20(2):177
31. Prehn A, Ohrt A, Sojka B, Ferstl R, Pause BM (2006) Chemosensory anxiety signals augment the startle reflex in humans. Neurosci Lett 394(2):127–130
32. Prehn-Kristensen A, Wiesner C, Bergmann TO, Wolff S, Jansen O, Mehdorn HM, Ferstl R, Pause BM (2009) Induction of empathy by the smell of anxiety. PLoS One 4(6):e5987
33. Mujica-Parodi LR, Strey HH, Frederick B, Savoy R, Cox D, Botanov Y, Tolkunov D, Rubin D, Weber J (2009) Chemosensory cues to conspecific emotional stress activate amygdala in humans. PLoS One 4(7):113–123
34. Magnetic Resonance Neuroimaging: Methods and Protocols (2011). Magn Reson Neuroimaging Methods Protoc 711:1–598. doi:10.1007/978-1-61737-992-5
35. Analgesia: Methods and Protocols (2010). Analgesia Methods Protoc 617:1–560. doi:10.1007/978-1-60327-323-7
36. fMRI Techniques and Protocols (2009). Fmri Tech Protoc 41:1–843. doi:10.1007/978-1-60327-919-2
37. Boulton Aa, Baker Gb, Hiscock M (1991). Neuromethods, vol 17 Neuropsychology. Perceptual and motor skills 72(2):703
38. Johnson BN, Sobel N (2007) Methods for building an olfactometer with known concentration outcomes. J Neurosci Methods 160(2):231–245,doi:S0165-0270(06)00461-4 [pii]10.1016/j.jneumeth.2006.09.008
39. Sobel N, Prabhakaran V, Desmond JE, Glover GH, Sullivan EV, Gabrieli JD (1997) A method for functional magnetic resonance imaging of olfaction. J Neurosci Methods 78(1–2):115–123
40. Lundstrom JN, Gordon AR, Alden EC, Boesveldt S, Albrecht J (2010) Methods for building an inexpensive computer-controlled olfactometer for temporally-precise experiments. Int J Psychophysiol 78(2):179–189, doi:S0167-8760(10)00666-5[pii]10.1016/j.ijpsycho.2010.07.007
41. Yantis S, Johnson DN (1990) Mechanisms of attentional priority. J Exp Psychol Hum Percept Perform 16(4):812–825

42. Levenson RW, Ekman P, Friesen WV (1990) Voluntary facial action generates emotion-specific autonomic nervous system activity. Psychophysiology 27(4):363–384
43. Baylis GC, Driver J (1993) Visual attention and objects: evidence for hierarchical coding of location. J Exp Psychol Hum Percept Perform 19(3):451–470
44. Willis CM, Church SM, Guest CM, Cook WA, McCarthy N, Bransbury AJ, Church MRT, Church JCT (2004) Olfactory detection of human bladder cancer by dogs: proof of principle study. Br Med J 329(7468):712
45. Mathwroks Matlab. http://www.mathworks.com/products/matlab/. Accessed 16 Sep 2012
46. National Instruments LabVIEW. http://www.ni.com/labview/. Accessed 16 Sep 2012
47. AD Instruments. http://www.adinstruments.com/. Accessed 16 Sep 2012
48. Entran—Measurment specialists. http://www.meas-spec.com/entran.aspx. Accessed 16 Sep 2012
49. Hauser MD, Kralik J (1997) Life beyond the mirror: a reply to Anderson & Gallup. Anim Behav 54(6):1568–1571
50. Swartz KB (1997) What is mirror self-recognition in nonhuman primates, and what is it not? Ann N Y Acad Sci 818:64–71
51. Salimetrics (2011) High sensitivity salivary cortisol enzyme immunassay kit protocol. http://www.salimetrics.com/documents/Cortisol_Kit_Insert.pdf. Accessed 16 Sep 2012
52. Salimetrics (2011) Salivary testosterone enzyme immunassay kit protocol. http://www.salimetrics.com/documents/Testo_Kit_Insert.pdf. Accessed 16 Sep 2012
53. Anderson JR, Gallup GG (1997) Self-recognition in Saguinus? A critical essay. Anim Behav 54:1563–1567
54. Delfour F, Marten K (2001) Mirror image processing in three marine mammal species: killer whales (Orcinus orca), false killer whales (Pseudorca crassidens) and California sea lions (Zalophus californianus). Behav Processes 53(3):181–190
55. Drea CM, Vignieri SN, Cunningham SB, Glickman SE (2002) Responses to olfactory stimuli in spotted hyenas (Crocuta crocuta): I. Investigation of environmental odors and the function of rolling. J Comp Psychol 116(4):331–341
56. Bekoff M, Sherman PW (2004) Reflections on animal selves. Trends Ecol Evol 19(4):176–180
57. Calderone JB, Reese BE, Jacobs GH (2003) Topography of photoreceptors and retinal ganglion cells in the spotted hyena (Crocuta crocuta). Brain Behav Evol 62(4):182–192
58. Drea CM, Vignieri SN, Kim HS, Weldele ML, Glickman SE (2002) Responses to olfactory stimuli in spotted hyenas (Crocuta crocuta): II. Discrimination of conspecific scent. J Comp Psychol 116(4):342–349
59. Jones DT, Monroy D, Pflugfelder SC (1997) A novel method of tear collection: comparison of glass capillary micropipettes with porous polyester rods. Cornea 16(4):450–458
60. Stuchell RN, Feldman JJ, Farris RL, Mandel ID (1984) The effect of collection technique on tear composition. Invest Ophthalmol Vis Sci 25(3):374–377
61. Hummer TA, McClintock MK (2009) Putative human pheromone androstadienone attunes the mind specifically to emotional information. Horm Behav 55(4):548–559, doi:S0018-506X(09)00005-1[pii] 10.1016/j.yhbeh.2009.01.002
62. Jacob S, McClintock MK (2000) Psychological state and mood effects of steroidal chemosignals in women and men. Horm Behav 37(1):57–78, doi:10.1006/hbeh.1999.1559S0018-506X(99)91559-3 [pii]
63. Lundstrom JN, Hummel T, Olsson MJ (2003) Individual differences in sensitivity to the odor of 4,16-androstadien-3-one. Chem Senses 28(7):643–650
64. Bensafi M, Brown WM, Khan R, Levenson B, Sobel N (2004) Sniffing human sex-steroid derived compounds modulates mood, memory and autonomic nervous system function in specific behavioral contexts. Behav Brain Res 152(1):11–22, doi:10.1016/j.bbr.2003.09.009 S0166432803003309[pii]
65. Bensafi M, Tsutsui T, Khan R, Levenson RW, Sobel N (2004) Sniffing a human sex-steroid derived compound affects mood and autonomic arousal in a dose-dependent manner. Psychoneuroendocrinology 29(10):1290–1299, doi:10.1016/j.psyneuen.2004.03.007 S0306453004000460[pii]
66. Frey WH II, DeSota-Johnson D, Hoffman C, McCall JT (1981) Effect of stimulus on the chemical composition of human tears. Am J Ophthalmol 92(4):559–567
67. Van Haeringen NJ (1981) Clinical biochemistry of tears. Surv Ophthalmol 26(2):84–96
68. Wyart C, Webster WW, Chen JH, Wilson SR, McClary A, Khan RM, Sobel N (2007) Smelling a single component of male sweat alters levels of cortisol in women. J Neurosci 27(6):1261–1265

INDEX

A

B

C

Kazushige Touhara (ed.), *Pheromone Signaling: Methods and Protocols*, Methods in Molecular Biology, vol. 1068, DOI 10.1007/978-1-62703-619-1, © Springer Science+Business Media, LLC 2013

H

I

L

M

MIX
Papier aus verantwortungsvollen Quellen
Paper from responsible sources
FSC® C105338

If you have any concerns about our products,
you can contact us on
ProductSafety@springernature.com

In case Publisher is established outside the EU,
the EU authorized representative is:
Springer Nature Customer Service Center GmbH
Europaplatz 3, 69115 Heidelberg, Germany

Printed by Libri Plureos GmbH
in Hamburg, Germany